Physics of Mass

Physics of Mass

Edited by

Behram N. Kursunoglu

Global Foundation, Inc.
Coral Gables, Florida

Stephan L. Mintz

Florida International University
Miami, Florida

and

Arnold Perlmutter

University of Miami
Coral Gables, Florida

Springer Science+Business Media, LLC

Library of Congress Cataloging-in-Publication Data

Physics of mass / edited by Behram N. Kursunoglu, Stephan L. Mintz and Arnold Perlmutter.
p. cm.
Proceedings of the International Conference on Orbis Scientiae 1997 II: Physics of Mass, held December 12-15, 1997, in Miami Beach, Florida --T.p. verso.
Includes bibliographical references and index.
ISBN 978-1-4419-3305-8 ISBN 978-0-306-47085-1 (eBook)
DOI 10.1007/978-0-306-47085-1

1. Atomic mass--Congresses. 2. Particles (Nuclear physics)--Congresses. 3. Astrophysics--Congresses. I. Kurşunoğlu, Behram, 1922- . II. Mintz, Stephan L. III. Perlmutter, Arnold, 1928- . IV. International Conference on Orbis Scientiae (1997 : Miami Beach, Florida)
QC172.P49 1998
539.7'2--dc21 98-42065
CIP

Proceedings of the International Conference on Orbis Scientiae 1997 II: Physics of Mass, held December 12 – 15, 1997, in Miami Beach, Florida

This volume was taken from a series of conferences sponsored by Global Foundation, Inc., Coral Gables, Florida

ISBN 978-1-4419-3305-8

Originally published by Kluwer Academic/Plenum Publishers, New York in 1998
Softcover reprint of the hardcover 1st edition 1998

10 9 8 7 6 5 4 3 2 1

A C.I.P. record for this book is available from the Library of Congress.

PREFACE

The 26th Conference on Orbis Scientiae 1997 is the second in this series of high energy physics and cosmology that took place in the month of December instead of the well established tradition where the month of January was the conference date. This change was due to the increased hotel rates in South Florida. Another change in the organization of these conferences is the choice of a core topic to take half of the conference time. The remaining time will be devoted to subjects of direct interest to participants. In the 1997 Orbis Scientiae we chose "Physics of Mass" as the core topic of the conference, which took over five sessions. The remaining five sessions were devoted to those presentations of direct interest to some of the participants.

The Orbis Scientiae 1998 will cover "The Physics of Spin" as a core topic. In anticipation of the ascending importance of gravity research in the 21st century, for the core topics for the Orbis Scientiae 1999 and 2000 we suggest "The Status of An Evolving Cosmological Parameter" and "The Nature of Gravity from Big Bang to Flat Universe", respectively. These two topics may be expected to lead on to the first conference in the 21st century in which the core topic should be chosen for the Orbis Scientiae 2001 by the participants. We are pleased to invite the conference participants to send us their choices of core topics.

The Trustees and the Chairman of the Global Foundation, Inc., wish to extend special thanks to Edward Bacinich of Alpha Omega Research Foundation for his continuing generous support including the 1997 Orbis Scientiae.

Behram N. Kursunoglu
Stephan L. Mintz
Arnold Perlmutter
Coral Gables, Florida
April 1998

ABOUT THE GLOBAL FOUNDATION, INC.

The Global Foundation, Inc., which was established in 1977, utilizes the world's most important resource... people. The Foundation consists of great senior men and women of science and learning, and of outstanding achievers and entrepreneurs from industry, governments, and international organizations, along with promising and enthusiastic young people. These people form a unique and distinguished interdisciplinary entity, and the Foundation is dedicated to assembling all the resources necessary for them to work together. The distinguished senior component of the Foundation transmits its expertise and accumulated experience, knowledge, and wisdom to the younger membership on important global issues and frontier problems in science.

Our work, therefore, is a common effort, employing the ideas of creative thinkers with a wide range of experience and viewpoints.

GLOBAL FOUNDATION BOARD OF TRUSTEES

GLOBAL FOUNDATION'S RECENT CONFERENCE PROCEEDINGS

Making the Market Right for the Efficient Use of Energy
Edited by: Behram N. Kursunoglu
Nova Science Publishers, Inc., New York, 1992

Unified Symmetry in the Small and in the Large
Edited by: Behram N. Kursunoglu, and Arnold Perlmutter
Nova Science Publishers, Inc., New York, 1993

Unified Symmetry in the Small and in the Large - 1
Edited by: Behram N. Kursunoglu, Stephan Mintz, and Arnold Perlmutter
Plenum Press, 1994

Unified Symmetry in the Small and in the Large - 2
Edited by: Behram N. Kursunoglu, Stephan Mintz, and Arnold Perlmutter
Plenum Press, 1995

Global Energy Demand in Transition: The New Role of Electricity
Edited by: Behram N. Kursunoglu, Stephan Mintz, and Arnold Perlmutter
Plenum Press, 1996

Economics and Politics of Energy
Edited by: Behram N. Kursunoglu, Stephan Mintz, and Arnold Perlmutter
Plenum Press, 1996

Neutrino Mass, Dark Matter, Gravitational Waves, Condensation of Atoms and Monopoles, Light Cone Quantization
Edited by: Behram N. Kursunoglu, Stephan Mintz, and Arnold Perlmutter
Plenum Press, 1996

Technology for the Global Economic, Environmental Survival and Prosperity
Edited by: Behram N. Kursunoglu, Stephan Mintz, and Arnold Perlmutter
Plenum Press, 1997

25th Coral Gables Conference on High Energy Physics and Cosmology
Edited by: Behram N. Kursunoglu, Stephan Mintz, and Arnold Perlmutter
Plenum Press, 1997

Environment and Nuclear Energy
Edited by: Behram N. Kursunoglu, Stephan Mintz, and Arnold Perlmutter
Plenum Press, 1998

26th Coral Gables Conference on Physics of Mass
Edited by: Behram N. Kursunoglu, Stephan Mintz, and Arnold Perlmutter
Plenum Press, 1998

CONTRIBUTING CO-SPONSORS OF THE GLOBAL FOUNDATION CONFERENCES

Electric Power Research Institute, Palo Alto, California
Gas Research Institute, Washington, DC
General Electric Company, San Jose, California
Northrop Grumman Aerospace Company, Bethpage, New York
Martin Marietta Astronautics Group, Denver, Colorado
Black and Veatch Company, Kansas City, Missouri
Bechtel Power Corporation, Gaithersburg, Maryland
ABB Combustion Engineering, Windsor, Connecticut
BellSouth Corporation, Atlanta, Georgia
National Science Foundation
United States Department of Energy

ORBIS SCIENTIAE 1997 - II
PROGRAM

FRIDAY, December 12, 1997

8:00 AM - Noon REGISTRATION

1:30 PM	**SESSION I:**	**GRAVITATIONAL MASS**
	Moderators:	**BEHRAM N. KURSUNOGLU,** Global Foundation, Inc., Coral Gables, Florida
	Dissertators:	**BEHRAM N. KURSUNOGLU** "Origin of Mass" **SYDNEY MESHKOV,** CALTECH "Current Status of LIGO" **CHRIS POPE,** Texas A & M, College Station, Texas "Gauge Dyonic Strings in Six Dimensions"
	Annotators:	**JANNA LEVIN,** University of California, Berkeley
	Session Organizer:	**BEHRAM N. KURSUNOGLU**
3:15 PM	**COFFEE BREAK**	
3:30 PM	**SESSION II:**	**NEUTRINO MASSES**
	Moderator:	**PETER ROSEN,** Department of Energy, Washington DC **GEOFFREY MILLS,** Los Alamos National Laboratory
	Dissertators:	**JONNY KLEINFELLER,** Forschungszentrum Karlsruhe "KARMEN-Upgrade: Improvement In The Search For Neutrino Oscillations And First Results" **GEOFFREY MILLS** "Results from the LSND Experiment" **SANJIB MISHRA,** Harvard University "Results from the NOMAD Experiment" **CHARLES LANE,** Drexel University, Philadelphia, Pennsylvania "First Results from the CHOOZ Experiment" **LAWRENCE WAI,** University of Washington, Seattle "Atmospheric Neutrino Flux Mesurements with the Super-Kamiokande Neutrino Obervatory"
	Annotators:	**STEPHAN MINTZ,** Florida International University
	Session Organizer:	**GEOFF MILLS**

6:00 PM - ***Welcoming Cocktails, Fontainebleau Cabana Area***
7:00 PM ***Courtesy of Maria and Edward Bacinich***

7:00 PM Orbis Scientiae Adjourns For The Day

SATURDAY, December 13, 1997

8:30 AM	**SESSION III:**	**PROGRESS ON NEW AND OLD IDEAS - I**
	Moderators:	**PAUL FRAMPTON,** University of North Carolina at Chapel Hill
	Dissertators:	**PAUL FRAMPTON** "Leptoquarks" **SHELDON GLASHOW,** Harvard University "Tests of Lorentz Invariance" **O.W. GREENBERG,** University of Maryland "Spin-Statistics, Spin-Locality and TCP: Three Distinct Theorems" **GERALD GURALNIK,** Brown University "Numerical Field Theory on the Continuum" **ALAN KOSTELECKY,** Indiana University "Testing CPT Symmetry"
	Annotators:	**SYDNEY MESHKOV,** CALTECH
	Session Organizer:	**PAUL FRAMPTON**
10:30 AM	**COFFEE BREAK**	
10:45 AM	**SESSION IV:**	**PARTICLE MASSES**
	Moderators:	**SYDNEY MESHKOV** **HARALD FRITZSCH,** Universität München
	Dissertators:	**HARALD FRITZSCH** "Quark Masses and the Description and Dynamics of Flavor Mixing" **RAJAN GUPTA,** Los Alamos National Laboratory "Quark Masses From Lattice and Sum Rules & Implications for Epsilon/Epsilon" **KENNETH LANE,** Boston University "Recent Development in Technicolor" **RICH PARTRIDGE,** Brown University "Top Mass Measurement" "Leptoquark Search" "Direct and Indirect Measurements of the Higgs Mass at the Tevatron (Present and Future" **S.G. RAJEEV,** University of Rochester, New York "Matrix Models of Particle Masses"
	Annotators:	**RICHARD P. WOODARD,** University of Florida
	Session Organizer:	**SYDNEY MESHKOV**
12:45 PM	Orbis Scientiae Adjourns for the Day	

7:30 PM - ***Conference Banquet - Fontainebleau Ballroom A***
10:30 PM ***Courtesy of Maria and Edward Bacinich***

SUNDAY, December 14, 1997

8:30 AM	**SESSION V:**	**COSMOLOGICAL MASSES (BLACK HOLES, NEUTRON STARS, ETC.)**
	Moderators:	**EDWARD KOLB,** FERMI Laboratory, Chicago **KATHERINE FREESE,** University of Michigan
	Dissertators:	**SCOTT DODELSON,** FERMI Laboratory "The CMB and Cosmological Parameters" **KATHERINE FREESE** "Massive Astrophysical Compact Halo Objects" **EDWARD KOLB** "Who is the Inflation?" **VIGDOR L. TEPLITZ,** Southern Methodist University "Mass of the Kuiper Belt" **EDWIN TURNER,** Princeton University "Gravitational Lens Determinations of the Mass and Age of the Universe"
	Annotators:	**FRANCESCO ANTONUCCIO,** The Ohio State University
	Session Organizer:	**EDWARD KOLB,** FERMI Laboratory, Chicago
10:30 AM	**COFFEE BREAK**	
10:45 AM	**SESSION VI:**	**PROGRESS ON NEW AND OLD IDEAS - II**
	Moderator:	**GEOFFREY WEST**, LANL
	Dissertators:	**ROBERT BLUHM,** Colby College "Testing CPT in Atomic Physics" **GERALD B. CLEAVER,** University of Pennsylvania "Mass Heirarchy and Flat Directions in String Models" **ZACHARY GURALNIK,** Princeton University "Branes, Torons, and Yang-Mills Dualities" **PAOLO PRIVITERA,** University of Rome, Italy "Measurement of the Mass of the Intermediate Vector Bosons" **RICHARD P. WOODARD** "Particles As Bound States In Their Own Potential"
	Annotators:	**KATSUMI TANAKA,** Ohio State University
	Session Organizer:	**STEPHAN MINTZ**
12:45 PM	**LUNCH BREAK**	
1:30 PM	**SESSION VII:**	**SUSY MASSES**
	Moderators:	**RICHARD ARNOWITT,** Texas A&M University
	Dissertators:	**RICHARD ARNOWITT** "Non-Universal Soft SUSY Breaking" **STEPHANE FICHET,** LPNHE, Paris France "Status of Higgses and SUSY Mass Limits at LEP" **CHUNG KAO,** University of Wisconsin "Infrared Fixed Point Results in mSUGRA Models" **STEPHAN MINTZ** "Observing the Weak Interaction in Polarized Parity Violating Electron Scattering"

SUNDAY, December 14, 1997

		JAMES WHITE, Texas A&M University "Supersymmetry at the Tevatron - Current Limits and Future Prospects"
	Annotators:	**RICH PARTRIDGE**
	Session Organizer:	**VERNON BARGER,** University of Wisconsin - Madison
3:30 PM	**COFFEE BREAK**	
3:45 PM	**SESSION VIII:**	**PROGRESS ON NEW AND OLD IDEAS - III**
	Moderators:	**LOUISE DOLAN,** University of North Carolina at Chapel Hill **KATSUMI TANAKA**
	Dissertators:	**LOUISE DOLAN** "The Tube Metric Conformal Field Theory as a String Vacuum" **PRAN NATH,** Northeastern University, Boston, Massachusetts "Nucleon Stability and Dark Matter Constraints on SUSY Unification" **KELLY STELLE,** Imperial College, London "Charge Quantization and P-Brane Spectra" **KATSUMI TANAKA,** Ohio State University "Quark Masses in Dual Theories"
	Annotators:	**STEPHANE FICHET**
	Session Organizer:	**LOUISE DOLAN**

5:30 PM Conference Adjourns for the Day

MONDAY, December 15, 1997

8:30 AM	**SESSION IX:**	**PROGRESS ON NEW AND OLD IDEAS - IV**
	Moderators:	**STEPHEN PINSKY,** Ohio State University **FRED ZACHARIASEN,** CALTECH
	Dissertators:	**FRANCESCO ANTONUCCIO** "DLCQ of Supersymmetric Matrix Models" **BRIAN MEADOWS,** University of Cincinnati "BaBar Experiment at SLAC" **TERRY WALKER,** Ohio State University "Primordial Lithium and the Baryonic Mass of the Universe"
	Annotators:	**ARNOLD PERLMUTTER,** University of Miami
	Session Organizer:	**STEPHEN PINSKY,** Ohio State University
10:15 AM	**COFFEE BREAK**	
10:30 AM	**SESSION X:**	**BIG BANG CREATION OF MASS**
	Moderators:	**EDWARD KOLB**
	Dissertators:	**JOSH FRIEMAN,** FERMI Laboratory "The Distribution of Mass in the Universe" **CHRIS HILL,** FERMI Laboratory "A Brief History of Mass" **PIERRE RAMOND** "The Issue of Mass"
	Annotators:	**BEHRAM KURSUNOGLU**
	Session Organizer:	**EDWARD KOLB**

12:30 PM ORBIS SCIENTIAE 1997 - II ADJOURNS

CONTENTS

SECTION I
Gravitational Mass

SECTION II
Neutrino Masses

SECTION III
Progress on New and Old Ideas

SECTION IV
Particle Masses

SECTION V
Progress on New and Old Ideas II

Physics of Mass

SECTION I
Gravitational Mass

A NEW COSMOLOGICAL PARAMETER SPANNING THE MICROCOSM AND MACROCOSM

Behram N. Kursunoglu
Global Foundation, Inc., P.O. Box 249055
Coral Gables, Florida 33124-9055

INTRODUCTION. The five most important advances, amongst others, in cosmology in the past five decades after Hubble's observation of the expanding universe, and after George Gamow's theory of the Big-Bang creation of the universe, include, in chronological order: (1) Ralph A. Alpher and Robert Herman's[1] theoretical prediction in the 1940's of the cosmic microwave background radiation (CMBR) left over from the Big-Bang; (2) Arno A. Penzias and Robert W. Wilson's observation in 1964 in the residual heat detected as the CMBR; (3) Alan H. Guth's[2] hypothesis in 1979 of an inflationary universe; (4) the observation in 1992 of microwave anisotropies in the CMBR as seen through COBE by physicists and cosmologists led by George Smoot[3,4]; (5) The recent data on distant light from stars, exploded before the sun was born, gives support to the idea of accelerated expansion of the universe which may be destined to continue forever.

There is, amongst physicists and cosmologists, a consensus that the description of the universe and the elementary particles before, during, and after Planck-time of 10^{-43} sec. requires a unified theory of the large and the small, i.e., of cosmology and elementary particle physics. Aside from the recent developments of quantum gravity, the problem of overproduction of monopoles in grand unified theories, accounting for the observed flatness of the universe, the problem of "horizon" were amongst the issues that inspired Alan Guth in 1979 to hypothesize the inflationary behavior of the universe as an exceedingly brief glitch to precede all the events in the Big-Bang cosmology.

EQUATION OF STATE. In this paper we show that the theory resulting from the non-symmetrization of general relativity does in fact predict a flat universe and the neutral magnetic charge confinements of monopoles as the constituents of integral spin particles (bosons of spins 0,1,and 2) and half-integral spin particles like, for example, quarks and leptons. It was a serendipitous observation of the physical significance of a relation, which is now more than forty years old[5],

$$q_o^2\ r_o^2 = c^4/2G, \tag{1}$$

that caused the writing of this paper, where c and G represent speed of light and gravitational constant, respectively, and where r_o is a "fundamental length" and q_o^2 is the

Physics of Mass
Edited by Kursunoglu *et al.*, Kluwer Academic / Plenum Publishers, New York, 1999

energy density, with q_o having the dimensions of an electric field. The relation (1) was the result of correspondence requirement, where in the $r_o=0$ limit the generalized theory of gravitation reduces to general relativity in the presence of the electromagnetic field without the electric and magnetic charges. The energy density q_o^2 in the region of size r_o (Planck-length size) is of the order of 10^{114}ergs.cm^{-3} and it represents the energy density at the beginning of time when the size of the universe was of the order of r_o. Thus, what was obtained more than 40 years ago as a fundamental length r_o only now turned out to be the size, as shown below, of the universe as it evolved from the time of the Big-Bang to the present time. The fundamental relation (1) does in fact govern the general behavior of the universe including the process of nucleosynthesis and the elementary particles as described by the field equations of the theory[6] for the 16 non-symmetric hermitian field variables $\hat{g}_{\mu\nu}= g_{\mu\nu}+ iq_o^{-1}\Phi_{\mu\nu}$ where $g_{\mu\nu}$ and $\Phi_{\mu\nu}$ represent generalized gravitational and

generalized electromagnetic fields, respectively. The theory includes also the *supersymmetric* transition of the hermitian field variables into the 16 non-symmetric non-hermitian field variables $\hat{g}_{\mu\nu}= g_{\mu\nu}+ q_o^{-1}\Phi_{\mu\nu}$, both sets of corresponding field equations reducing in the limit $r_o=0$, to general relativity .

The relation (1) can be written as an analogue of $E = mc^2$ in its density form by

$$q_o^2=\rho_o c^2, \tag{2}$$

where $\rho_o = (c^2/2G)r_0^{-2}$, not to be confused with the function ρ appearing in the field equations below, says that mass density is proportional to the new cosmological parameter r_o^{-2} $(=\lambda)$ of the space time. The quantity λ is not related to Einstein's cosmological constant which excludes the early universe. At the instant of creation of the universe from a region of size r_o with a curvature r_o^{-2} ($\sim10^{66}$cm^{-2}), mass density ρ_o ($\sim10^{95}$gr.cm^{-3}), and energy density (10^{114}erg.cm^{-3}), the corresponding Big-Bang values, as follows by setting $r_o=0$, are infinite. For the present universe, by replacing r_o by r_t and q_t, the same relation (1) applies,

$$q_t^2\, r_t^2 = c^4/2G, \quad q_t^2= \rho_t\, c^2, \tag{3}$$

which is satisfied for the present values of $q_t^2 \sim 10^{-8}$erg.cm^{-3}, $r_t \sim 2.58\times10^{10}$ light years. The total mass of the universe, as follows from (3), $M_u c^2= \frac{1}{4\pi} \int q_t^2\, dV=c^2\int \rho_t dV = (c^4/2G) \int_0^{r_t} (r^2/r^2)dr = (c^4/2G)r_t$ which, with $r_t \sim 2.58\times10^{10}$ light years, yields the value $M_u \sim8\times10^{22}$ solar mass. The relation (1) states that as the universe expands, its average energy density decreases. Furthermore, the relation (2), for two conjugate pairs q_t, r_t, yields the ratio Ω in the form

$$\Omega=q_t^2/q_{t_o}^2 = \rho_t/\rho_{t_o} = (r_{t_o}^2 / r_t^2)\leq 1, \tag{4}$$

where ρ_t and ρ_{t_o} represent the actual and critical densities, respectively. The size r_{t_o} of the universe corresponding to the critical density may be expected to be less than, or at most equal to the size r_t of the universe for the actual density. The theoretical justification of the result (4) requires the use of the standard cosmological model of the present theory, i.e., one that results from the study of the time-dependent spherically symmetric field equations. However, for the present considerations the time-independent spherically symmetric field equations are quite adequate since the equation (1) is valid for all the solutions of the field equations.

SPHERICALLY SYMMETRIC CASE. The field equations for the 16 hermitian field variables for the spherically symmetric case are given[7] by

$$\tfrac{1}{2} r_o^2 \frac{d}{d\beta}[S_B\, \Phi \exp(\rho)] = R_o^2 \sinh \Phi - (-1)^s\, \ell_o^2 \cosh\Phi, \qquad (5)$$

$$\tfrac{1}{2} r_o^2 \frac{d}{d\beta}[S_B \rho \exp(\rho)] = R_o^2 \cosh \Phi - (-1)^s\, \ell_o^2 \sinh\Phi - \exp(\rho), \qquad (6)$$

$$\tfrac{1}{2} r_o^2 \frac{d}{d\beta}[S_B \exp(\rho)] = \exp(\rho)\left[\frac{\cosh\Phi}{\cos\Gamma} - 1\right], \qquad (7)$$

$$\left[\frac{d^2}{d\beta^2} - \tfrac{1}{4}\dot{\Phi}^2\right] \exp(\tfrac{1}{2}\rho) = 0 \qquad (8)$$

where $\dot{\Phi} = \frac{d\Phi}{d\beta}$, $\dot{\rho} = \frac{d\rho}{d\beta}$, $\dot{S}_B = \frac{dS_B}{d\beta}$, and where

$$s = 0,1, \quad R_o^2 - r_o^2 = [\exp(2\rho) - \lambda_o^4]^{1/2}, \; S_B = \exp(u)/\cos^2\Gamma, \; 0 \le \Gamma \le \frac{\pi}{2} \qquad (9)$$

$$\cos\Gamma = \exp(-\rho)\,(R_o^2 - r_o^2) = \lambda_o^2 \exp(-\rho)\,[\lambda^{-4}\exp(2\rho) - 1]^{1/2}, \; -\infty < \Phi < \infty, \qquad (10)$$

$$\exp(\rho) \sin\Gamma = \pm\, \lambda_o^2, \quad \ell_o^2 = g\, q^{-1}, \quad \lambda_o^2 = e\, q^{-1} \qquad (11)$$

and where the constants of integration, λ_o^2 and ℓ_o^2, are identified as functions of electric charge and magnetic charge to obtain the Coulomb's law of force. The approximate solutions of the field equations yield, for the various lengths the results

$$r_o^2 = \frac{2G}{c^4}(e^2 + g^2), \quad \lambda_o^2 = \frac{2G}{c^4} e\sqrt{(e^2 + g^2)}, \quad \ell_o^2 = \frac{2G}{c^4} g\sqrt{(e^2 + g^2)}. \qquad (12)$$

The singularity at $\Gamma = \frac{\pi}{2}$ or $\exp(\rho) = \lambda_o^2$ for the function S_B is invariant and cannot be removed by a coordinate transformation, and the definition of λ_o^2 in (12) shows that $\lambda_o^2 < r_o^2$ and, therefore, the surface of singularity lies inside the region whose horizon has the dimension of r_o. Furthermore, the magnetic charge g differs from the electric charge e in sourcing a short-range interaction only and that it is a *running coupling constant* with a *magnetic horizon* defined by $g(r_c)=0$, where r_c is the short distance beyond which there is no magnetic charge sourced field, no magnetic charge distribution. The field equations (5) - (8) at the origin are solved by $\exp(\rho) = 0$, $\Phi(0) = 0$, $\lambda_o^2 = 0$, $\ell_o^2 = 0$, $r_o^2 = 0$. With the function S_B assuming arbitrary values. At the current epoch of the universe the parameter r_o can be assigned the value of the size of the universe, and therefore, dividing the field equations by r_o and using its evolved value of 10^{10} light years, we find that the field equations are satisfied by the flat space-time values

$$v = \pm 1, \quad \Phi(r_t) = 0, \; \exp(\rho) = \beta^2, \quad \Gamma = 0, \; S = 1 \;, \qquad (13)$$

where the remaining term $-(-1)^s\, \ell_o^2\, r_o^{-2} = -(-1)^s\, g\, q\, \frac{2G}{c^4}$ on the right hand side of the equation (5), because of the small factor $\frac{2G}{c4}$ (~10^{-49}) and the small energy density q^2 (~ 10^{-8}erg.cm^{-3}), is not affected by the size of the magnetic charge and, therefore, is negligibly small. Thus, *the universe is flat*. The solution (13) if used in the line element

$$ds_B{}^2 = S_B \cos^2\Gamma\, dx_o{}^2 - S_B{}^{-1} d\beta^2 - \exp(\rho) \cosh\Phi\, d\Omega_0{}^2, \tag{14}$$

where $d\Omega_0{}^2 = d\theta^2 + \sin^2\theta\, d\varphi^2$, and the variable β is related to the usual radial coordinate by the definition $dr = f\, d\beta$, $f = v \cos\Gamma$, yields the flat space-time metric $ds^2 = dx_0^2 - dr^2 - r^2\, d\Omega_0^2$. The function S_B is defined by

$$S_B = \frac{\lambda_0{}^{-4} \exp(u + 2\rho)}{[\lambda_0{}^{-4} \exp(2\rho) - 1]} \tag{15}$$

The field equations (5)-(8) being invariant under the transformations

$$\Phi \rightarrow -\Phi, \quad g \rightarrow -g, \tag{16}$$

have only magnetic dipole solutions with equal and opposite signs of magnetic charges, i.e., solutions with magnetic charge distributions, where there is no limitation on the amount of magnetic charge g. There are in fact no free monopole solutions. A more general proof of the absence of free monopole is given in reference 6, equation (184). However, prior to the time 10^{-43} sec. the primordial *Bose-Einstein fluid* of monopoles of all fractional spins and sizes with neutral distribution in an equilibrium state at a temperature of the order of 10^{30}K could have deviated from the equilibrium state by a *transient gravitational repulsion* arising from a singularity, as seen from the definition (15) of S_B, inside the region of the dimension of r_o, which could have energized the explosive creation of the universe, where $-\lambda_0^2 \leq \exp(\rho) \leq \lambda_0^2$. The interior singularity is at $\exp(\rho)=\lambda_0^2$, and S_B is the negative in the interval of the singularity.

In the hot region of dimension r_o the *Bose-Einstein fluid* of monopoles and electric charges is not the only equilibrium state but a *super-symmetric transition* to acquire a different neutral distribution of monopoles with a different statistics can be obtained uniquely by the transformations,

$$\Phi \rightarrow i\Phi + \frac{\pi}{2}, \quad \Gamma \rightarrow i\Gamma, \quad r_0 \rightarrow ir_0 \tag{17}$$

$$q^{-1} \rightarrow iq^{-1}, \quad \ell_o^2 \rightarrow i\ell_o^2, \quad \lambda_o{}^2 \rightarrow i\lambda_o{}^2 \tag{18}$$

leading to the field equations

$$\tfrac{1}{2} r_o{}^2 \frac{d}{d\beta}[\, S_F \exp(\rho)\, \Phi\,] = R_o{}^2 \cos\Phi + \ell_o^2 \sin\Phi, \tag{19}$$

$$\tfrac{1}{2} r_o{}^2 \frac{d}{d\beta}\, [\, S_F \exp(\rho)\ \rho\,] = -R_o{}^2 \sin\Phi + \ell_o^2 \cos\Phi + \exp(\rho), \tag{20}$$

$$\tfrac{1}{2} r_o{}^2 \frac{d}{d\beta}[\, S_F \exp(\rho)\,] = \exp(\rho)\, [\, 1 - \frac{\sin\Phi}{\cosh\Gamma}], \tag{21}$$

$$[\frac{d^2}{d\beta^2} + \tfrac{1}{4}\dot{\Phi}^2] \exp(\tfrac{1}{2}\rho) = 0, \tag{22}$$

where $f = v \cosh\Gamma$, $\cosh\Gamma = \exp(-\rho)(R_0^2 + r_0^2)$, $R_0^2 + r_0^2 = [\exp(2\rho) + \lambda_0^4]^{1/2}$ and where Φ is now an angle restricted by $0 \le \Phi \le \frac{\pi}{2}$. The new gravitational potential

$$S_F = \frac{\lambda_0^{-4} \exp(u+2\rho)}{\lambda_0^{-4} \exp(2\rho) + 1} \tag{23}$$

is a positive function and differs from S_B, defined in (15), by the +1 sign in the denominator versus -1 in that of S_B. In this case the neutral magnetic charge distribution is synthesized in an infinitely layered form of infinite number of monopoles of decreasing magnitudes with distance and of alternating signs, where $\sum_0^\infty g_n = 0$ with g_n representing the positive or negative magnetic charge in the n-th. layer. The *change of statistics* resulting from the cooling of the *Bose-Einstein fluid* and synthesizing of the monopoles creates either quarks with spin ½, based on an Ansatz $\sum_0^\infty g_n^2 = \frac{1}{2}\hbar c$, or leptons, will depend on the interpretation of the *generalized Dirac Wave equation* obtained from this theory[8]. The field equations (19)-(22) also in the limit $r_0 = 0$ reduce to the *spherically symmetric* field equations of general relativity. In the limit of r_0 evolving into the value at the present cosmological epoch we find that the field equations (19)-(22) have the flat space-time solutions,

$$\phi = \frac{\pi}{2}, \ \exp(\rho) = r^2, \ \Gamma = 0, \ S_F = 1 \tag{24}$$

The fact that the universe began with a structure as described by its primordial size r_0 and finite energy density q_0^2 could be related to temperature fluctuations in CMBR and to the structure formation in the large. Finally, in view of the present theory's prediction of the magnetic charge structure of all matter, luminous or dark, the assumption of the ratio of dark matter to luminous matter being proportional to $\ell_0^2 / \lambda_0^2 = g/e = \frac{eg}{e^2} = \frac{1}{2} n \frac{\hbar c}{e^2}$ may not be too far off the actual reality of the dark matter cosmology, where we used Dirac's relation $eg = \frac{1}{2} n\hbar c$ for a free monopole. Alternatively we can obtain the same result from summing over all fractional spins in

$$\sum_0^\infty (g_n^2/e^2) = \frac{1}{2} \frac{\hbar c}{e^2} \tag{25}$$

The author wishes to thank Ralph A. Alpher and the late Robert Herman for many discussions that have greatly improved this paper.

REFERENCES

[1] Ralph A. Alpher and Robert Herman "Remembrance of Things Past" in proceedings of the Coral Gables Conference "*Unified Symmetry: In the small and in the large*,"1993.Nova Science Publishers, New York eds. Behram N. Kursunoglu and Arnold Perlmutter.

[2] Alan A. Guth, Phys. Rev. D.23, 347 (1981).

[3] George F. Smoot et al., in "After the First Three Minutes" eds. S.S. Holt, C.L. Bennet, and Virginia Trimble (New York: AIP Conference Proc. 222), 95 (1991 C). See also J.C. Mather et al. In the same proceedings.

[4] J.C. Mather in the proceedings of the Coral Gables Conference in "*Unified Symmetry: In the small and in the large*," 1993, Nova Science Publishers, New York, eds. Behram N. Kursunoglu and Arnold Perlmutter.

[5] For more references, advances, and recent developments, see Behram N. Kursunoglu, Journal of Physics Essays Vol. 4, No.4, pp 439-518, 1991. Behram N. Kursunoglu, Phys Rev. 88, 1369 (1952); proceedings of Les Theories Relativistes de la Gravitation, edited by M.S. Lichnerowicz (Royaumont, 1959), p.359; Rev. Mod. Phys. 29, 412 (1957); Nuovo Cimento 15, 729 (1960); Phys. Rev. D9, 2723 (1974); ibid. 12, 1850(E) (1975); Phys. Rev. D13 1538 (1976); ibid D14, 1518 (1974).

[6] See in reference 4 above, the generalized Dirac wave equation in B.N. Kursunoglu's paper, "The Ascent of Gravity."

TOP QUARK AND ELECTROWEAK MASS

Christopher T. Hill
Fermi National Accelerator Laboratory
P.O. Box 500, Batavia, Illinois, 60510
The Department of Physics and Enrico Fermi Institute
The University of Chicago
Chicago, Illinois

Abstract. We describe a class of models of electroweak symmetry breaking that involve strong dynamics and top quark condensation. A new scheme based upon a seesaw mechanism appears particularly promising. Various implications for the first-stage muon collider are discussed.

TOPCOLOR I

The top quark mass may be large because it is a combination of a *dynamical condensate component*, $(1-\epsilon)m_t$, generated by a new strong dynamics [1], together with a small *fundamental component*, ϵm_t, i.e, $\epsilon << 1$, generated by something else. The most obvious "handle" on the top quark for new dynamics is the color index. Invoking new dynamics involving the top quark color index leads directly to a class of Technicolor-like models incorporating "Topcolor". We expect in such schemes that the new strong dynamics occurs primarily in interactions that involve $\bar{t}t\bar{t}t$, $\bar{t}t\bar{b}b$, and $\bar{b}b\bar{b}b$.

In Topcolor I the dynamics at the ~ 1 TeV scale involves the following structure at the TeV scale (or a generalization thereof) [2]:

$$SU(3)_1 \times SU(3)_2 \times U(1)_{Y1} \times U(1)_{Y2} \times SU(2)_L \rightarrow SU(3)_{QCD} \times U(1)_{EM} \quad (1)$$

where $SU(3)_1 \times U(1)_{Y1}$ ($SU(3)_2 \times U(1)_{Y2}$) generally couples preferentially to the third (first and second) generations. The $U(1)_{Yi}$ are just strongly rescaled versions of electroweak $U(1)_Y$.

The fermions are then assigned $(SU(3)_1, SU(3)_2, Y_1, Y_2)$ quantum numbers in the following way:

$$\begin{array}{ll} (t,b)_L \sim (3,1,1/3,0) & (t,b)_R \sim (3,1,(4/3,-2/3),0) \\ (\nu_\tau,\tau)_L \sim (1,1,-1,0) & \tau_R \sim (1,1,-2,0) \end{array} \quad (2)$$

$$\begin{array}{ll} (u,d)_L,\ (c,s)_L \sim (1,3,0,1/3) & (u,d)_R,\ (c,s)_R \sim (1,3,0,(4/3,-2/3)) \\ (\nu,\ell)_L\ \ \ell = e,\mu \sim (1,1,0,-1) & \ell_R \sim (1,1,0,-2) \end{array}$$

Topcolor must be broken, which we describe by an (effective) scalar field:

$$\Phi \sim (3, \bar{3}, y, -y) \tag{3}$$

When Φ develops a VEV, it produces the simultaneous symmetry breaking

$$SU(3)_1 \times SU(3)_2 \rightarrow SU(3)_{QCD} \qquad \text{and} \qquad U(1)_{Y1} \times U(1)_{Y2} \rightarrow U(1)_Y \tag{4}$$

$SU(3)_1 \times U(1)_{Y1}$ is assumed to be strong enough to form chiral condensates which will be "tilted" in the top quark direction by the $U(1)_{Y1}$ couplings. The theory is assumed to spontaneously break down to ordinary QCD $\times U(1)_Y$ at a scale of ~ 1 TeV, before it becomes confining. The isospin splitting that permits the formation of a $\langle \bar{t}t \rangle$ condensate but disables the $\langle \bar{b}b \rangle$ condensate is due to the $U(1)_{Yi}$ couplings. The b-quark mass in this scheme can arise from a combination of ETC effects and instantons in $SU(3)_1$. The θ-term in $SU(3)_1$ may manifest itself as the CP-violating phase in the CKM matrix. Above all, the new spectroscopy of such a system should begin to materialize indirectly in the third generation, perhaps at the Tevatron in top and bottom quark production, or possibly in a muon collider.

The symmetry breaking pattern outlined above will generically give rise to three (pseudo)-Nambu-Goldstone bosons $\tilde{\pi}^a$, or "top-pions", near the top mass scale. *This is the smoking gun of Topcolor.* [We were led to Topcolor by considering how strong dynamics might produce the analog of the decay $t \rightarrow H^+ + b$, considered to be a SUSY signature for a charged Higgs-boson H. This is an example of "SUSY-Technicolor/Topcolor duality".] If the Topcolor scale is of the order of 1 TeV, the top-pions will have a decay constant of $f_\pi \approx 50$ GeV, and a strong coupling given by a Goldberger-Treiman relation, $g_{tb\pi} \approx m_t/\sqrt{2} f_\pi \approx 2.5$, potentially observable in $\tilde{\pi}^+ \rightarrow t + \bar{b}$ if $m_{\tilde{\pi}} > m_t + m_b$.

We assume presently that ESB can be primarily driven by a Higgs sector or Technicolor, with gauge group G_{TC}[3] [4]. This gives the $\mathcal{O}(\epsilon)$ component of m_t. Technicolor can also provide condensates which generate the breaking of Topcolor to QCD and $U(1)_Y$.

The coupling constants (gauge fields) of $SU(3)_1 \times SU(3)_2$ are respectively h_1 and h_2 ($A^A_{1\mu}$ and $A^A_{2\mu}$) while for $U(1)_{Y1} \times U(1)_{Y2}$ they are respectively q_1 and q_2, ($B_{1\mu}, B_{2\mu}$). The $U(1)_{Yi}$ fermion couplings are then $q_i \frac{Y_i}{2}$, where Y_1, Y_2 are the charges of the fermions under $U(1)_{Y1}, U(1)_{Y2}$ respectively.

Topcolor I produces new gauge heavy bosons Z', and "colorons" B^A with couplings to fermions given by:

$$\mathcal{L}_{Z'} = g_1 (Z' \cdot J_{Z'}) \qquad \mathcal{L}_B = g_3 \cot\theta (B^A \cdot J^A_B) \tag{5}$$

where the currents $J_{Z'}$ and J_B in general involve all three generations of fermions

$$\begin{aligned} J_{Z'} &= -(J_{Z',1} + J_{Z',2}) \tan\theta' + J_{Z',3} \cot\theta' \\ J_B &= -(J_{B,1} + J_{B,2}) \tan\theta + J_{B,3} \cot\theta \end{aligned} \tag{6}$$

For example, for the third generation the currents read explicitly (in a weak eigenbasis):

$$J^\mu_{Z',3} = \frac{1}{6}\bar{t}_L \gamma^\mu t_L + \frac{1}{6}\bar{b}_L \gamma^\mu b_L + \frac{2}{3}\bar{t}_R \gamma^\mu t_R - \frac{1}{3}\bar{b}_R \gamma^\mu b_R$$

$$-\frac{1}{2}\bar{\nu}_{\tau L}\gamma^{\mu}\nu_{\tau L} - \frac{1}{2}\bar{\tau}_L\gamma^{\mu}\tau_L - \bar{\tau}_R\gamma^{\mu}\tau_R$$

$$J_{B,3}^{A,\mu} = \bar{t}\gamma^{\mu}\frac{\lambda^A}{2}t + \bar{b}\gamma^{\mu}\frac{\lambda^A}{2}b \tag{7}$$

where λ^A is a Gell-Mann matrix acting on color indices. We ultimately demand $\cot\theta \gg 1$ and $\cot\theta' \gg 1$ to select the top quark direction for condensation.

The attractive Topcolor interaction, for sufficiently large $\kappa = g_3^2 \cot^2\theta/4\pi$, would by itself trigger the formation of a low energy condensate, $\left\langle \bar{t}t + \bar{b}b \right\rangle$, which would break $SU(2)_L \times SU(2)_R \times U(1)_Y \to U(1) \times SU(2)_c$, where $SU(2)_c$ is a global custodial symmetry. On the other hand, the $U(1)_{Y1}$ force is attractive in the $\bar{t}t$ channel and repulsive in the $\bar{b}b$ channel. Thus, to make $\left\langle \bar{b}b \right\rangle = 0$ and $\langle \bar{t}t \rangle \neq 0$ we can have in concert critical and subcritical values of the combinations:

$$\kappa + \frac{2\,\kappa_1}{9N_c} > \kappa_{crit}; \qquad \kappa_{crit} > \kappa - \frac{\kappa_1}{9N_c}; \tag{8}$$

Here N_c is the number of colors and $\kappa_1 = g_1^2 \cot^2\theta'/4\pi$. (It should be mentioned that our analyses are performed in the context of a large-N_c approximation). This leads to "tilted" gap equations in which the top quark acquires a constituent mass, while the b quark remains massless. Given that both κ and κ_1 are large there is no particular fine tuning occuring here, only "rough-tuning" of the desired tilted configuration. Of course, the NJL approximation is crude, but as long as the associated phase transitions of the real strongly coupled theory are approximately second order, analogous rough-tuning in the full theory is possible.

TOPCOLOR II

If the above described "Topcolor I" is the analog of Weinberg's original version of the SM, incorporating standard fermions and the Z-boson, then Topcolor II is the analog of the original Georgi-Glashow model, which incorporated no new Z boson, but rather included additional fermions. [This is an example of "Weinberg– Georgi-Glashow" duality.] The strong $U(1)$ is present in the previous scheme to avoid a degenerate $\langle \bar{t}t \rangle$ with $\left\langle \bar{b}b \right\rangle$. However, we can give a model in which there is: (i) a Topcolor $SU(3)$ group but (ii) no strong $U(1)$ with (iii) an anomaly-free representation content. In fact the original model of [2] was of this form, introducing a new quark of charge $-1/3$. Let us consider a generalization of this scheme which consists of the gauge structure $SU(3)_Q \times SU(3)_1 \times SU(3)_2 \times U(1)_Y \times SU(2)_L$. We require an additional triplet of fermions fields (Q_R^a) transforming as $(3,3,1)$ and $Q_L^{\dot{a}}$ transforming as $(3,1,3)$ under the $SU(3)_Q \times SU(3)_1 \times SU(3)_2$.

The fermions are then assigned the following quantum numbers in $SU(2) \times SU(3)_Q \times SU(3)_1 \times SU(3)_2 \times U(1)_Y$:

$$\begin{aligned}
(t,b)_L \;\; (c,s)_L &\sim (2,1,3,1) \qquad Y = 1/3 \\
(t)_R &\sim (1,1,3,1) \qquad Y = 4/3; \\
(Q)_R &\sim (1,3,3,1) \qquad Y = 0 \\
(u,d)_L &\sim (2,1,1,3) \qquad Y = 1/3
\end{aligned}$$

$$
\begin{aligned}
(u,d)_R \ \ (c,s)_R &\sim (1,1,1,3) \qquad Y=(4/3,-2/3) \\
(\nu,\ell)_L \ \ \ell=e,\mu,\tau &\sim (2,1,1,1) \qquad Y=-1; \\
(\ell)_R &\sim (1,1,1,1) \qquad Y=-2 \\
b_R &\sim (1,1,1,3) \qquad Y=2/3; \\
(Q)_L &\sim (1,3,1,3) \qquad Y=0;
\end{aligned} \tag{9}
$$

Thus, the Q fields are electrically neutral. One can verify that this assignment is anomaly free.

The $SU(3)_Q$ confines and forms a $\langle \bar{Q}Q \rangle$ condensate which acts like the Φ field and breaks the Topcolor group down to QCD dynamically. We assume that Q is then decoupled from the low energy spectrum by its large constituent mass. There is a lone $U(1)$ Nambu–Goldstone boson $\sim \bar{Q}\gamma^5 Q$ which acquires a large mass by $SU(3)_Q$ instantons.

TRIANGULAR TEXTURES

The texture of the fermion mass matrices will generally be controlled by the symmetry breaking pattern of a horizontal symmetry. In the present case we are specifying a residual Topcolor symmetry, presumably subsequent to some initial breaking at some scale Λ, large compared to Topcolor, e.g., the third generation fermions in Model I have different Topcolor assignments than do the second and first generation fermions. Thus the texture will depend in some way upon the breaking of Topcolor [5] [3].

Let us study a fundamental Higgs boson, which ultimately breaks $SU(2)_L \times U(1)_Y$, together with an effective field Φ breaking Topcolor as in eq.(4). We must now specify the full Topcolor charges of these fields. As an example, under $SU(3)_1 \times SU(3)_2 \times U(1)_{Y1} \times U(1)_{Y2} \times SU(2)_L$ let us choose:

$$
\Phi \sim (3, \bar{3}, \frac{1}{3}, -\frac{1}{3}, 0) \qquad H \sim (1, 1, 0, -1, \frac{1}{2}) \tag{10}
$$

The effective couplings to fermions that generate mass terms in the up sector are of the form

$$
\begin{aligned}
\mathcal{L}_{M_U} = {} & m_0 \bar{t}_L t_R + c_{33}\bar{T}_L t_R H \frac{\det \Phi^\dagger}{\Lambda^3} + c_{32}\bar{T}_L c_R H \frac{\Phi}{\Lambda} + c_{31}\bar{T}_L u_R H \frac{\Phi}{\Lambda} \\
& + c_{23}\bar{C}_L t_R H \Phi^\dagger \frac{\det \Phi^\dagger}{\Lambda^4} + c_{22}\bar{C}_L c_R H + c_{21}\bar{C}_L u_R H \\
& + c_{13}\bar{F}_L t_R H \Phi^\dagger \frac{\det \Phi^\dagger}{\Lambda^4} + c_{12}\bar{F}_L c_R H + c_{11}\bar{F}_L u_R H + \text{h.c.}
\end{aligned} \tag{11}
$$

Here $T = (t,b)$, $C = (c,s)$ and $F = (u,d)$. The mass m_0 is the dynamical condensate top mass. Furthermore $\det \Phi$ is defined by

$$
\det \Phi \equiv \frac{1}{6}\epsilon_{ijk}\epsilon_{lmn}\Phi_{il}\Phi_{jm}\Phi_{kn} \tag{12}
$$

where in Φ_{rs} the first(second) index refers to $SU(3)_1$ ($SU(3)_2$). The matrix elements now require factors of Φ to connect the third with the first or second generation color indices. The down quark and lepton mass matrices are generated by couplings analogous to (11).

To see what kinds of textures can arise naturally, let us assume that the ratio Φ/Λ is small, $O(\epsilon)$. The field H acquires a VEV of v. Then the resulting mass matrix is approximately triangular:

$$\begin{pmatrix} c_{11}v & c_{12}v & \sim 0 \\ c_{21}v & c_{22}v & \sim 0 \\ c_{31}O(\epsilon)v & c_{32}O(\epsilon)v & \sim m_0 + O(\epsilon^3)v \end{pmatrix} \tag{13}$$

where we have kept only terms of $\mathcal{O}(\epsilon)$ or larger.

This is a triangular matrix (up to the c_{12} term). When it is written in the form $U_L \mathcal{D} U_R^\dagger$ with U_L and U_R unitary and $\mathcal{D}$ positive diagonal, there automatically result restrictions on U_L and U_R. In the present case, the elements $U_L^{3,i}$ and $U_L^{i,3}$ are vanishing for $i \neq 3$, while the elements of U_R are not constrained by triangularity. Analogously, in the down quark sector $D_L^{i,3} = D_L^{3,i} = 0$ for $i \neq 3$ with D_R unrestricted. The situation is reversed when the opposite corner elements are small, which can be achieved by choosing $H \sim (1, 1, -1, 0, \frac{1}{2})$.

These restrictions on the quark mass rotation matrices have important phenomenological consequences. For instance, in the process $B^0 \to \overline{B^0}$ there are potentially large contributions from top-pion and coloron exchange. However, these contributions are proportional to the product $D_L^{3,2} D_R^{3,2}$. The same occurs in $D^0 - \bar{D}^0$ mixing, where the effect goes as products involving U_L and U_R off-diagonal elements. Therefore, triangularity can naturally select these products to be small.

The precise selection rules depend upon the particular symmetry breaking that occurs. This example is merely illustrative of the systematic effects that can occur in such schemes.

TOP-PIONS; INSTANTONS; THE B-QUARK MASS

Since the top condensation is a spectator to the TC (or Higgs) driven ESB, there must occur a multiplet of top-pions. A chiral Lagrangian can be written:

$$L = i\overline{\psi}\partial\!\!\!/ \psi - m_t(\overline{\psi}_L \Sigma P \psi_R + h.c.) - \epsilon m_t \overline{\psi} P \psi, \qquad P = \begin{pmatrix} 1 & 0 \\ 0 & 0 \end{pmatrix} \tag{14}$$

and $\psi = (t, b)$, and $\Sigma = \exp(i\tilde{\pi}^a \tau^a / \sqrt{2} f_\pi)$. With $\epsilon = 0$ this is invariant under $\psi_L \to e^{i\theta^a \tau^a / 2} \psi_L$, $\tilde{\pi}^a \to \tilde{\pi}^a + \theta^a f_\pi / \sqrt{2}$. Hence, the relevant currents are left-handed, $j_\mu^a = \psi_L \gamma_\mu \frac{\tau^a}{2} \psi_L$, and $< \tilde{\pi}^a | j_\mu^b | 0 >= \frac{f_\pi}{\sqrt{2}} p_\mu \delta^{ab}$. The Pagels-Stokar relation, eq.(1), then follows by demanding that the $\tilde{\pi}^a$ kinetic term is generated by integrating out the fermions. The top-pion decay constant estimated from eq.(1) using $\Lambda = M_B$ and $m_t = 175$ GeV is $f_\pi \approx 50$ GeV. The couplings of the top-pions take the form:

$$\frac{m_t}{\sqrt{2} f_\pi} \left[i\bar{t}\gamma^5 t \tilde{\pi}^0 + \frac{i}{\sqrt{2}} \bar{t}(1 - \gamma^5) b \tilde{\pi}^+ + \frac{i}{\sqrt{2}} \bar{b}(1 + \gamma^5) t \tilde{\pi}^- \right] \tag{15}$$

and the coupling strength is governed by the relation $g_{bt\tilde{\pi}} \approx m_t / \sqrt{2} f_\pi$.

The small ETC mass component of the top quark implies that the masses of the top-pions will depend upon ϵ and Λ. Estimating the induced top-pion mass from the fermion loop yields:

$$m_{\tilde{\pi}}^2 = \frac{N \epsilon m_t^2 M_B^2}{8\pi^2 f_\pi^2} = \frac{\epsilon M_B^2}{\log(M_B / m_t)} \tag{16}$$

where the Pagels-Stokar formula is used for f_π^2 (with $k = 0$) in the last expression. For $\epsilon = (0.03,\ 0.1)$, $M_B \approx (1.5,\ 1.0)$ TeV, and $m_t = 180$ GeV this predicts $m_{\tilde{\pi}} = (180,\ 240)$ GeV. The bare value of ϵ generated at the ETC scale Λ_{ETC}, however, is subject to very large radiative enhancements by Topcolor and $U(1)_{Y1}$ by factors of order $(\Lambda_{ETC}/M_B)^p \sim 10^1$, where the $p \sim O(1)$. Thus, we expect that even a bare value of $\epsilon_0 \sim 0.005$ can produce sizeable $m_{\tilde{\pi}} > m_t$. Note that $\tilde{\pi}$ will generally receive gauge contributions to it's mass; these are at most electroweak in strength, and therefore of order ~ 10 GeV.

Top-pions can be as light as ~ 150 GeV, in which case they would emerge as a detectable branching fraction of top decay [6]. However, there are dangerous effects in $Z \to b\bar{b}$ with low mass top pions and decay constamnts as small as ~ 60 GeV [8]. A more comfortable phenomenological range is slightly larger than our estimates, $m_{\tilde{\pi}} \gtrsim 300$ GeV and $f_\pi \gtrsim 100$ GeV.

The b quark receives mass contributions from ETC of $O(1)$ GeV, but also an induced mass from instantons in $SU(3)_1$. The instanton effective Lagrangian may be approximated by the 't Hooft flavor determinant (we place the cut-off at M_B):

$$L_{eff} = \frac{k}{M_B^2} e^{i\theta_1} \det(\bar{q}_L q_R) + h.c. = \frac{k}{M_B^2} e^{i\theta_1} [(\bar{b}_L b_R)(\bar{t}_L t_R) - (\bar{t}_L b_R)(\bar{b}_L t_R)] + h.c. \tag{17}$$

where θ_1 is the $SU(3)_1$ strong CP-violation phase. θ_1 cannot be eliminated because of the ETC contribution to the t and b masses. It can lead to induced scalar couplings of the neutral top pion [5], and an induced CKM CP-phase, however, we will presently neglect the effects of θ_1.

We generally expect $k \sim 1$ to 10^{-1} as in QCD. Bosonizing in fermion bubble approximation $\bar{q}_L^i t_R \sim \frac{N}{8\pi^2} m_t M_B^2 \Sigma_1^i$, where $\Sigma_j^i = \exp(i\tilde{\pi}^a \tau^a/\sqrt{2} f_\pi)_j^i$ yields:

$$L_{eff} \to \frac{Nkm_t}{8\pi^2} e^{i\theta} [(\bar{b}_L b_R)\Sigma_1^1 + (\bar{t}_L b_R)\Sigma_1^2 + h.c.] \tag{18}$$

This implies an instanton induced b-quark mass:

$$m_b^\star \approx \frac{3km_t}{8\pi^2} \sim 6.6\ k\ GeV \tag{19}$$

This is not an unreasonable estimate of the observed b quark mass, as we might have feared it would be too large.

TOP SEE-SAW

EWSB may occur via the condensation of the top quark in the presence of an extra vectorlike, weak-isoscalar quark [7]. The mass scale of the condensate is large, of order 0.6 TeV corresponding to the electroweak scale $f_\pi \approx 175$ GeV. The vectorlike iso-scalar then naturally admits a seesaw mechanism, yielding the physical top quark mass, which is then adjusted to the experimental value. The choice of a natural ~TeV scale for the topcolor dynamics then determines the mass of the weak-isoscalar see-saw partner. The scheme is economical, requiring no additional weak isodoublets, and therefore easily satisfies the constraints upon the S parameter using estimates made in the large-N approximation. The constraints on custodial symmetry violation, i.e., the value of the $\delta\rho$ or equivalently, T parameter, are easily satisfied, being principally the usual m_t contribution, plus corrections that are suppressed by the see-saw mechanism.

The dynamical fermion masses that are induced can be written as:

$$\mathcal{L} = -\,(\bar{t}_L\ ,\ \overline{\chi}_L)\begin{pmatrix} 0 & m_{t\chi} \\ m_{\chi t} & m_{\chi\chi} \end{pmatrix}\begin{pmatrix} t_R \\ \chi_R \end{pmatrix} + \text{h.c.} \tag{20}$$

Typically $\overline{\chi}_L \chi_R$ is the most attractive channel, and it is possible to arrange the $\langle \overline{\chi}_L \chi_R \rangle$ condensate to be significantly larger than the other ones, such that $m^2_{\chi\chi} \gg m^2_{\chi t} > m^2_{t\chi}$. As a result the physical top mass is suppressed by a seesaw mechanism:

$$m_t \approx \frac{m_{\chi t} m_{t\chi}}{m_{\chi\chi}}\left[1 + O\left(m^2_{\chi t,\, t\chi}/m^2_{\chi\chi}\right)\right] \ . \tag{21}$$

The electroweak symmetry is broken by the $m_{t\chi}$ dynamical mass. Therefore, the electroweak scale is estimated to be given by

$$v^2 \approx \frac{3}{16\pi^2} m^2_{t\chi} \ln\left(\frac{M}{m_{t\chi}}\right) \ . \tag{22}$$

Thus, $v \approx 174$ GeV requires a dynamical mass $m_{t\chi} \sim 620$ GeV for $M \sim 5$ TeV (and $m_{t\chi} = 520$ GeV for $M \sim 10$ TeV). From eq. (21) follows then that a top mass of 175 GeV requires $m_{\chi t}/m_{\chi\chi} \approx 0.29$. The electroweak T parameter can be estimated in fermion–bubble large-N approximation as:

$$T \approx \frac{3m_t^2}{16\pi^2 \alpha(M_Z^2) v^2} \frac{m^2_{t\chi}}{m^2_{\chi t}}\left[1 + O\left(m^2_{\chi t,\, t\chi}/m^2_{\chi\chi}\right)\right] \ , \tag{23}$$

where α is the fine structure constant. Moreover, we obtain the usual Standard Model result for the S parameter. Requiring that our model does not exceed the 1σ upper bound on S and T, we obtain $m_{t\chi}/m_{\chi t} \leq 0.55$.

It should be emphasized that these results do not require excessive fine-tuning. The top-seesaw is therefore a plausible natural theory of dynamical EWSB with a minimal number of new degrees of freedom. This model also implies the existence of pseudo–Nambu-Goldstone bosons (pNGB's). A cursory discussion of that is given in ref.[7].

OBSERVABLES

There are several classes of possible experimental implications of the kinds of models we described above that may be relevant to the muon collider. We will describe them here briefly as lines to be developed further. These may be enumerated as follows:

1. $\mu\bar{\mu} \to Z'$: this is the province of high energy machine, since we expect $M_{Z'} \gtrsim 0.5$ TeV.

2. $\mu\bar{\mu} \to \pi_{top}$: the notion that the muon collider can see technipions, or other PNGB's, such as top-pions has emerged from discussions in this workshop, prompted by MacKenzie and myself. Lane has presented the multi-scale technicolor signal [4].

3. Effects in Z physics involving the third generation, such as $Z \to b\bar{b}$ [8].

4. Effects in top-quark pair production at threshold, e.g., see [11] for analogous case in e^+e^- and $p\bar{p}$ collider physics.

5. Induced GIM violation in low energy processes such as $K^+ \to \pi^+ \nu\bar{\nu}$; we discuss this below as an example of a potential signature that can be enhanced by Topcolor wrt the Standard Model (this result was anticipated in ref[5] before the observation of the single event at Brookhaven E787).

6. Induced lepton family number violation, e.g. $\mu\bar{\mu} \to \tau\bar{\mu}$.

7. Flavor dependent production effects, e.g. anomalous $\mu\bar{\mu} \to b\bar{s}$, etc.

8. New physics in e.g. μp collisions, such as $d(u) + \bar{\mu} \to b(t) + \bar{\tau}$.

GIM and lepton family number violation arise because of the generational structure of topcolor. (It is actually more general than topcolor; the mere statement than the top mass is largely dynamical implies effects like this) In going to the mass eigenbasis, quark (and lepton) fields are rotated, e.g., by the matrices U_L, U_R (for the up-type left and right handed quarks) and D_L, D_R (for the down-type left and right handed quarks). For example, for the b-quark we make the replacement

$$b_L \to D_L^{bb} b_L + D_L^{bs} s_L + D_L^{bd} d_L \tag{24}$$

and analogously for b_R. Thus there will be induced FCNC interactions. This provides constraints and opportunities. Thus, induced effects like $\mu\bar{\mu} \to b\bar{s}$ may be enhanced, and effects like $\mu\bar{\mu} \to \tau\bar{\mu}$ may occur. Since the muon is presumably closer in affiliation to the third generation than is the electron, such effects may show up in muon collider physics, but be inaccessible in electron linear colliders! Similarly, induced effects like $\mu\bar{\mu} \to b\bar{s}$ may be enhanced.

For the FMC, sensitive probes arise in e.g., K-physics. there is a Z' induced contact term at low energies of the form $b\bar{b}\nu_\tau\bar{\nu}_\tau$ (this assumes that the τ is associated with the third generation; nothing fundamentally compels this, but we shall assume it to be true in the following). The above mass rotation induces a $s\bar{d}\nu_\tau\bar{\nu}_\tau$ which contributes to $K^+ \to \pi^+\nu\bar{\nu}$. The ratio of the Topcolor amplitude to the SM is then

$$\frac{\mathcal{A}^{TC}}{\mathcal{A}^{SM}} = -\left(\frac{g_1 \cot\theta'}{M_{Z'}}\right)^2 \frac{\sqrt{2}\pi \sin^2\theta_W}{24\alpha\, G_F} \frac{\delta_{ds}}{\sum_j V_{js}^* V_{jd} D_j(x_j)} \sim -3 \times 10^9\, \delta_{ds} \frac{\kappa_1}{M_{Z'}^2} \tag{25}$$

where $\delta_{ds}^* = D_L^{bs} D_L^{bd*} - 2 D_R^{bs} D_R^{bd*}$. The form-factor $f_+(q^2)$ is experimentally well known. We expect, $|\delta_{ds}| \sim \lambda^{10}$ where λ is the Wolfenstein CKM parameter. For $M_{Z'} = 500$ GeV and $\kappa_1 = 1$ the ratio of amplitudes is about ~ 4.0, and the branching ratio is between 0.3 to $O(10)$, times the SM result, depending on the sign of the interference. The recent observation of one event by the Brookhaven E787 Collaboration [10] makes this an exciting channel in which to search for new physics. High sensitivity experiments are possible at the front-end muon collider with its copious K-meson yields.

REFERENCES

1. Y. Nambu, "BCS Mechansim, Quasi-Supersymmetry, and Fermion Mass Matrix," Talk presented at the Kasimirz Conference, EFI 88-39 (July 1988); "Quasi-Supersymmetry, Bootstrap Symmetry Breaking, and Fermion Masses,"

EFI 88-62 (August 1988); a version of this work appears in "*1988 International Workshop on New Trends in Strong Coupling Gauge Theories*," Nagoya, Japan, ed. Bando, Muta and Yamawaki; "Bootstrap Symmetry Breaking in Electroweak Unification," EFI Preprint, 89–08 (1989). V. A. Miransky, M. Tanabashi, K. Yamawaki, *Mod. Phys. Lett.* **A4**, 1043 (1989); *Phys. Lett.* **221B** 177 (1989); W. J. Marciano, *Phys. Rev. Lett.* **62**, 2793 (1989). W. A. Bardeen, C. T. Hill, M. Lindner *Phys. Rev.* **D41**, 1647 (1990).

2. C. T. Hill, *Phys. Lett.* **B266** 419 (1991); C. T. Hill, *Phys. Lett.*; **B345** 483, (1995); S. P. Martin, *Phys. Rev.* **D46**, 2197 (1992); *Phys. Rev.* **D45**, 4283 (1992), *Nucl. Phys.* **B398**, 359 (1993); M. Lindner and D. Ross, *Nucl. Phys.* **B 370**, 30 (1992); R. Bönisch, *Phys. Lett.* **B268** 394 (1991); C. T. Hill, D. Kennedy, T. Onogi, H. L. Yu, *Phys. Rev.* **D47** 2940 (1993); B. Pendleton, G.G. Ross, *Phys. Lett.* **98B** 291, (1981); C.T. Hill, *Phys. Rev.* **D24**, 691 (1981); C. T. Hill, C. N. Leung, S. Rao, *Nucl. Phys.* **B262**, 517 (1985).
3. K. Lane, E. Eichten, *Phys. Lett.* **B352**, 382, (1995); K. Lane, *Phys. Rev.* **D54** 2204, (1996).
4. See the talk by K. Lane, "Technicolor and the First Muon Collider" in these proceedings, or hep-ph/9801385.
5. G. Buchalla, G. Burdman, C. T. Hill, D. Kominis, *Phys. Rev.* **D53** 5185, (1996);
6. B. Balaji, *Phys. Lett.* **B393** 89 (1997).
7. "Electroweak Symmetry Breaking via a Top Seesaw," B. Dobrescu, C. T. Hill FERMILAB-PUB-97-409-T, hep-ph/9712319
8. G. Burdman, D. Kominis, *Phys. Lett.* **B403** 101 (1997). C. T. Hill, Xinmin Zhang, *Phys.Rev.* **D51** 3563, (1995).
9. R. S. Chivukula, B. Dobrescu, J. Terning, *Phys. Lett.* **B353**, 289 (1995); R. S. Chivukula, J. Terning, *Phys. Lett.* **B385** 209 (1996).
10. S. Adler *et al.* , *Phys. Rev. Lett.* **79** 2204 (1997).
11. M. Strassler and M. Peskin, *Phys. Rev.* **D43** 1500 (1991); see also C. T. Hill and S. Parke, *Phys. Rev.* **D49**, 4454 (1994).

ANTI-DE SITTER BLACK HOLES AND THEIR SUPERPARTNERS IN 2+1 DIMENSIONS

Sharmanthie Fernando and Freydoon Mansouri
Physics Department
University of Cincinnati
Cincinnati, Ohio 45221

1 INTRODUCTION

The main objective of this work is to study the anti-de Sitter black holes in 2+1 dimensions and their supersymmetric generalization. To see that it is related to the main theme of this Conference, which is *Physics of Mass*, let us recall our theoretical framework for the description of *mass* : Following Wigner, we identify a particle state with an irreducible representation of the Poincaré group. In the same sense, a typical black hole may be viewed, asymptotically, as a Poincaré state characterized by its mass and spin. So, as long as the Poincaré group retains its role as the asymptotic symmetry group of space-time, the notion of *mass* will retain its invariant meaning.

The asymptotic symmetry group of space-time must, ultimately, be determined by experiments. In particular, if the cosmological constant turns out to be not identically zero, no matter how small, then the asymptotic symmetry group will change to, say, anti-de Sitter (AdS) group. One immediate consequence of such a change is that mass, as a Casimir invariant of the Poincaré group, will no longer retain its invariant meaning. One must then either replace the notion of Poincaré mass with one of the invariants of, say, the AdS group or consider it as a quantity which changes under AdS boosts. In 2+1 dimensions, the issues discussed above acquire more immediate relevance because the only known black hole solution with finite "mass" and "angular momentum" is of anti-de Sitter variety [1].

2 STATEMENT OF THE PROBLEM

The traditional way of searching for signs of supersymmetry in black hole solutions has been to look for Killing spinors. Some examples of the searches of this kind can be seen in, e.g., references [2-6]. The rationale underlying this approach is to proceed in analogy with the situation in the bosonic case: Just as the presence of a Killing vector signals the presence of an observable associated with an asymptotic symmetry group, one might argue that a Killing spinor could signal the presence of some sort of asymptotic supersymmetry. One would have to be careful, however, in pushing this analogy too far. By definition, a supersymmetric charge transforms as a spinor under Lorentz group and is not an observable whereas the corresponding bosonic charge

Physics of Mass
Edited by Kursunoglu *et al.*, Kluwer Academic / Plenum Publishers, New York, 1999

could be. More generally, to have an invariant meaning, all asymptotic observables must be functions of the Casimir invariants of the corresponding asymptotic (super) group. So, if one stops at the level of identifying a Killing spinor, many issues such as the supersymmetric quantum numbers carried by the (super)black hole and the superpartners of a given black hole, if any, will remain obscure. One way to overcome these limitations, is to make use of the notion of a *supersymmetric space-time* [7].

Before giving its supersymmetric generalization, we must understand the BTZ black hole from the point of view of the Chern Simons gauge theory [8,9] coupled to sources [9-11]. Steps have already been taken in this direction [12,13]. However, a number of issues need further clarification. Among these are the determination of the asymptotic observables of the black hole in terms of the properties of the source(s). Our general view is that the Chern Simons theory is an explicit realization of the Mach Principle, so that in the absence of sources only trivial solutions are possible. Any changes in topology must come about due to the presence of sources. To implement this idea, we must have a localized source (particle) carrying an irreducible representation of the gauge symmetry group [7,11]. This means that for the BTZ black hole, the source is an irreducible representation of the anti-de Sitter group labeled by its Casimir invariants. It is this formulation of the BTZ solution which lends itself to generalization to the supersymmetric case.

3 ANTI-DE SITTER SPACE AND ALGEBRA

The anti-de Sitter space in 2+1 dimensions can be viewed as a subspace of a flat 4-dimensional space with the line element

$$ds^2 = dX_A dX^A = dX_0^2 - dX_1^2 - dX_2^2 + dX_3^2 \tag{1}$$

It is determined by the constraint

$$(X_0)^2 - (X_1)^2 - (X_2)^2 + (X_3)^2 = l^2 \tag{2}$$

where l is a real constant . The set of transformations which leave the line element invariant form the anti-de Sitter group $SO(2,2)$. It is locally isomorphic to $SL(2,R) \times SL(2,R)$ or $SU(1,1) \times SU(1,1)$. From here on by anti-de Sitter group we shall mean its universal covering group.

The AdS algebra consists of the elements M_{AB} satisfying the commutation relations

$$[M_{AB}, M_{CD}] = i\left(\eta_{AD}M_{BC} + \eta_{BC}M_{AD} - \eta_{AC}M_{BD} - \eta_{BD}M_{AC}\right) \tag{3}$$

With $A = (a,3)$ and $a = 0,1,2$, we can write the algebra in two more convenient forms:

$$\begin{aligned} M^{ab} &= \epsilon^{abc}J_c = \epsilon^{abc}(J_c^+ + J_c^-) \\ M^{a3} &= l\Pi^a = (J^{+a} - J^{-a}) \end{aligned} \tag{4}$$

where

$$\epsilon^{012} = 1; \qquad \eta^{ab} = (1,-1,-1) \tag{5}$$

Then, the commutation relations in these bases take the form, respectively,

$$\left[J^a, J^b\right] = -i\epsilon^{abc}J_c; \quad \left[J^a, \Pi^b\right] = -i\epsilon^{abc}\Pi_c; \quad \left[\Pi^a, \Pi^b\right] = -il^{-2}\epsilon^{abc}J_c. \tag{6}$$

$$\left[J_a^{\pm}, J_b^{+}\right] = -i\epsilon_{ab}^{c}J_c^{\pm}; \qquad \left[J_a^{+}, J_b^{-}\right] = 0 \tag{7}$$

The Casimir operators look simplest in the latter basis:

$$j_{\pm}^2 = \frac{1}{l^2}\eta^{ab} J_a^{\pm} J_b^{\pm} \tag{8}$$

In the (J, Π) basis, they have the form,

$$\begin{aligned} M &= \Pi^a \Pi_a + l^{-2} J^a J_a = 2(j_+^2 + j_-^2) \\ J/l &= 2\Pi_a J^a / l = 2(j_+^2 - j_-^2) \end{aligned} \tag{9}$$

We will use the same symbols for operators and their eigenvalues.

An irreducible representation of AdS group can be labeled by the eigenvalues of either the pair (M, J) or the pair (j_+, j_-). For our applications, it is often advantageous to use a third set of labels which we denote by (H,S). They correspond to the maximal compact subgroup $SO(2) \times SO(2)$ of $SO(2,2)$, which is generated by J^0 and Π^0. The labels (H, S) are a natural choice from the point of view of the theory of induced representations. This can be seen from the comparison with the more familiar situation in the Poincaré group which can be obtained from anti-de Sitter group in the limit $l \to \infty$. From here on, we will use the labels, (j_+, j_-), (M, J), and (H, S) interchangeably. The last two are related to each other according to

$$M = H^2 + (S/l)^2; \qquad J = 2HS \tag{10}$$

Note that in order for M to assume negative values, H and S must, in general, be complex.

To see the relevance of H and S to the BTZ solution, let us express H and S in terms of the labels M and J by inverting Eqs. (10). We obtain

$$H^2 = \frac{1}{2} M \left[1 + \sqrt{1 - (\frac{J}{lM})^2} \right] \tag{11}$$

$$S^2 = \frac{l^2}{2} M \left[1 - \sqrt{1 - (\frac{J}{lM})^2} \right] \tag{12}$$

For $M > 0$ and $|J| \leq lM$, H and S are thus proportional to the horizon radii, $r_{\pm}$, of the BTZ black hole [1]:

$$r_+ = lH; \qquad r_- = S \tag{13}$$

4 CONNECTION AND THE CHERN SIMONS ACTION

We begin by writing the connection in $SL(2, R) \times SL(2, R)$ basis

$$A_\mu = \omega_\mu^{AB} M^{AB} = \omega_\mu^a J_a + e_\mu^a \Pi_a = A_\mu^{+a} J_a^+ + A_\mu^{-a} J_a^- \tag{14}$$

where

$$A_\mu^{\pm a} = \omega_\mu^a \pm l^{-1} e_\mu^a \tag{15}$$

The covariant derivative will have the form

$$D_\mu = \partial_\mu - iA_\mu = \partial_\mu - iA_\mu^{+a} J_a^+ - iA_\mu^{-a} J_a^- \tag{16}$$

Then the components of the field strength are given by

$$[D_\mu, D_\nu] = -iF_{\mu\nu}^{+a} J_a^+ - iF_{\mu\nu}^{-a} J_a^- = -iF_{\mu\nu}^+[A^+] - iF_{\mu\nu}^-[A^-] \tag{17}$$

For a simple or a semi-simple group, the Chern Simons action has the form

$$I_{cs} = \frac{1}{2}Tr\int_M A\wedge\left(dA+\frac{2}{3}A\wedge A\right) \tag{18}$$

where Tr stands for trace and

$$A = A_\mu dX^\mu = A^+ + A^- \tag{19}$$

We require the 2+1 dimensional manifold M to have the topology $R\times\Sigma$, with Σ a two-manifold. So, in our $SL(2,R)\times SL(2,R)$ basis we get

$$I_{cs} = \frac{1}{2}Tr\int_M\left[A^+\wedge\left(dA^+ + \frac{2}{3}A^+\wedge A^+\right) + A^-\wedge\left(dA^- + \frac{2}{3}A^-\wedge A^-\right)\right] \tag{20}$$

In this work, we take the point of view that in a free Chern Simons theory the field strengths vanish everywhere.

Under infinitesimal gauge transformations

$$u_\pm = \theta^{\pm\, a}J^\pm_a \tag{21}$$

the gauge fields transform as

$$\delta A_\mu = -\partial_\mu u - i[A_\mu, u] \tag{22}$$

More specifically,

$$\delta A^{\pm\, a}_\mu = -\partial_\mu\theta^{\pm\, a} - \epsilon^a_{\ bc}A^{\pm\, b}\theta^{\pm\, c} \tag{23}$$

As we have stated, the manifold M has the topology $R\times\Sigma$ with R representing x^0. Then subject to the constraints

$$F^\pm_a[A^\pm] = \frac{1}{2}\eta_{ab}\epsilon^{ij}(\partial_i A^{\pm\, b}_j - \partial_j A^{\pm\, b}_i + \epsilon^b_{\ cd}A^{\pm\, c}_i A^{\pm\, d}_j) = 0 \tag{24}$$

the Chern Simons action for $SO(2,2)$ will take the form

$$\begin{aligned} I_{cs} = &\int_R dx^0\int_\Sigma d^2x\left(-\epsilon^{ij}\eta_{ab}A^{+a}_i\partial_0 A^{+b}_j + A^{+a}_0 F^+_a\right)\\ &+\int_R dx^0\int_\Sigma d^2x\left(-\epsilon^{ij}\eta_{ab}A^{-a}_i\partial_0 A^{-b}_j + A^{-a}_0 F^-_a\right) \end{aligned} \tag{25}$$

5 INTERACTION WITH SOURCES

Following the approach which has been successful in coupling sources to Poincaré Chern Simons theory [11], we take a source for the present problem to be an irreducible representation of anti-de-Sitter group characterized by Casimir invariants M and J (or H and S). Within the representation, the states are further specified by the phase space variables of the source Π^A and q^A, $A = 0,1,2,3$, subject to anti-de Sitter constraints.

For illustrative purposes, let us consider first the interaction term for a special case which is the analog of the Poincaré case [11] with the intrinsic spin set to zero.

$$\begin{aligned} I_1 = &\int_C d\tau\left[\Pi_A D_\tau q^A + \lambda\left(q^A q_A - l^2\right)\right]\\ &+\int_C d\tau\left[\lambda_+\left(J^{+a}J^+_a - l^2 j^2_+\right) + \lambda_-\left(J^{-a}J^-_a - l^2 j^2_-\right)\right] \end{aligned} \tag{26}$$

where C is a path in M, τ is a parameter along C, and the covariant derivative D_τ is given by

$$D_\tau = \partial_\tau - i\omega^{AB} M_{AB} \tag{27}$$

The first term in this action is the same as that given in reference [13]. The second term ensures that $q^A(\tau)$ satisfy the AdS constraint. It is not the manifold M over which the gauge theory is defined but the space of $q's$ which give rise to the classical space-time. The last two constraints identify the source being coupled to the Chern Simons theory as an anti-de Sitter state with invariants j_+ and j_-. These constraints are crucial in relating the invariants of the source to the asymptotic observable of the coupled theory. In this respect, our action differs from that given in reference [13]. Although the word "constraints" was mentioned there in connection with this action, they were not explicitly stated or made use of in the sequel.

Using the standard (orbital) representation of the generators

$$M_{AB} = i(q_A\partial_B - q_B\partial_A) \tag{28}$$

we have

$$\Pi_C \omega^{AB} M_{AB} q^C = \omega^{AB}(q_A\Pi_B - q_B\Pi_A) = \omega^{AB} L_{AB} \tag{29}$$

Here L_{AB} are c-number quantities transforming like M_{AB}. Breaking up this expression into $SL(2,R) \times SL(2,R)$ form just as was done M_{AB}, we get

$$\omega^{AB} L_{AB} = A^{+a} L^+_a + A^{-a} L_a \tag{30}$$

So, the action I_1 can be written as

$$I_1 = \int_C d\tau \left[\Pi_A \partial_\tau q^A - i\left(A^{+a}L^+_a + A^{-a}L^-_a\right) + \lambda\left(q^A q_A - l^2\right)\right]$$

$$+ \int_C d\tau \left[\lambda_+\left(J^{+a}J^+_a - l^2 j^2_+\right) + \lambda_-\left(J^{-a}J^-_a - l^2 j^2_-\right)\right] \tag{31}$$

In this expression $L^\pm_a$ play the role of (c-number) generalized orbital angular momenta. If, in addition, the representation carries generalized intrinsic (spin) angular momenta, then $L^\pm_a$ would have to be replaced by $J^\pm_a$, respectively, where

$$J^\pm_a = L^\pm_a \oplus S^\pm_a \tag{32}$$

It is now clear how the interaction term I_1 can be generalized to the case when $S^\pm_a \neq 0$. We simply replace $L^\pm_a$ with $J^\pm_a$ in I_1 to get

$$I_s = \int_C d\tau \left[\Pi_A D_\tau q^A + \lambda\left(q^A q_A - l^2\right)\right]$$

$$+ \int_C d\tau \left[\lambda_+\left(J^{+a}J^+_a - l^2 j^2_+\right) + \lambda_-\left(J^{-a}J^-_a - l^2 j^2_-\right)\right] \tag{33}$$

This action can be expressed in a form in which the $SL(2,R) \times SL(2,R)$ structure of the gauge group is transparent:

$$I_s = \int_C d\tau \left[\Pi_A \partial_\tau q^A - (A^{+a}J^+_a + A^{-a}J^-_a) + \lambda\left(q^A q_A - l^2\right)\right]$$

$$+ \int_C d\tau \left[\lambda_+\left(J^{+a}J^+_a - l^2 j^2_+\right) + \lambda_-\left(J^{-a}J^-_a - l^2 j^2_-\right)\right] \tag{34}$$

In this expression $J^\pm_a$ play the role of c-number generalized angular momenta which transform in the same way as the corresponding generators and which label the source.

If there are several sources, an interaction of the form (35) must be written down for each source.

It is well known that for a Poincaré state with mass $m^2 > 0$, there is a (rest) frame in which, e.g., the momentum vector takes the form

$$p^a = (p^0, \vec{p}) \rightarrow (m, 0) \tag{35}$$

Similarly, in the present case, there is a frame such that when, e.g., $J^{\pm a} J^{\pm}_a > 0$, we have

$$J^{\pm a} = (J^{\pm\,0}, \vec{J^{\pm}}) \rightarrow (l j_{\pm}, 0) \tag{36}$$

Combining, the interaction term I_s with the Chern Simons action I_{cs}, we get the total action for the theory;

$$I = I_{cs} + I_s \tag{37}$$

In this theory, the components of the field strength still vanish everywhere except at the location of the sources. So, the analog of Eqs.(24) becomes

$$\epsilon^{ij} F^{\pm\,a}_{ij} = J^{\pm\,a} \delta^2(\vec{x}, \vec{x}_0) \tag{38}$$

In particular, when $\eta^{ab} J^{\pm}_a J^{\pm}_b > 0$, we get in the special (rest) frame

$$\epsilon^{ij} F^{\pm\,0}_{ij} = l j_{\pm} \delta^2(\vec{x}, \vec{x^0}) \tag{39}$$

All other components of the field strength vanish. We thus see that because of the constraints appearing in the action (34), the strength of the sources become identified with their Casimir invariants. These invariants, in turn, determine the asymptotic observables of the theory. Since such observables must be gauge invariant, they are expressible in terms of Wilson loops, and a Wilson loop about our source can only depend on, e.g., j_+ and j_-.

6 THE BLACK HOLE SPACE-TIME

To see how the space-time structure emerges from our anti-de Sitter gauge theory, we follow the same procedure which led to the emergence of space-time from Poincaré [11] and super Poincaré [7] Chern Simons gauge theories. For the AdS Chern Simons gauge theory, we note from Eq. (1) that the manifold M_q is now an anti-de Sitter space satisfying the constraint

$$q_0^2 - q_1^2 - q_2^2 + q_3^2 = l^2 = -\Lambda^{-1} \tag{40}$$

where Λ = cosmological constant. In fact, our $SL(2,R) \times SL(2,R)$ formulation allows us to take M_q to be the universal covering space of the AdS space. Moreover, the source coupled to the Chern Simons action is an AdS state characterized by the Casimir invariants (M, J) or, equivalently, (H, S). To parametrize M_q consistent with the above constraint, consider a pair of 2-vectors,

$$\vec{q}_\phi = (q^1, q^2) = (f cos\phi, f sin\phi) \tag{41}$$

$$\vec{q}_t = (q^0, q^3) = \left(\sqrt{f^2 + l^2} cos(t/l), \sqrt{f^2 + l^2} sin(t/l)\right) \tag{42}$$

where $f = f(r)$, with r a radial coordinate which for an appropriate $f(r)$ will become the radial coordinate appearing in the line element for the BTZ black hole. As far the constraint (40) is concerned, the functional form of $f(r)$ is irrelevant. The parameters ϕ and t/l are both periodic. We will keep ϕ periodic throughout. However, since we are taking M_q to be universal covering space of AdS space, we do not have to , and we will not, identify t with $t + 2\pi l$. Computing the line element in terms of the parameters $(t/l, r, \phi)$, we get

$$ds^2 = (1 + \frac{f^2}{l^2})dt^2 - \frac{f'^2 dr^2}{(1 + \frac{f^2}{l^2})} - f^2 d\phi^2 \tag{43}$$

where "prime" indicates differentiation with respect to r.

Anticipating the results to be given below, we note that if we compare this line element with that of BTZ, we see that it corresponds to an irreducible representation of the AdS group with $J = 0$ and $M = -1$. As we have noted in connection with Eqs. (11) and (12), for these values of J and M, the invariant H is pure imaginary. This, in turn, implies that the quantities $r_\pm$ will also be imaginary. Thus, we can interpret the line element (43) as a special form of the BTZ line element which has been "Wick rotated" into the imaginary axis in the complex H space. In this form, the consequences of the residual qauge transformations involving H and S, or $r_\pm$, which we will perform below on $q^A(\tau)$ become very similar to those performed in the Poincaré [10,11] Chern Simons gravity. We must keep in mind, however, that in the end, we must Wick rotate the results back to the real $r_\pm$ axes.

With this in mind, let us now consider *local* gauge transformations. Although the original theory was invariant under $SL(2,R) \times SL(2,R)$ gauge transformations, we have already reduced this symmetry by choosing to work in a gauge in which equation (39) holds. In fact, the left over symmetry is just $SO(2) \times SO(2)$ generated, respectively, by J_0 and Π_0, or, equivalently, by $J^{\pm 0}$. So, identifying the parameters ϕ and t/l, respectively, with each $SO(2)$, consider the local gauge transformation

$$\vec{q'}_\phi(\phi) = e^{i\frac{r_+}{l}\phi J^0} \vec{q}_\phi(\phi) \tag{44}$$

It leaves $\vec{q}_t(t/l)$ invariant. Then, since ϕ is 2π periodic,

$$\vec{q}_{\phi'}(\phi + 2\pi) = e^{i2\pi\frac{r_+}{l}J^0} \vec{q}_\phi(\phi) \tag{45}$$

Similarly, consider the gauge transformation

$$\vec{q'}_t(t/l, \phi) = e^{i\frac{r_-}{l}\phi\Pi^0} \vec{q}_t(t/l) \tag{46}$$

It leaves $\vec{q}_\phi$ invariant and leads to

$$\vec{q}_{t'}(t/l, \phi + 2\pi) = e^{i\frac{r_-}{l}\phi\Pi^0} \vec{q}_t(t/l, \phi) \tag{47}$$

Thus, the periodicity of ϕ has led to a discrete subgroup group of isometries in the universal covering space of the AdS space. The parameters $\frac{2\pi}{l}r_\pm$ for these transformations were chosen to demonstrate the ease with which one can obtain the identifications necessary for the BTZ black hole. We note, however, that in contrast to the situation for the Poincaré group, the residual symmetry $SO(2) \times SO(2)$ assigns symmetrical roles to the invariants (H, S) or (r_+, r_-) as well as the parameters ϕ and t/l. To reflect this symmetrical role, we can perform our gauge transformations on $\vec{q}_\phi$ and $\vec{q}_t$ in the following more symmetrical manner:

$$\begin{aligned} \vec{\hat{q}}_{\phi'}(\phi, t/l) &= e^{i\left(\frac{r_+}{l}\phi - \frac{r_- t}{l^2}\right)J^0} \vec{q}_\phi(\phi) \\ \vec{\hat{q}}_{t'}(t/l, \phi) &= e^{i\left(\frac{r_-}{l}\phi - \frac{r_+ t}{l^2}\right)\Pi^0} \vec{q}_t(t/l) \end{aligned} \tag{48}$$

It then follows that

$$
\begin{aligned}
\vec{\hat{q}}_{\phi'}(\phi+2\pi, t/l) &= e^{i2\pi \frac{r_+}{l} J^0} \vec{\hat{q}}_{\phi'}(\phi, t/l) \\
\vec{\hat{q}}_{t'}(t/l, \phi+2\pi) &= e^{i2\pi \frac{r_-}{l} \Pi^0} \vec{\hat{q}}_{t}(t/l, \phi) \\
\vec{\hat{q}}_{\phi'}(\phi+2\pi, t/l+2\pi) &= e^{i2\pi (\frac{r_+}{l} - \frac{r_-}{l}) J^0} \vec{\hat{q}}_{\phi'}(\phi, t/l) \\
\vec{\hat{q}}_{t'}(t/l+2\pi, \phi+2\pi) &= e^{i2\pi (\frac{r_-}{l} - \frac{r_+}{l}) \Pi^0} \vec{\hat{q}}_{t'}(\phi, t/l)
\end{aligned} \tag{49}
$$

Thus, given the previous identifications, the last two expressions do not lead to any new identifications. We can now write

$$\vec{\hat{q}}_{\phi'}(\phi, t/l) = \vec{\hat{q}}_{\phi'}(\phi'); \qquad \vec{\hat{q}}_{t'}(\phi, t/l) = \vec{\hat{q}}_{t'}(t'/l) \tag{50}$$

where

$$\phi' = \frac{r_+}{l}\phi - \frac{r_- t}{l^2}; \qquad t' = \frac{r_-}{l}\phi - \frac{r_+ t}{l^2} \tag{51}$$

It then follows from (48) that after these transformations, we obtain a manifold $M_{\hat{q}}$ which can be parametrized as follows:

$$
\begin{aligned}
\hat{q}^1 &= f cos\left(\frac{r_+}{l}\phi - \frac{r_- t}{l^2}\right) \\
\hat{q}^2 &= f sin\left(\frac{r_+}{l}\phi - \frac{r_- t}{l^2}\right) \\
\hat{q}^0 &= \sqrt{f^2 + l^2} cos\left(\frac{r_-}{l}\phi - \frac{r_+ t}{l^2}\right) \\
\hat{q}^3 &= \sqrt{f^2 + l^2} sin\left(\frac{r_-}{l}\phi - \frac{r_+ t}{l^2}\right)
\end{aligned} \tag{52}
$$

From these we can compute the line element. It is given by

$$ds^2 = \frac{f^2}{l^2}\left(r_+ d\phi - r_- \frac{dt}{l}\right)^2 - \frac{f'^2 dr^2}{(1 + \frac{f^2}{l^2})} - \left(\frac{f^2}{l^2} - 1\right)\left(r_- d\phi - r_+ \frac{dt}{l}\right)^2 \tag{53}$$

It will now be recalled that the quantities $r_\pm$ appearing in this expression are "Wick rotated" relative to the corresponding invariants which appear in the BTZ solution. We must, therefore, rotate them back to the Re $r_\pm$ axes by letting

$$r_\pm \rightarrow -i r_\pm \tag{54}$$

Then, we get

$$ds^2 = -\frac{f^2}{l^2}\left(r_+ d\phi - r_- \frac{dt}{l}\right)^2 - \frac{f' \, 2dr^2}{(\frac{f^2}{l^2} + 1)} + \left(\frac{f^2}{l^2} - 1\right)\left(r_- d\phi - r_+ \frac{dt}{l}\right)^2 \tag{55}$$

Finally, to put this expression in a form identical to that given by BTZ [1], let

$$\frac{f^2}{l^2} = \frac{r_-^2 - r^2}{r_+^2 - r_-^2}; \qquad r < r_- \tag{56}$$

The result is

$$ds^2 = [\frac{r^2}{l^2} - M + \frac{J^2}{4r^2}]dt^2 - \frac{dr^2}{[\frac{r^2}{l^2} - M + \frac{J^2}{4r^2}]} - r^2[d\phi - \frac{J^2}{2r^2}dt]^2 \tag{57}$$

It can be seen from Eqs. (52) and (56) that the parametrization leading to this expression is valid for $r < r^-$ and any value of the parameter l. The simplest way of obtaining suitable parametrizations for all values of l is to observe that parametrization in terms of circular functions are Wick rotated relative to the BTZ solution. Then, as can be seen from Eq. (54), when we rotate the Casimir invariants $r_\pm$ back to their real axes in their respective complex $r_\pm$ planes, as we did in the above example, we are effectively replacing trigonometric functions by their corresponding hyperbolic functions. We emphasize that this replacement leaves the periodicity of the angle ϕ intact since the Wick rotation occurs not in ϕ but in complex r_+ and r_- spaces. This means that we do not need to impose periodicity on ϕ "by hand" if we wish to use a hyperbolic parametrization [1,14] which is advantageous in many instances.

It is, nevertheless, of interest to see if a parametrization in terms of circular functions works for $r > r^+$. Consider the following expressions:

$$\begin{aligned}
\hat{q}^1 &= f\cos\left(\frac{r_-}{l}\phi - \frac{r_+ t}{l^2}\right) \\
\hat{q}^2 &= f\sin\left(\frac{r_-}{l}\phi - \frac{r_+ t}{l^2}\right) \\
\hat{q}^0 &= \sqrt{f^2 + l^2}\cos\left(\frac{r_+}{l}\phi - \frac{r_- t}{l^2}\right) \\
\hat{q}^0 &= \sqrt{f^2 + l^2}\sin\left(\frac{r_+}{l}\phi - \frac{r_- t}{l^2}\right)
\end{aligned} \tag{58}$$

Then we can get back the BTZ metric of Eq. (57) by computing the line element in terms of these parameters, using (54) for inverse Wick rotation, and setting f to

$$\frac{f^2}{l^2} = \frac{r^2 - r_+^2}{r_+^2 - r_-^2}; \qquad r > r_+ \tag{59}$$

7 SUPER ANTI-DE SITTER GROUP AND ITS REPRESENTATIONS

The simplest way of obtaining the supersymmetric extension of the anti-de Sitter group is to begin with the AdS group in its $SL(2,R) \times SL(2,R)$ basis. The $N = 1$ supersymmetric form of each $SL(2,R)$ factor is the supergroup $OSp(1|2;R)$. Thus, one arrives at the (1,1) form of the $N = 2$ super AdS group. Its algebra is given by

$$\begin{gathered}
[J_a^\pm, J_b^\pm] = -i\epsilon_{ab}{}^c J_c^\pm; \quad [J_a^\pm, Q_\alpha^\pm] = -\sigma_\alpha^{a\,\beta} Q_\alpha^\pm; \quad \{Q_\alpha^\pm, Q_\beta^\pm\} = -\sigma_{\alpha\beta}^a J_a^\pm \\
\{Q_\alpha^+, Q_\beta^-\} = 0; \qquad [J^+, J^-] = 0
\end{gathered} \tag{60}$$

The Casimir invariants are given by

$$C_\pm = j_\pm^2 + \epsilon^{\alpha\beta} Q_\alpha^\pm Q_\beta^\pm \tag{61}$$

The spinor indices are raised and lowered by antisymmetric metric $\epsilon^{\alpha\beta}$ defined by $\epsilon^{12} = -\epsilon_{12} = 1$. The matrices $(\sigma^a)_\alpha^\beta$, $(a = 0,1,2)$, form a representation of $SL(2,R)$ and satisfy the Clifford algebra

$$\{\sigma^a, \sigma^b\} = \frac{1}{2}\eta^{ab} \tag{62}$$

More explicitly, we can take them to be:

$$\sigma^0 = \frac{1}{2}\begin{pmatrix} 1 & 0 \\ 0 & -1 \end{pmatrix}; \quad \sigma^1 = \frac{1}{2}\begin{pmatrix} 0 & i \\ i & 0 \end{pmatrix}; \quad \sigma^2 = \frac{1}{2}\begin{pmatrix} 0 & 1 \\ -1 & 0 \end{pmatrix} \tag{63}$$

Since super AdS group is semi-simple, we can construct its irreducible representations by first constructing the irreducible representations of $OSp(1|2,R)$. Depending on which $OSp(1|2,R)$ we are considering, the states within any such supermultiplet are the corresponding irreducible representations of $SL(2,R)$ Characterized by the Casimir invariants j_+ and j_-, respectively. To construct the supermultiplet corresponding to the "plus" generators in Eq. (60), let us define a Clifford vacuum state $|\Omega^+>$ by the requirement that

$$Q_1^+|\Omega^+>=0 \tag{64}$$

Without loss of generality, we can work in a frame in which the analog of Eq. (40) holds. Then, the Casimir invariant C_+ acting on this state will give

$$C_+|\Omega_+>=j_+|\Omega_+> \tag{65}$$

Thus, as indicated above, the Clifford vacuum $|\Omega^+>$ is an $SL(2,R)$ state with eigenvalue $j_+ = C_+$. The superpartner of this state is the state

$$|\Omega_1^+>=Q_2^+|\Omega^+> \tag{66}$$

It is easy to verify that this is an $SL(2,R)$ state with Casimir eigenvalue $j_+ + 1/2$.

The supermultiplet for the second $OSp(1|2,R)$ can be constructed in a similar way. In this case, the Clifford vacuum $|\Omega^->$ is an $SL(2,R)$ state such that

$$C_-|\Omega^->=j_-|\Omega^-> \tag{67}$$

The corresponding superpartner state is given by

$$|\Omega_1^->=Q_2^-|\Omega^-> \tag{68}$$

This is an $SL(2,R)$ state with eigenvalue $j_- + 1/2$.

We are now in a position to construct the (1,1) super AdS supermultiplet as a direct product of the two $OSp(1|2,R)$ doublets. Altogether, there will be four states in the supermultiplet. They will have the following labels:

$$|j_+,j_->; \quad |j_+,j_-+1/2>; \quad |j_++1/2,j_->; \quad |j_++1/2,j_-+1/2> \tag{69}$$

We can also label the states in terms of the quantities (H,S) by noting from section 3 that

$$H = j_+ + j_-; \qquad S/l = j_+ - j_- \tag{70}$$

Thus, the eigenvalues of $(H, S/l)$ for members of the above supermultiplet are, respectively, $(H, S/l)$, $(H+1/2, S/l+1/2)$, $(H+1/2, S/l-1/2)$, and $(H+1, S/l)$. From these, we can also obtain the expressions for the eigenvalues (M,J) of various states within the supermultiplet. The corresponding expressions are not as simple or intuitive as in the previous two bases.

8 CHERN SIMONS AND SOURCE ACTIONS FOR THE SUPER ADS GROUP

The Chern Simons term for simple and semisimple supergroups has the same structure as that for Lie groups. The only difference is that the trace operation is replaced by super trace (Str) operation. So, in the $OSp(1|2,R) \times OSp(1|2,R)$ basis the Chern Simons action has the same form as in (20). But now

$$A^{\pm} = \left[A_{\mu}^{\pm\,a} J_a^{\pm} + \chi_{\mu}^{\pm\alpha} Q_{\alpha}^{\pm} \right] dx^{\mu} \tag{71}$$

Just as in the non-supersymmetric case, to have a nontrivial theory, we must couple sources to the Chern Simons action. To do this in a gauge invariant and locally supersymmetric fashion, we must take a source to be an irreducible representation of the super AdS group. As we saw in the previous section, such a supermultiplet consists of four AdS states. To couple it to the gauge fields, we must first extend the AdS canonical variables we used in section 5 to their supersymmetric form:

$$\Pi_A \rightarrow (\Pi_A, \Pi_\alpha) \qquad q_A \rightarrow (q_A, q_\alpha) \tag{72}$$

Then, the source coupling can be written as

$$I_s = \int_C \left[\Pi_A dq^A + \Pi_\alpha dq^\alpha + (A^+ + A^-) + constraints \right] \tag{73}$$

The constraints here include those discussed for the AdS group in section 5, and, in addition, those which relate the AdS labels of the Clifford vacuum to the Casimir eigenvalues of the super AdS group. The combined action

$$I = I_{cs} + I_s \tag{74}$$

leads to the constraint equations

$$\epsilon^{ij} F_{ij}^{\pm\,a} = J^{\pm}\, \delta^2(\vec{x}, \vec{x^0}); \qquad \epsilon^{ij} F_{ij}^{\pm\alpha} = Q^{\pm\alpha} \delta^2(\vec{x}, \vec{x^0}) \tag{75}$$

where $i, j = 1, 2$. Just as in the AdS case in section 5, we can choose a gauge (frame) in which

$$\epsilon^{ij} F^{\pm\,0} = l j_{\pm} \delta^2(\vec{x}, \vec{x^0}) \tag{76}$$

9 THE EMERGING SUPERSYMMETRIC SPACE-TIME

To display the space-time structure which emerges from our super AdS Chern Simons theory, we follow the same procedure which led us to the structure of space time in the AdS theory in section 6. There we first parametrized the field space M_q in terms of the quantities (t, r, ϕ). Then, by fixing the gauge via specific gauge transformations and performing a Wick rotation, we obtained the BTZ solution. These transformations involved the Casimir invariants of the AdS state which represented our source. In the super AdS case, to have a source coupling which was invariant under $(1, 1)$ supersymmetry transformations, it was necessary that the source be not an AdS state but a super AdS state consisting of four AdS states given by Eq. (69). Each one of these AdS states is labeled by its own set of invariants (r_+, r_-) or, equivalently, (M, J). As a result, a gauge transformation involving $r_{\pm}$ for one of these AdS states will not be appropriate for gauge fixing of all the states of the supermultiplet. Moreover, the BTZ line element itself carries the labels (M, J) of an AdS state:

$$ds^2 = -(\frac{r^2}{l^2} - M + \frac{J^2}{4l^2})dt^2 + \frac{dr^2}{(\frac{r^2}{l^2} - M + \frac{J^2}{4r^2})} + r^2(d\phi - \frac{J^2}{2r^2}dt)^2 \tag{77}$$

This makes it impossible for a single c-number line element of this type to describe all the AdS states of a supermultiplet.

The situation here runs parallel to what was encountered in connection with super Poincaré Chern Simons theory [7]. There it was pointed out that standard classical geometries were not capable describing various features inherent in these geometries and that one must make use of nonclassical geometries. Such geometries can be based on three elements: 1. An algebra such as a Lie or a super Lie algebra. 2. A line element operator with values in this algebra. 3. A Hilbert space on which the algebra acts linearly. Consider, e.g., the BTZ line element given above. We replace the c-number quantities (M, J) with Casimir operators and obtain a line element operator. When this operator acts on one of the AdS states in the Hilbert space of the supermultiplet, it creates a classical line element. In this way, we obtain a supermultiplet of space-times the dimension of which is equal to that of source supermultiplet. Supersymmetry transformations are the messengers linking different layers of this multilayered space-time which we will refer to as a *supersymmetric black hole.*

This work was supported, in part by the Department of Energy under the contract number DOE-FGO2-84ER40153.

REFERENCES

1. M. Bañados, C. Teitelboim and J. Zanelli, Phys. Rev. Lett. **69** (1992) 1849; M. Bañados, M. Henneaux, C. Teitelboim and J. Zanelli, Phys. Rev. D **48** (1993) 1506.

2. O. Coussaert, M. Henneaux, eprint no. hep- th/9310194; Phys. Rev. Lett. **72** (1993) 183

3. J.M. Izquierdo and P.K. Townsend, eprint no. gr-qc/9501018; Clas. Quan. Grav. **12** (1995) 895

4. R. Kallosh, eprint no. hep-th/9503029; Phys. Rev. **D52** (1995) 1234

5. T. Ortin, eprint no. hep-th/9705095

6. A.R. Steif, Phys. Rev. Lett. **69** (1995) 1849

7. Sunme Kim, F. Mansouri, Phys. Lett. **B397** (1997) 81; F. Ardalan, S. Kim, F. Mansouri, Int. Jour. Mod. Phys., **A12** (1997) 1183

8. A. Achucarro, P. Townsend, Phys. Lett.B **180** (1986) 35

9. E. Witten, Nucl. Phys **B311** (1988) 46; **B323** (1989) 113

10. K. Koehler *et al*, Nucl. Phys. **B348** (1990) 373

11. F. Mansouri, M.K. Falbo-Kenkel, Mod. Phys Lett. **A8** (1993) 2503; F. Mansouri, Proceedings of XIIth Johns Hopkins Workshop, ed. Z. Horwath, World Scientific, 1994

12. D.Cangemi, M. Leblanc and R.B. Mann, Phys. Rev. Lett. **38** (1993) 739

13. C. Vaz and L. Witten, Phys. Lett. **B327** (1994) 29

14. S. Carlip, eprint no. gr-qc/9506079; Clas. Quan. Grav. **12** (1995) 2853

THE CASE FOR A STANDARD MODEL WITH ANOMALOUS $U(1)$

Pierre Ramond
Institute for Fundamental Theory
Department of Physics, University of Florida
Gainesville, FL 32611

Abstract

A gauged phase symmetry with its anomalies cancelled by a Green-Schwarz mechanism, broken at a large scale by an induced Fayet-Iliopoulos term, is a generic feature of a large class of superstring theories. It induces many desirable phenomenological features: Yukawa coupling hierarchy, the emergence of a small Cabibbo-like expansion parameter, relating the Weinberg angle to $b - \tau$ unification, and the linking of R-parity conservation to neutrino masses. Some are discussed in the context of a three-family model which reproduces all quark and lepton mass hierarchies as well as the solar and atmospheric neutrino oscillations.

1 Introduction

The commonly accepted lore that string theories do not imply robust relations among measurable parameters is challenged in a large class of effective low energy theories derived from string models contain an anomalous $U(1)$ with anomalies cancelled by the Green-Schwarz mechanism [1] at cut-off. As emphasized by 't Hooft long ago, anomalies provide a link between infrared and ultraviolet physics. In these theories, this yields relations between the low-energy parameters of the standard model and those of the underlying theory through anomaly coefficients. An equally important feature is that as the dilaton gets a vacuum value, it generates a Fayet-Iliopoulos that triggers the breaking [2] of the anomalous gauged symmetry at a large computable scale.

Through the anomalous $U(1)$, the Weinberg angle at cut-off is related to anomaly coefficients [3]. A simple model [4] with one family-dependent anomalous $U(1)$ beyond the standard model was the first to exploit these features to produce Yukawa hierarchies through the Froggatt-Nielsen mechanism [5], and determine the Weinberg angle. It was soon realized that some features could be abstracted from the presence of the anomalous $U(1)$: expressing the ratio of down-like quarks to charged lepton masses in terms of the Weinberg angle [6, 7, 8], the suppression of the bottom to the top quark masses [9], relating [10] the uniqueness of the vacuum to Yukawa hierarchies and the presence

Physics of Mass
Edited by Kursunoglu *et al.*, Kluwer Academic / Plenum Publishers, New York, 1999

of MSSM invariants in the superpotential, and finally relating the seesaw mechanism [11] to R-parity conservation [12].

These theories are expressed as effective low-energy supersymmetric theories with a cut-off scale M. The anomalous symmetry implies:

- A Cabibbo-like expansion parameter for the mass matrices.
- Quark and charged lepton Yukawa hierarchies, and mixing, including the bottom to top Yukawa suppression.
- The value of the Weinberg angle at unification.
- Natural R-parity conservation linked to massive neutrinos.
- A hidden sector that contains strong gauge interactions with chiral matter.

An important theoretical requirement is that the vacuum, in which the anomalous symmetry is broken by stringy effects, be free of flat directions associated with the MSSM invariants, and preserve supersymmetry.

The anomalous $U(1)$ also provides a possible explanation of supersymmetry breaking. Since the hidden sector contains a gauge theory with strong coupling, the Green-Schwarz mechanism requires that it have a mixed anomaly as well. This implies that the hidden matter is chiral with respect to the anomalous symmetry. As shown by Binétruy and Dudas [13], any strong coupling gauge theory with X-chiral fermions breaks supersymmetry (even QCD). In the context of special free-fermion models [14], this mechanism can, with several Abelian symmetries, produce flavor-independent squark masses [15]. A recent analysis [16] improves the BD mechanism by showing that the dilaton F-term does not vanish, providing for gaugino masses and possibly solving the FCNC problem.

In the following, we present the generic features of this type of model, and illustrate some in the context of a realistic three-family model.

2 Applications to the standard model

We consider models which have a gauge structure broken in two sectors: a visible sector, and a hidden sector, linked by the anomalous symmetry and possibly other Abelian symmetries (as well as gravity).

$$G_{\text{visible}} \times U(1)_X \times U(1)_{Y^{(1)}} \cdots \times U(1)_{Y^{(M)}} \times G_{\text{hidden}} , \tag{2.1}$$

where G_{hidden} is the hidden gauge group, and

$$G_{\text{visible}} = SU(3) \times SU(2) \times U(1)_Y . \tag{2.2}$$

is the standard model. Of the $M+1$ extra symmetries,one which we call X, is anomalous in the sense of Green-Schwarz.

The symmetries, X, $Y^{(a)}$ are spontaneously broken at a high scale by the Fayet-Iliopoulos term generated by the dilaton vacuum. This DSW vacuum [2] is required by phenomenology to preserve both supersymmetry and the standard model symmetries.

We assume the smallest matter content needed to reproduce the observed quark and charged lepton hierarchy, cancel the anomalies associated with the extra gauge symmetries, and has a unique vacuum structure:

- Three chiral families
- One standard-model vector-like pair of Higgs weak doublets.
- Three right-handed neutrinos $\overline{N}_i$,
- Standard model vector-like pairs,
- Chiral fields that are needed to break the three extra $U(1)$ symmetries in the DSW vacuum. We denote these fields by θ_a.
- Hidden sector gauge interactions and their matter, together with singlet fields, needed to cancel the remaining anomalies.

3 Anomalies

When viewed from the infrared, the anomaly constraints put strong restrictions on the low energy theory. In a four-dimensional theory, the Green-Schwarz anomaly compensation mechanism occurs through a dimension-five term that couples an axion to all the gauge fields. As a result, any anomaly linear in the X-symmetry must satisfy the Green-Schwarz relations

$$(XG_iG_j) = \delta_{ij}C_i \ , \tag{3.1}$$

where G_i is any gauge current. The anomalous symmetry must have a mixed gravitational anomaly, so that

$$(XTT) = C_{\text{grav}} \neq 0 \ , \tag{3.2}$$

where T is the energy-momentum tensor. In addition, the anomalies compensated by the Green-Schwarz mechanism satisfy the universality conditions

$$\frac{C_i}{k_i} \equiv \frac{C_{\text{grav}}}{12} \quad \text{for all } i \ . \tag{3.3}$$

A similar relation holds for $C_X \equiv (XXX)$, the self-anomaly coefficient of the X symmetry. These result in important numerical constraints, which restrict the matter content of the model. All other anomalies must vanish:

$$(G_iG_jG_k) = (XXG_i) = 0 \ . \tag{3.4}$$

In terms of the standard model, the vanishing anomalies are therefore of the following types:

- The first involve only standard-model gauge groups G_{SM}, with coefficients $(G_{\text{SM}}G_{\text{SM}}G_{\text{SM}})$, which cancel for each chiral family and for vector-like matter. Also the hypercharge mixed gravitational anomaly (YTT) vanishes.
- The second type is where the new symmetries appear linearly, of the type $(Y^{(i)}G_{\text{SM}}G_{\text{SM}})$. If we assume that the $Y^{(i)}$ are traceless over the three chiral families, these vanish over the three families of fermions with standard-model charges. Hence they must vanish on the Higgs fields: with $G_{\text{SM}} = SU(2)$, it implies the Higgs pair is vector-like with respect to the

$Y^{(i)}$. It also follows that the mixed gravitational anomalies $(Y^{(i)}TT)$ are zero over the fields with standard model quantum numbers.

- The third type involve the new symmetries quadratically, of the form $(G_{\rm SM}Y^{(i)}Y^{(j)})$. These vanish by group theory except for those of the form $(YY^{(i)}Y^{(j)})$. In general two types of fermions contribute: the three chiral families and standard-model vector-like pairs.

- The remaining vanishing anomalies involve the anomalous charge X.
  - With X family-independent, and $Y^{(i)}$ family-traceless, the vanishing of the $(XYY^{(i)})$ anomaly coefficients over the three families is assured: so they must also vanish over the Higgs pair. This means that X is vector-like on the Higgs pair. This is an important result, as it implies that the standard-model invariant H_uH_d (the μ term) has zero X and $Y^{(i)}$ charges; it can appear by itself in the superpotential, but we are dealing with a string theory, where mass terms do not appear in the superpotential: it can appear only in the Kähler potential. This results, after supersymmetry-breaking in an induced μ-term, of weak strength, as suggested by Giudice and Masiero [17].
  - The coefficients $(XY^{(i)}Y^{(j)})$, $i \neq j$. Since standard-model singlets can contribute to these anomalies, we expect cancellation to come about through a combination of hidden sector and singlet fields.
  - The coefficient (XXY). This imposes an important constraint on the X charges on the chiral families.
  - The coefficients $(XXY^{(i)})$; with family-traceless symmetries, they vanish over the three families of fermions with standard-model charges, but contributions are expected from other sectors of the theory.

The building of models in which these anomaly coefficients vanish is highly non-trivial. Finding a set of charges which satisfy all anomaly constraints, and reproduce phenomenology is highly constrained. In the three-family model it will even prove predictive in the neutrino sector.

3.1 Standard Model Anomalies

In the standard model, we consider three anomalies associated with its three gauge groups,

$$C_{\rm color} = (XSU(3)SU(3)) \; ; \qquad C_{\rm weak} = (XSU(2)SU(2)) \; ; \qquad C_Y = (XYY) \; , \tag{3.5}$$

when () stands for the trace. They can be expressed [9] in terms of the X-charges of the invariants of the MSSM

$$C_{\rm color} = \frac{1}{2}\sum_i \left[X^{[u]}_{ii} + X^{[d]}_{ii}\right] - 3X^{[\mu]} \; , \tag{3.6}$$

$$C_Y + C_{\rm weak} - \frac{8}{3}C_{\rm color} = 2\sum_i \left[X^{[e]}_{ii} - X^{[d]}_{ii}\right] + 2X^{[\mu]} \; , \tag{3.7}$$

where $X^{[u]}_{ij}$ is the X-charge of $\mathbf{Q}_i\overline{\mathbf{u}}_jH_u$, $X^{[d]}_{ij}$ that of $\mathbf{Q}_i\overline{\mathbf{d}}_jH_d$, $X^{[e]}_{ij}$ that of $L_i\overline{e}_jH_d$, and finally $X^{[\mu]}$ that of the μ-term H_uH_d, where i,j are the family indices. Also the mixed gravitational anomaly over the three chiral families is given by

$$C_{\rm grav}^{[\rm fam]} = 3C_{\rm color} + \sum_i X_{ii}^{[e]} - X^{[\mu]} \ . \tag{3.8}$$

In theories derived from superstrings, the integer level numbers $k_{\rm color}$ and $k_{\rm weak}$ are equal, resulting in the equality

$$C_{\rm weak} = C_{\rm color} \ . \tag{3.9}$$

These imply that, as long as these anomaly coefficients do not vanish, the MSSM Yukawa invariants cannot all appear at tree level, as their X-charges are necessarily non-zero. This means that not all Yukawa couplings can be of the same order of magnitude, resulting in some sort of Yukawa hierarchy.

More specific conclusions can be reached by assuming that the X charges are family-independent and the $Y^{(i)}$ are family-traceless. As we have seen, the μ-term has vector-like charges, $X^{[\mu]} = 0$.

By further assuming that the top quark Yukawa mass coupling occurs at tree-level, we have $X_{33}^{[u]} = X^{[u]} = 0$. This implies that the X-charge of the down quark Yukawa is proportional to the color anomaly, and thus cannot vanish: the down Yukawa is *necessarily* smaller than the top Yukawa, leading to the suppression of m_b over m_t, after electroweak breaking! The non-vanishing of the color anomaly implies the (observed) suppression of the bottom mass relative to the top mass.

The second anomaly equation simplifies to

$$C_{\rm Y} - \frac{5}{3}C_{\rm weak} = 6\left[X^{[e]} - X^{[d]}\right] \ , \tag{3.10}$$

stating that the relative suppression of the down to the charged lepton sector is proportional to the difference of two anomaly coefficients. The data, extrapolated to near unification scales indicates that there is no relative suppression between the two sectors, suggesting that difference should vanish. Remarkably, the vanishing [3] of that combination fixes the value of the Weinberg angle through the string of relations

$$\frac{3}{5} = \frac{C_{\rm weak}}{C_{\rm Y}} = \frac{k_{\rm weak}}{k_{\rm Y}} = \frac{g_{\rm Y}^2}{g_{\rm weak}^2} = \tan^2\theta_w \ . \tag{3.11}$$

This happens exactly at the phenomenologically preferred value of the Weinberg angle: the $b-\tau$ unification is related to the value of the Weinberg angle [6]!

The application of the Green-Schwarz structure to the standard model is consistent with many of its phenomenological patterns. However, more can be said through a careful study of the DSW vacuum.

4 The DSW vacuum

When the dilaton acquires its vacuum value, an anomalous Fayet-iliopoulos D-term is generated through the gravitational anomaly. In the weak coupling limit of the string, it is given by

$$\xi^2 = \frac{g^2}{192\pi^2} M_{\rm Planck}^2 C_{\rm grav} \ , \tag{4.1}$$

where g is the string coupling constant. This induces the breaking of X and $Y^{(i)}$ below the cut-off.

Phenomenology require that neither supersymmetry nor any of the standard model symmetries be broken at that scale. This puts severe restrictions on the form of the superpotential and the matter fields [10].

The analysis of the vacuum structure of supersymmetric theories is greatly facilitated by the fact that the solutions of the vacuum equations for the D-terms are in one-to-one correspondance with holomorphic invariants. This analysis has been recently generalized to include an anomalous Fayet-Iliopoulos term.

In order to get a unique determination of the DSW vacuum, we need as many singlet superfields, θ_a, as there are broken symmetries. Only they assume vev's as a result of the FI term. They are standard model singlets, but not under X and $Y^{(a)}$. If more fields than broken symmetries assume non-zero values in the DSW vacuum, we would have undetermined flat directions and hierarchies.

We assemble the charges in a $(M+1)\times(M+1)$ matrix $\mathbf{A}$, whose rows are the X, $Y^{(i)}$ charges of the θ fields, respectively. Assuming the existence of a supersymmetric vacuum where only the θ fields have vacuum values, implies from the vanishing of the $M+1$ D-terms

$$\mathbf{A}\begin{pmatrix} |\theta_1|^2 \\ \cdot \\ |\theta_{M+1}|^2 \end{pmatrix} = \begin{pmatrix} \xi^2 \\ \cdot \\ 0 \end{pmatrix} . \tag{4.2}$$

For this vacuum solution to exist, the matrix $\mathbf{A}$ must have an inverse and the entries in the first row of its inverse must be positive. The solution to these equations naturally provide computably small expansion parameters $\lambda_a =< |\theta_a| >_0 /M$. In the case when all expansion parameters are the same we can relate their value in terms of standard model quantities

$$\lambda_a = \frac{\alpha}{4\pi} C_{\text{color}} , \tag{4.3}$$

where α is the unified gauge coupling at unification.

Another important consequence is that there is no holomorphic invariant polynomial involving the θ fields alone. Another is that the θ sector is necessarily anomalous. Indeed, let us assume that it has no mixed gravitational anomalies. This means that all the charges are traceless over the θ fields. now the $(M+1)$ θ fields form a representation of $SU(M+1)$, and the tracelessness of the charges insures that they be members of $SU(M+1)$. So we are looking for M non-anomalous symmetries in $SU(M+1)$, which is impossible except for $M=1$. If two or more of the charges are the same on the θ's, we could have anomaly cancellation, but then the matrix A would be singular, contrary to the assumption of the DSW vacuum. Hence this sector will in general be anomalous.

For a thorough analysis of the vacuum with FY term, we refer the reader to Ref. [18, 10]. Here, we simply note two striking generic facts of phenomenological import. Consider any invariant I of the MSSM. It corresponds [19] to a possible flat direction of the non-anomalous supersymmetric vacuum. For that configuration, all its fields are aligned to the same vacuum value, as required by the vanishing of the non-anomalous D-terms of the standard model symmetries. It follows that the contribution of these terms of the anomalous D_X will be proportional to its X-charge [20]. In order to forbid this flat direction to appear alone in the vacuum, it is therefore necessary to require that its charge be of the wrong sign to forbid a solution of $D_X = \xi^2$. This implies a holomorphic invariant of the form $I\mathcal{P}(\theta_a)$, where $\mathcal{P}$ is a holomorphic polynomial in the θ's. The D-term equations are not sufficient to forbid this flat direction together with θ fields. We have to rely on the F-terms associated with that invariant polynomial, and its presence is needed in the superpotential. Fortunately, phenomenology **also** requires such terms to appear in the superpotential. This is the first of several curious links between phenomenology and the vacuum structure near unification scales! One can see that the existence of this invariant is predicated on the invertibility of **A**, the same condition for the DSW vacuum.

The second point addresses singlet fields that do not get vev's in the DSW vacuum. To implement the seesaw mechanism, there must be right-handed neutrinos, $\overline{N}_i$. Since they have no vev, their X-charge must also be of the wrong sign, which allows for holomorphic invariants of the form $\overline{N}^A\mathcal{P}(\theta)$, where A is a positive integer. The case $A = 1$ is forbidden as it breaks supersymmetry. Thus $A \geq 2$. The case $A = 2$ generates Majorana masses for these fields in the DSW vacuum. W scale. To single out $A = 2$ we need to choose the X charges of the $\overline{N}_i$ to be a negative half-odd integers. To implement the seesaw, the right-handed neutrinos couple to the standard model invariants L_iH_u, which requires that $X_{L_iH_u}$ is also a half-odd integer, while all other MSSM invariants have positive or zero integers X-charges.

5 A Three-Family Model

We can see how some of the features we have just discussed lead to phenomenological consequences in the context of a three-family model [21, 22], with three Abelian symmetries broken in the DSW vacuum. The matter content of the theory is inspired by E_6, which contains two Abelian symmetries outside of the standard model: the first $U(1)$, which we call V', appears in the embedding

$$E_6 \subset SO(10) \times U(1)_{V'} \tag{5.1}$$

The second $U(1)$, called V, appears in

$$SO(10) \subset SU(5) \times U(1)_V \ . \tag{5.2}$$

The two non-anomalous symmetries are

$$Y^{(1)} = \frac{1}{5}(2Y + V)\begin{pmatrix} 2 & 0 & 0 \\ 0 & -1 & 0 \\ 0 & 0 & -1 \end{pmatrix} \tag{5.3}$$

$$Y^{(2)} = \frac{1}{4}(V + 3V')\begin{pmatrix} 1 & 0 & 0 \\ 0 & 0 & 0 \\ 0 & 0 & -1 \end{pmatrix} , \tag{5.4}$$

The family matrices run over the three chiral families, so that $Y^{(1,2)}$ are family-traceless. Since $\mathrm{Tr}(YY^{(i)}) = 0$, there is no appreciable kinetic mixing between the non-anomalous $U(1)$s.

The X charges on the three chiral families in the **27** are of the form

$$X = (\alpha + \beta V + \gamma V')\begin{pmatrix} 1 & 0 & 0 \\ 0 & 1 & 0 \\ 0 & 0 & 1 \end{pmatrix} , \tag{5.5}$$

where α, β, γ are expressed in terms of the X-charges of $\overline{N}_i$ (=-3/2), that of $\mathbf{Q}\overline{\mathbf{d}}H_d$ (=-3), and that of the vector-like pair ,mass term $\overline{E}E$ (=-3).
The matter content of this model is the smallest that reproduces the observed quark and lepton hierarchy while cancelling the anomalies associated with the extra gauge symmetries:

- Three chiral families each with the quantum numbers of a **27** of E_6. This means three chiral families of the standard model, $\mathbf{Q}_i$, $\overline{\mathbf{u}}_i$, $\overline{\mathbf{d}}_i$, L_i, and $\overline{e}_i$, together with three right-handed neutrinos $\overline{N}_i$, three vector-like pairs denoted by $E_i + \overline{\mathbf{D}}_i$ and $\overline{E}_i + \mathbf{D}_i$, with the quantum numbers of the $\mathbf{5} + \overline{\mathbf{5}}$ of $SU(5)$, and finally three real singlets S_i.
- One standard-model vector-like pair of Higgs weak doublets.
- Chiral fields that are needed to break the three extra $U(1)$ symmetries in the DSW vacuum. We denote these fields by θ_a. In our minimal model with three symmetries that break through the FI term, we just take $a = 1, 2, 3$. The θ sector is necessarily anomalous.
- Other standard model singlet fields.
- Hidden sector gauge interactions and their matter.

Finally, the charges of the θ fields is given in terms of the matrix

$$\mathbf{A} = \begin{pmatrix} 1 & 0 & 0 \\ 0 & -1 & 1 \\ 1 & -1 & 0 \end{pmatrix} . \tag{5.6}$$

Its inverse

$$\mathbf{A}^{-1} = \begin{pmatrix} 1 & 0 & 0 \\ 1 & 0 & -1 \\ 1 & 1 & -1 \end{pmatrix} , \tag{5.7}$$

shows all three fields acquire the same vacuum value.

In the following, we will address only the features of the model which are of more direct phenomenological interest. For more details, the interested reader is referred to the original references [21, 22].

5.1 Quark and Charged Lepton Masses

The Yukawa interactions in the charge 2/3 quark sector are generated by operators of the form

$$\mathbf{Q}_i\overline{\mathbf{u}}_j H_u \left(\frac{\theta_1}{M}\right)^{n_{ij}^{(1)}} \left(\frac{\theta_2}{M}\right)^{n_{ij}^{(2)}} \left(\frac{\theta_3}{M}\right)^{n_{ij}^{(3)}} , \tag{5.8}$$

in which the exponents must be positive integers or zero. Assuming that only the top quark Yukawa coupling appears at tree-level, a straighforward computation of their charges yields in the DSW vacuum the charge 2/3 Yukawa matrix

$$Y^{[u]} \sim \begin{pmatrix} \lambda^8 & \lambda^5 & \lambda^3 \\ \lambda^7 & \lambda^4 & \lambda^2 \\ \lambda^5 & \lambda^2 & 1 \end{pmatrix} , \tag{5.9}$$

where $\lambda = |\theta_a|/M$ is the common expansion parameter.

A similar computation is now applied to the charge $-1/3$ Yukawa standard model invariants $\mathbf{Q}_i\bar{\mathbf{d}}_jH_d$. The difference is that $X^{[d]}$, its X-charge does not vanish. As long as $X^{[d]} \leq -3$, we deduce the charge $-1/3$ Yukawa matrix

$$Y^{[d]} \sim \lambda^{-3X^{[d]}-6} \begin{pmatrix} \lambda^4 & \lambda^3 & \lambda^3 \\ \lambda^3 & \lambda^2 & \lambda^2 \\ \lambda & 1 & 1 \end{pmatrix} . \tag{5.10}$$

Diagonalization of the two Yukawa matrices yields the CKM matrix

$$\mathcal{U}_{CKM} \sim \begin{pmatrix} 1 & \lambda & \lambda^3 \\ \lambda & 1 & \lambda^2 \\ \lambda^3 & \lambda^2 & 1 \end{pmatrix} . \tag{5.11}$$

This shows the expansion parameter to be of the same order of magnitude as the Cabibbo angle λ_c.

The eigenvalues of these matrices reproduce the geometric interfamily hierarchy for quarks of both charges

$$\frac{m_u}{m_t} \sim \lambda_c^8 , \qquad \frac{m_c}{m_t} \sim \lambda_c^4 . \tag{5.12}$$

$$\frac{m_d}{m_b} \sim \lambda_c^4 , \qquad \frac{m_s}{m_b} \sim \lambda_c^2 , \tag{5.13}$$

while the quark intrafamily hierarchy is given by

$$\frac{m_b}{m_t} = \cot\beta\lambda_c^{-3X^{[d]}-6} . \tag{5.14}$$

implying the relative suppression of the bottom to top quark masses, without large $\tan\beta$. These quark-sector results are the same as in a previously published model [21], but our present model is different in the lepton sector.

The analysis in the charged lepton sector proceeds in similar ways. No dimension-three term appears and the standard model invariant $L_i\bar{e}_jH_d$ have X-charge $X^{[e]}$. For $X^{[e]} = -3$, there are supersymmetric zeros in the (21) and (31) position, yielding

$$Y^{[e]} \sim \lambda_c^3 \begin{pmatrix} \lambda_c^4 & \lambda_c^5 & \lambda_c^3 \\ 0 & \lambda_c^2 & 1 \\ 0 & \lambda_c^2 & 1 \end{pmatrix} . \tag{5.15}$$

Its diagonalization yields the lepton interfamily hierarchy

$$\frac{m_e}{m_\tau} \sim \lambda_c^4 , \qquad \frac{m_\mu}{m_\tau} \sim \lambda_c^2 . \tag{5.16}$$

Our choice of X insures that $X^{[d]} = X^{[e]}$, which guarantees through the anomaly conditions the correct value of the Weinberg angle at cut-off, since

$$\sin^2\theta_w = \frac{3}{8} \quad \leftrightarrow \quad X^{[d]} = X^{[e]} ; \tag{5.17}$$

it sets $X^{[d]} = -3$, so that

$$\frac{m_b}{m_\tau} \sim 1 ; \qquad \frac{m_b}{m_t} \sim \cot\beta\lambda_c^3 . \tag{5.18}$$

It is a remarkable feature of this type of model that both inter- and intrafamily hierarchies are linked not only with one another but with the value of the Weinberg angle as well. In addition, the model predicts a natural suppression of m_b/m_τ, which suggests that $\tan\beta$ is of order one.

5.2 Neutrino Masses

Neutrino masses are naturally generated by the seesaw mechanism [11] if the three right-handed neutrinos $\overline{N}_i$ acquire a Majorana mass in the DSW vacuum. The flat direction analysis indicates that their X-charges must be negative half-odd integers, with $X_{\overline{N}} = -3/2$ preferred by the vacuum analysis. Their standard-model invariant masses are generated by terms of the form

$$M\overline{N}_i\overline{N}_j\left(\frac{\theta_1}{M}\right)^{p^{(1)}_{ij}}\left(\frac{\theta_2}{M}\right)^{p^{(2)}_{ij}}\left(\frac{\theta_3}{M}\right)^{p^{(3)}_{ij}} , \tag{5.19}$$

where M is the cut-off of the theory. In the (ij) matrix element. The Majorana mass matrix is computed to be

$$M\lambda_c^7\begin{pmatrix} \lambda_c^6 & \lambda_c^5 & \lambda_c \\ \lambda_c^5 & \lambda_c^4 & 1 \\ \lambda_c & 1 & 0 \end{pmatrix} . \tag{5.20}$$

Its diagonalization yields three massive right-handed neutrinos with masses

$$m_{\overline{N}_e} \sim M\lambda_c^{13} ; \qquad m_{\overline{N}_\mu} \sim m_{\overline{N}_\tau} \sim M\lambda_c^7 . \tag{5.21}$$

By definition, right-handed neutrinos are those that couple to the standard-model invariant L_iH_u, and serve as Dirac partners to the chiral neutrinos. In our model,

$$X(L_iH_u\overline{N}_j) \equiv X^{[\nu]} = 0 . \tag{5.22}$$

The superpotential contains the terms

$$L_iH_u\overline{N}_j\left(\frac{\theta_1}{M}\right)^{q^{(1)}_{ij}}\left(\frac{\theta_2}{M}\right)^{q^{(2)}_{ij}}\left(\frac{\theta_3}{M}\right)^{q^{(3)}_{ij}} \tag{5.23}$$

resulting, after electroweak symmetry breaking, in the orders of magnitude (we note $v_u = \langle H_u^0\rangle$)

$$v_u\begin{pmatrix} \lambda_c^8 & \lambda_c^7 & \lambda_c^3 \\ \lambda_c^5 & \lambda_c^4 & 1 \\ \lambda_c^5 & \lambda_c^4 & 1 \end{pmatrix} \tag{5.24}$$

for the neutrino Dirac mass matrix. The actual neutrino mass matrix is generated by the seesaw mechanism. A careful calculation yields the orders of magnitude

$$\frac{v_u^2}{M\lambda_c^3}\begin{pmatrix} \lambda_c^6 & \lambda_c^3 & \lambda_c^3 \\ \lambda_c^3 & 1 & 1 \\ \lambda_c^3 & 1 & 1 \end{pmatrix} . \tag{5.25}$$

A characteristic of the seesaw mechanism is that the charges of the $\overline{N}_i$ do not enter in the determination of these orders of magnitude as long as there are no massless right-handed neutrinos. Hence the structure of the neutrino mass matrix depends only on the charges of the invariants L_iH_u, already fixed by phenomenology and anomaly cancellation. In particular, the family structure is that carried by the lepton doublets L_i. In our model, since L_2 and L_3 have the same charges, it ias not surprising that we have no flavor distinction between the neutrinos of the second and third family. In models with two non-anomalous flavor symmetries based on E_6 the matrix (5.25) is a very stable prediction of our model. Its diagonalization yields the neutrino mixing matrix [23]

$$\mathcal{U}_{\text{MNS}} = \begin{pmatrix} 1 & \lambda_c^3 & \lambda_c^3 \\ \lambda_c^3 & 1 & 1 \\ \lambda_c^3 & 1 & 1 \end{pmatrix}, \tag{5.26}$$

so that the mixing of the electron neutrino is small, of the order of λ_c^3, while the mixing between the μ and τ neutrinos is of order one. Remarkably enough, this mixing pattern is precisely the one suggested by the non-adiabatic MSW [24] explanation of the solar neutrino deficit and by the oscillation interpretation of the reported anomaly in atmospheric neutrino fluxes (which has been recently confirmed by the Super-Kamiokande [25] and Soudan [26] collaborations).

Whether the present model actually fits both the experimental data on solar and atmospheric neutrinos or not depends on the eigenvalues of the mass matrix (5.25). A naive order of magnitude diagonalization gives a μ and τ neutrinos of comparable masses, and a much lighter electron neutrino:

$$m_{\nu_e} \sim m_0\,\lambda_c^6\,; \qquad m_{\nu_\mu}, m_{\nu_\tau} \sim m_0\,; \qquad m_0 = \frac{v_u^2}{M\lambda_c^3}, \tag{5.27}$$

The overall neutrino mass scale m_0 depends on the cut-off M. Thus the neutrino sector allows us, in principle, to measure it.

At first sight, this spectrum is not compatible with a simultaneous explanation of the solar and atmospheric neutrino problems, which requires a hierarchy between m_{ν_μ} and m_{ν_τ}. However, the estimates (5.27) are too crude: since the (2,2), (2,3) and (3,3) entries of the mass matrix all have the same order of magnitude, the prefactors that multiply the powers of λ_c in (5.25) can spoil the naive determination of the mass eigenvalues. A more careful analysis shows that even with factors of order one, it is possible to fit the atmospheric neutrino anomaly as well. A welcome by-product of the analysis is that the mixing angle is actually driven to its maximum value. We refer the reader to Ref.[22] for more details. The main point of this analysis is that maximal mixing between the second and third family in the neutrino sector occurs naturally as it is determined from the structure of the quark and charged lepton hierarchies.

6 R-Parity

The invariants of the minimal standard model and their associated flat directions have been analyzed in detail in the literature [27]. In models with an anomalous $U(1)$, these invariants carry in general X-charges, which, as we have seen, determines their suppression in the effective Lagrangian. Just as there is a basis of invariants, proven long ago by Hilbert, the charges of these invariants are not all independent; they can in fact be expressed in terms of the charges of the lowest order invariants built out of the fields of the minimal standard model, and some anomaly coefficients.

The X-charges of the three types of cubic standard model invariants that violate R-parity as well as baryon and/or lepton numbers can be expressed in terms of the X-charges of the MSSM invariants and the R-parity violating invariant

$$X^{[\not R]} \equiv X(LH_u) , \tag{6.1}$$

through the relations

$$X_{LQ\bar{d}} = X^{[d]} - X^{[\mu]} + X^{[\not R]} , \tag{6.2}$$

$$X_{LL\bar{e}} = X^{[e]} - X^{[\mu]} + X^{[\not R]} . \tag{6.3}$$

$$X_{\bar{u}\bar{d}\bar{d}} = X^{[d]} + X^{[\not R]} + \frac{1}{3}(C_{\rm color} - C_{\rm weak}) - \frac{2}{3}X^{[\mu]} . \tag{6.4}$$

Although they vanish in our model, we still display $X^{[u]}$ and $X^{[\mu]} = 0$, since these sum rules are more general.

In the analysis of the flat directions, we have seen how the seesaw mechanism forces the X-charge of $\overline{N}$ to be half-odd integer. Also, the Froggatt-Nielsen [5] suppression of the minimal standard model invariants, and the holomorphy of the superpotential require $X^{[u,d,e]}$ to be zero or negative integers, and the equality of the Kác-Moody levels of $SU(2)$ and $SU(3)$ forces $C_{\rm color} = C_{\rm weak}$, through the Green-Schwarz mechanism. Thus we conclude that the X-charges of these operators are half-odd integers, and thus they cannot appear in the superpotential unless multiplied by at least one $\overline{N}$. This reasoning can be applied to the higher-order $\not R$ operators since their charges are given by

$$\begin{aligned}
X_{\mathbf{QQQ}H_d} &= X^{[u]} + X^{[d]} - \frac{1}{3}X^{[\mu]} - X^{[\not R]} , &(6.5)\\
X_{\bar{d}\bar{d}\bar{d}LL} &= 2X^{[d]} - X^{[u]} - \frac{5}{3}X^{[\mu]} + 3X^{[\not R]} , &(6.6)\\
X_{\mathbf{QQQ}\bar{u}} &= 2X^{[u]} + X^{[d]} - \frac{4}{3}X^{[\mu]} - X^{[\not R]} , &(6.7)\\
X_{\bar{u}\bar{u}\bar{u}\bar{e}\bar{e}} &= 2X^{[u]} - X^{[d]} + 2X^{[e]} - \frac{2}{3}X^{[\mu]} - X^{[\not R]} , &(6.8)
\end{aligned}$$

It follows that **there are no R-parity violating operators, whatever their dimensions** : through the right-handed neutrinos, R-parity is linked to half-odd integer charges, so that charge invariance results in R-parity invariance. Thus **none** of the operators that violate R-parity can appear in holomorphic invariants: even after breaking of the anomalous X symmetry, the remaining interactions all respect R-parity, leading to an **absolutely stable superpartner**. This is a general result deduced from the uniqueness of the DSW vacuum, the Green-Schwarz anomaly cancellations, and the seesaw mechanisms.

7 Conclusion

The case for an anomalous $U(1)$ extension to the standard model is particularly strong. We have presented many of its phenomenological consequences. In a very unique model, we detailed how the neutrino matrices are predicted. However much remains to be done: the nature of the hidden sector, and supersymmetry breaking. Our model only predicts orders of magnitude of Yukawa couplings. To calculate the prefactors, a specific theory is required. Many of the features we have discussed are found in the context of free fermion theories [14], which arise in the context of perturbative string theory. It is hoped that since they involve anomalies, these features can also be derived under more general assumptions. A particularly difficult problem is that of the cut-off scale. From the point of view of the low energy, there is only one scale of interest, that at which the couplings unify, and the Green-Schwarz mechanism, by fixing the weak and color anomalies, identifies the cut-off as the unification scale. On the other hand, another mass scale appears in the theory through the size of the anomalous FI term, and the two values do not coincide, the usual problem of string unification. It is hoped that the calculation of the Fayet-Iliopoulos term in other regimes will throw some light on this problem.

Our simple model has too many desirable phenomenological features to be set aside, and we hope that a better understanding of fundamental theories will shed light on this problem.

Acknowledgements

I would like to thank Professor B. Kursunoglu for his kind hospitality and for giving me the opportunity to speak at this pleasant conference, as well as my collaborators, N. Irges and S. Lavignac, on whose work much of the above is based. This work was supported in part by the United States Department of Energy under grant DE-FG02-97ER41029.

References

[1] M. Green and J. Schwarz, Phys. Lett. B149 (1984) 117.

[2] M. Dine, N. Seiberg and E. Witten, Nucl. Phys. B289 (1987) 589; J. Atick, L. Dixon and A. Sen, Nucl. Phys. B292 (1987) 109.

[3] L. Ibáñez, Phys. Lett. B303 (1993) 55.

[4] L. Ibáñez and G. G. Ross, Phys. Lett. B332 (1994) 100.

[5] C. Froggatt and H. B. Nielsen Nucl. Phys. B147 (1979) 277.

[6] P. Binétruy and P. Ramond, Phys. Lett. B350 (1995) 49; P. Binétruy, S. Lavignac, and P. Ramond, Nucl. Phys. B477 (1996) 353.

[7] Y. Nir, Phys. Lett. B354 (1995) 107.

[8] V. Jain and R. Shrock, Phys. Lett. B352 (1995) 83.

[9] P. Ramond, Kikkawa Proceedings, 1996.hep-ph/9604251

[10] P. Binétruy, S. Lavignac, Nikolaos Irges, and P. Ramond, Phys. Lett. B403 (1997) 38.

[11] M. Gell-Mann, P. Ramond, and R. Slansky in Sanibel Talk, CALT-68-709, Feb 1979, and in *Supergravity* (North Holland, Amsterdam 1979). T. Yanagida, in *Proceedings of the Workshop on Unified Theory and Baryon Number of the Universe*, KEK, Japan, 1979.

Since super AdS group is semi-simple, we can construct its irreducible representations by first constructing the irreducible representations of $OSp(1|2,R)$. Depending on which $OSp(1|2,R)$ we are considering, the states within any such supermultiplet are the corresponding irreducible representations of $SL(2,R)$ Characterized by the Casimir invariants j_+ and j_-, respectively. To construct the supermultiplet corresponding to the "plus" generators in Eq. (60), let us define a Clifford vacuum state $|\Omega^+>$ by the requirement that

$$Q_1^+|\Omega^+>=0 \tag{64}$$

Without loss of generality, we can work in a frame in which the analog of Eq. (40) holds. Then, the Casimir invariant C_+ acting on this state will give

$$C_+|\Omega_+>=j_+|\Omega_+> \tag{65}$$

Thus, as indicated above, the Clifford vacuum $|\Omega^+>$ is an $SL(2,R)$ state with eigenvalue $j_+=C_+$. The superpartner of this state is the state

$$|\Omega_1^+>=Q_2^+|\Omega^+> \tag{66}$$

It is easy to verify that this is an $SL(2,R)$ state with Casimir eigenvalue $j_++1/2$.

The supermultiplet for the second $OSp(1|2,R)$ can be constructed in a similar way. In this case, the Clifford vacuum $|\Omega^->$ is an $SL(2,R)$ state such that

$$C_-|\Omega^->=j_-|\Omega^-> \tag{67}$$

The corresponding superpartner state is given by

$$|\Omega_1^->=Q_2^-|\Omega^-> \tag{68}$$

This is an $SL(2,R)$ state with eigenvalue $j_-+1/2$.

We are now in a position to construct the (1,1) super AdS supermultiplet as a direct product of the two $OSp(1|2,R)$ doublets. Altogether, there will be four states in the supermultiplet. They will have the following labels:

$$|j_+,j_->;\quad |j_+,j_-+1/2>;\quad |j_++1/2,j_->;\quad |j_++1/2,j_-+1/2> \tag{69}$$

We can also label the states in terms of the quantities (H,S) by noting from section 3 that

$$H=j_++j_-;\qquad\qquad S/l=j_+-j_- \tag{70}$$

Thus, the eigenvalues of $(H,S/l)$ for members of the above supermultiplet are, respectively, $(H,S/l)$, $(H+1/2,S/l+1/2)$, $(H+1/2,S/l-1/2)$, and $(H+1,S/l)$. From these, we can also obtain the expressions for the eigenvalues (M,J) of various states within the supermultiplet. The corresponding expressions are not as simple or intuitive as in the previous two bases.

8 CHERN SIMONS AND SOURCE ACTIONS FOR THE SUPER ADS GROUP

The Chern Simons term for simple and semisimple supergroups has the same structure as that for Lie groups. The only difference is that the trace operation is replaced by

p-FORM CHARGES AND p-BRANE SPECTRA

K.S. Stelle

The Blackett Laboratory, Imperial College
London, SW7, England
and
TH Division, CERN
CH-1211 Geneva 23, Switzerland

ABSTRACT

We discuss the character of the p-form charges carried by p-brane solutions to supergravity theories. The antisymmetric tensorial nature of these charges gives rise to a new feature of the Dirac quantisation condition, that there are "Dirac-insensitive" configurations. Although they constitute only a measure-zero set, the existence of these insensitive configurations is important because they are needed to understand the pattern of Dirac conditions in lower dimensions obtained by dimensional reduction.

1. – INTRODUCTION

Let us start from the bosonic sector of $D = 11$ supergravity,

$$I_{11} = \int d^{11}x \left\{ \sqrt{-g}(R - \tfrac{1}{48}F_{[4]}^2) + \tfrac{1}{6}F_{[4]} \wedge F_{[4]} \wedge A_{[3]} \right\} . \tag{1}$$

In addition to the metric, one has a 3-form antisymmetric-tensor gauge potential $A_{[3]}$ with a gauge transformation $\delta A_{[3]} = d\Lambda_{[2]}$ and a field strength $F_{[4]} = dA_{[3]}$. The third term in the Lagrangian is invariant under the $A_{[3]}$ gauge transformation only up to a total derivative, so the action (1) is invariant under gauge transformations that are continuously connected to the identity.

The equation of motion for the $A_{[3]}$ gauge potential is

$$d\,{}^*F_{[4]} + \tfrac{1}{2}F_{[4]} \wedge F_{[4]} = 0 \; ; \tag{2}$$

this equation of motion gives rise to the conservation of an "electric" type charge [2]

$$U = \int_{\partial\mathcal{M}_8} ({}^*F_{[4]} + \tfrac{1}{2}A_{[3]} \wedge F_{[4]}) \; . \tag{3}$$

Another conserved charge relies on the Bianchi identity $dF_{[4]} = 0$ for its conservation,

$$V = \int_{\partial\widetilde{\mathcal{M}}_5} F_{[4]} \; . \tag{4}$$

Physics of Mass
Edited by Kursunoglu *et al.*, Kluwer Academic / Plenum Publishers, New York, 1999

Charges such as (3,4) can occur on the right-hand side of the supersymmetry algebra,

$$\{Q,Q\} = C(\Gamma^A P_A + \Gamma^{AB} U_{AB} + \Gamma^{ABCDE} V_{ABCDE}) \ . \tag{5}$$

Note that since the supercharge Q in $D = 11$ supergravity is a 32-component Majorana spinor, the LHS of (5) has 528 components. The symmetric spinor matrices $C\Gamma^A$, $C\Gamma^{AB}$ and $C\Gamma^{ABCDE}$ on the RHS of (5) also have a total of 528 independent components: 11 for the momentum P_A, 55 for the "electric" charge U_{AB} and 462 for the "magnetic" charge V_{ABCDE}.

Now the question arises as to the relation between the charges U and V in (3,4) and the 2-form and 5-form charges appearing in (5). One thing that immediately stands out is that the Gauss' law integration surfaces in (3,4) are the boundaries of integration volumes $\mathcal{M}_8$, $\widetilde{\mathcal{M}}_5$ that do not fill out a whole spacelike hypersurface in spacetime, unlike the more familiar situation for charges in ordinary electrodynamics. Nonetheless, this does not impede the conservation of (3,4), which only requires that no electric or magnetic currents are present at the boundaries $\partial\mathcal{M}_8$, $\partial\widetilde{\mathcal{M}}_5$. Before we can discuss such currents, we need to consider the supergravity solutions that carry charges like (3,4).

An important family of supergravity solutions has the character of static $(p+1)$-dimensional hyperplanes where the U or V charge resides. These may be viewed as static histories of infinite p-dimensional objects, *i.e.* p-branes. For solutions of $D = 11$ supergravity (1), the metric takes the form

$$\begin{aligned} ds^2 &= H^{-\tilde{d}/4} dx^\mu dx^\nu \eta_{\mu\nu} + H^{d/9} dy^m dy^m \\ \mu &= 0, 1, \ldots, p \qquad m = p+1, \ldots, D-1 = 10 \ . \end{aligned} \tag{6}$$

The function $H(y)$ is a harmonic function in the $(11-d)$-dimensional transverse space, $\nabla^2 H = \partial_m \partial_m H = 0$. Completing the specification of the solution involves making a choice between two further refinements of the p-brane ansatz: an "electric" case which follows closely the analogue of the electric field set up by an electrically charged particle in $D = 4$ Maxwell theory, or a "magnetic" case which is analogous to the field from a $D = 4$ magnetic monopole. The forms of these two ansätze are best compared in terms of the $F_{[4]}$ field strength,

$$F_{m\mu_1\mu_2\mu_3} = \epsilon_{\mu_1\mu_2\mu_3} \partial_m (H^{-1}) \quad m = 3, \ldots, 10 \quad \text{electric 2-brane} \tag{7a}$$
$$F_{m_1 \ldots m_4} = -\epsilon_{m_1 \ldots m_4 r} \partial_r H \quad m = 7, \ldots, 10 \quad \text{magnetic 5-brane,} \tag{7b}$$

with all other independent components vanishing in either case. From (7), we may now identify the "worldvolume" dimension $d = p+1$ in the two cases: $d_{\rm el} = 3$; $d_{\rm mag} = 6$.

Letting the transverse-space harmonic function $H(y)$ take an isotropic SO$(11-d)$ invariant form in the transverse space,

$$H(r) = 1 + \frac{k}{r^{\tilde{d}}} \ , \qquad r = \sqrt{y^m y^m} \ , \qquad \tilde{d} = D - d - 2 = 9 - d \ , \tag{8}$$

where k is an integration constant, one obtains from (3,4) the corresponding electric/magnetic charges:

$$U = \int_{\partial\mathcal{M}_8} d\Sigma^n F_{n012} = \Omega_7 \tilde{d}_{\rm el} k = 6\Omega_7 k \qquad \text{electric} \tag{9a}$$
$$V = \frac{1}{4!} \int_{\partial\widetilde{\mathcal{M}}_5} d\Sigma^r \epsilon_{rm_1 \ldots m_4} F^{m_1 \ldots m_4} = \Omega_4 \tilde{d}_{\rm mag} k = 3\Omega_4 k \qquad \text{magnetic,} \tag{9b}$$

where Ω_{10-d} is the volume of the unit $(10-d)$ sphere.

Comparing ($9a,b$) to the ADM energy per unit p-volume,

$$\mathcal{E} = \Omega_{10-d} k \tilde{d} \ , \tag{10}$$

one sees that the charges *saturate* the Bogomol'ny-Prasad-Sommerfield bound on the energy density in either case: $\mathcal{E} = U$ or $\mathcal{E} = V$. Thus, the p-brane hyperplane solutions ($6;9a,b$) are BPS solutions of the supergravity equations. Any solution carrying a U or V charge must asymptotically approach the form of the corresponding flat static solution ($6;9a,b$), provided the energy difference with respect to ($6;9a,b$) is finite.

2. – p-FORM CHARGES

Now let us consider the inclusion of sources. The harmonic function (8) has a singularity which has for simplicity been placed at the origin of the transverse coordinates y^m. Whether or not this gives rise to a physical singularity in a solution depends on the global structure of that solution. In the electric 2-brane case, the solution does in the end have a singularity.[3] This singularity is unlike the Schwarzschild singularity, however, in that it is a *timelike* curve, and thus it may be considered to be the wordvolume of a δ-function source. The electric source that couples to $D = 11$ supergravity is the fundamental supermembrane action,[4] whose bosonic part is

$$\begin{aligned} I_{\text{source}} = Q_{\text{e}} \int_{\mathcal{W}_3} d^3\xi \big[& \sqrt{-\det(\partial_\mu x^M \partial_\nu x^N g_{MN}(x))} \\ & + \frac{1}{3!} \epsilon^{\mu\nu\rho} \partial_\mu x^M \partial_\nu x^N \partial_\rho x^R A_{MNR}(x) \big] \ . \end{aligned} \tag{11}$$

The source strength Q_{e} will shortly be found to be equal to the electric charge U upon solving the coupled equations of motion for the supergravity fields and a single source of this type. Varying $\delta/\delta A_{[3]}$ the source action (11), one obtains the δ-function current

$$J^{MNR}(z) = Q_{\text{e}} \int_{\mathcal{W}_3} \delta^3(z - x(\xi)) dx^M \wedge dx^N \wedge dx^R \ . \tag{12}$$

This current now stands on the RHS of the $A_{[3]}$ equation of motion:

$$d({}^*F_{[4]} + \tfrac{1}{2} A_{[3]} \wedge F_{[4]}) = {}^*J_{[3]} \ . \tag{13}$$

Thus, instead of the Gauss' law expression for the charge, one may instead rewrite it as a volume integral of the source,

$$U = \int_{\mathcal{M}_8} {}^*J_{[3]} = \frac{1}{3!} \int_{\mathcal{M}_8} J^{0MN} d^8 S_{MN} \ , \tag{14}$$

where $d^8 S_{MN}$ is the 8-volume element on $\mathcal{M}_8$, specified in a $D = 10$ spatial section of the supergravity spacetime by a 2-form. The charge derived in this way from a single 2-brane source is thus $U = Q_{\text{e}}$ as expected.

Now consider the effect of making different choices of the $\mathcal{M}_8$ integration volume within the $D = 10$ spatial spacetime section, as shown in Figure 1. Let the difference between the surfaces $\mathcal{M}_8$ and $\mathcal{M}'_8$ be infinitesimal and be given by a vector field $v^N(x)$. The difference in the electric charges obtained is then given by

$$\delta U = \int_{\mathcal{M}_8} \mathcal{L}_v {}^*J_{[3]} = \frac{1}{3!} \int_{\partial \mathcal{M}_8} J^{0MN} v^R d^7 S_{MNR} \ , \tag{15}$$

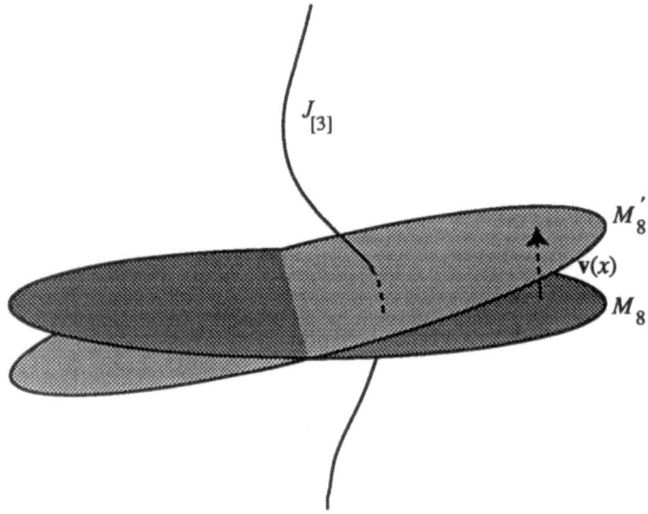

Figure 1. Different choices of charge integration volume "capturing" the current $J_{[3]}$.

where $\mathcal{L}_v$ is the Lie derivative along the vector field v. The second equality in (15) follows using Stokes' theorem and the conservation of the current $J_{[3]}$.

Now, a topological nature of the charge integral (3) becomes apparent; similar considerations apply to the magnetic charge (4). As long as the current $J_{[3]}$ vanishes on the boundary $\partial\mathcal{M}_8$, the difference (15) between the charges calculated using the integration volumes $\mathcal{M}_8$ and $\mathcal{M}_8'$ will vanish. This divides the electric-charge integration volumes into two topological classes distinguishing those for which $\partial\mathcal{M}_8$ "captures" the p-brane current, as shown in Figure 1 and giving $U = Q_e$, from those that do not capture the current, giving U=0.

The above discussion shows that the orientation-dependence of the U charges (3) is essentially topological. The topological classes for the charge integrals are naturally labeled by the asymptotic orientations of the p-brane spatial surfaces; an integration volume $\mathcal{M}_8$ extending out to infinity flips from the "capturing" class into the "non-capturing" class when $\partial\mathcal{M}_8$ crosses the δ-function surface defined by the current $J_{[3]}$. The charge thus naturally has a magnitude $|Q_{[p]}| = Q_e$ and a unit p-form orientation $Q_{[p]}/|Q_{[p]}|$ that is proportional to the ***asymptotic spatial volume form*** of the p=brane. Both the magnitude and the orientation of this p-form charge are conserved using the supergravity equations of motion.

The necessity of considering asymptotic p-brane volume forms arises because the notion of a p-form charge is not limited to static, flat p-brane solutions such as (6,7). Such charges can also be defined for any solution whose energy differs from that of a flat, static one by a finite amount. The charges for such solutions will also appear in the supersymmetry algebra (5) for such backgrounds, but the corresponding energy densities will not in general saturate the BPS bounds. For a finite energy difference with respect to a flat, static p-brane, the asymptotic orientation of the p-brane volume form must tend to that of a static flat solution, which plays the rôle of a "BPS vacuum" in a given p-form charge sector of the theory.

In order to have a non-vanishing value for a charge (3) or (4) occurring in the supersymmetry algebra (5), the p-brane must be either ***infinite*** or ***wrapped*** around a compact spacetime dimension. The case of a finite p-brane is sketched in Figure 2. Since the boundary $\partial\mathcal{M}$ of the infinite integration volume $\mathcal{M}$ does not capture the locus

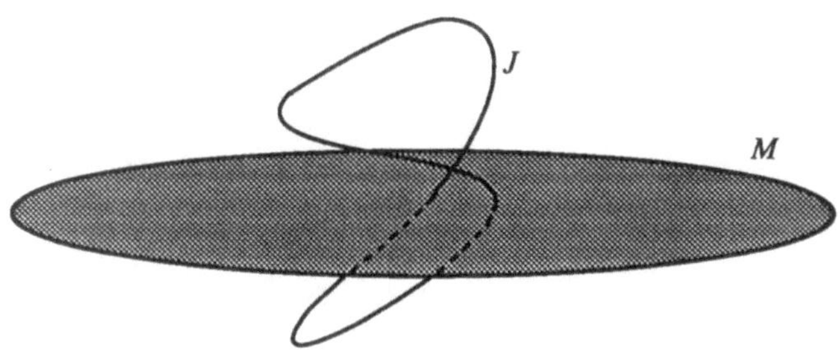

Figure 2. Finite p-brane not captured by $\partial\mathcal{M}$, giving zero charge.

where the p-brane current is non-vanishing, the current calculated using $\mathcal{M}$ will vanish as a result. Instead of an infinite p-brane, one may alternately have a p-brane wrapped around a compact dimension of spacetime, so that an integration-volume boundary $\partial\mathcal{M}_8$ is still capable of capturing the p-brane locus (if one considers this case as an infinite, but periodic, solution, this case may be considered simultaneously with that of the infinite p-branes). Only in such cases do the p-form charges occurring in the supersymmetry algebra (5) take non-vanishing values.[1]

3. – p-FORM CHARGE QUANTISATION CONDITIONS

Now we come to the question of what Dirac quantisation conditions arise for p-form charges. We shall first review a Wu-Yang style of argument,[6] (for a Dirac-string argument, see Ref.[7]) considering a closed sequence $\mathcal{W}$ of deformations of one p-brane, say the electric one, in the background fields set up by a dual, magnetic, $\hat{p}$-brane. After such a sequence of deformations, one sees from the supermembrane action (11) that the electric p-brane wavefunction picks up a phase factor

$$\exp\left(\frac{iQ_e}{(p+1)!}\oint_{\mathcal{W}} A_{M_1\ldots M_{p+1}}dx_1^M\wedge\ldots\wedge dx^{M_{p+1}}\right)\ , \tag{16}$$

where $A_{[p+1]}$ is the gauge potential set up (locally) by the magnetic $\hat{p}=D-p-4$ brane background.

A number of differences arise in this problem with respect to the ordinary Dirac quantisation condition for particles. One of these is that, as we have noted, objects carrying p-form charges appearing in the supersymmetry algebra (5) are necessarily either infinite or are wrapped around compact spacetime dimensions. For infinite p-branes, some deformation sequences $\mathcal{W}$ will lead to a divergent integral in the exponent in (16); such deformations also would require an infinite amount of energy, and so should be excluded from consideration. In particular, this excludes deformations that involve rigid rotations of an entire infinite brane. Thus, at least the *asymptotic* orientation of the electric brane must be preserved throughout the sequence of deformations. Another way of viewing this restriction on the deformations is to note that the asymptotic orientation of a brane is encoded into the electric p-form charge, and one should not consider changing this p-form in the course of the deformation any more than one should consider changing the magnitude of the electric charge in the ordinary $D=4$ Maxwell case.

[1]If one considers integration volumes that do not extend out to infinity, then one can construct integration surfaces that capture finite p-branes. Such charges do not occur in the supersymmetry algebra (5), but they are still of importance in determining the possible intersections of p-branes.[5]

We shall see shortly that another difference with respect to the ordinary Dirac quantisation of $D=4$ electrodynamics particles will be the existence of "Dirac-insensitive" configurations, for which the phase in (16) vanishes.

Restricting attention to deformations that give finite phases, one may use Stoke's theorem to rewrite the integral in (16):

$$\frac{Q_{\rm e}}{(p+1)!}\oint_{\mathcal{W}} A_{M_1\ldots M_{p+1}} dx_1^M \wedge \ldots \wedge dx^{M_{p+1}} = \\ \frac{Q_{\rm e}}{(p+2)!}\int_{\mathcal{M}_{\mathcal{W}}} F_{M_1\ldots M_{p+2}} dx^{M_1} \wedge \ldots \wedge dx^{M_{p+2}} = Q_{\rm e}\Phi_{\mathcal{M}_{\mathcal{W}}}\ , \qquad (17)$$

where $\mathcal{M}_{\mathcal{W}}$ is any surface "capping" the closed surface $\mathcal{W}$, *i.e.* a surface such that $\partial\mathcal{M}_{\mathcal{W}} = \mathcal{W}$; $\Phi_{\mathcal{M}_{\mathcal{W}}}$ is the flux through the cap $\mathcal{M}_{\mathcal{W}}$. Choosing the capping surface in two different ways, one can find a flux discrepancy $\Phi_{\mathcal{M}_1} - \Phi_{\mathcal{M}_2} = \Phi_{\mathcal{M}_1\cap\mathcal{M}_2} = \Phi_{\mathcal{M}_{\rm total}}$ (taking into account the orientation sensitivity of the flux integral). Then if $\mathcal{M}_{\rm total} = \mathcal{M}_1\cap\mathcal{M}_2$ "captures" the magnetic $\hat{p}$-brane, the flux $\Phi_{\mathcal{M}_{\rm total}}$ will equal the magnetic charge $Q_{\rm m}$ of the $\hat{p}$-brane; thus the discrepancy in the phase factor (16) will be $\exp({\rm i}Q_{\rm e}Q_{\rm m})$. Requiring this to equal unity gives,[6] in strict analogy to the ordinary case of electric and magnetic particles in $D=4$, the Dirac quantisation condition

$$Q_{\rm e}Q_{\rm m} = 2\pi n\ , \qquad n\in\mathbf{Z}\ . \qquad (18)$$

The charge quantisation condition (18) is almost, but not quite, the full story. In deriving (18), we have not taken into account the p-form character of the charges. Taking this into account shows that the phase in (16) vanishes for a measure-zero set of configurations of the electric and magnetic branes.[1] This is easiest to explain in a simplified situation where the electric and magnetic branes are in static flat configurations, with the electric p-brane oriented along the directions $\{x^{M_1}\ldots x^{M_p}\}$. The phase factor (16) then becomes $\exp({\rm i}Q_{\rm e}\oint_{\mathcal{W}} A_{M_1\ldots M_pR}\partial x^R/\partial\sigma)$, where σ is an ordering parameter for the closed sequence of deformations $\mathcal{W}$. In making this sequence, we recall from the above discussion that one should restrict the deformations so as to preserve the asymptotic orientation of the deformed p-brane. For simplicity, one may simply consider moving the electric p-brane by parallel transport around the magnetic $\hat{p}$-brane in a closed loop. The accrued phase factor is invariant under gauge transformations of the potential $A_{[p+1]}$. This makes it possible to simplify the discussion by making use of a specially chosen gauge. Note that magnetic $\hat{p}$-branes have purely transverse field strengths like (7b); there is accordingly a gauge in which the gauge potential $A_{[p+1]}$ is also purely transverse, *i.e.* it vanishes whenever any of its indices point along a worldvolume direction of the magnetic $\hat{p}$-brane.

Now one can see how Dirac-insensitive configurations arise: the phase in (16) vanishes whenever there is even a partial alignment between the electric and the magnetic branes, *i.e.* when there are shared worldvolume directions between the two branes. This measure-zero set of Dirac-insensitive configurations may be simply characterised in terms of the p and $\hat{p}$ charges themselves by the condition $Q^{\rm el}_{[p]}\wedge Q^{\rm mag}_{[\hat{p}]} = 0$. For such configurations, one learns no Dirac quantisation condition. To summarise the overall situation, one may incorporate this restriction into the Dirac quantisation condition (18) by replacing it by the $(p+\hat{p})$-form quantisation condition

$$Q^{\rm el}_{[p]}\wedge Q^{\rm mag}_{[\hat{p}]} = 2\pi n\frac{Q^{\rm el}_{[p]}\wedge Q^{\rm mag}_{[\hat{p}]}}{|Q^{\rm el}_{[p]}|\,|Q^{\rm mag}_{[\hat{p}]}|}\ , \qquad n\in\mathbf{Z}\ , \qquad (19)$$

which reduces to (18) for all except the Dirac-insensitive set of configurations.

4. – KALUZA-KLEIN DIMENSIONAL REDUCTION

The panoply of p-brane solutions to supergravity theories is evident in the form of the D-dimensional action obtained by Kaluza-Klein dimensional reduction from the $D = 11$ theory (1). The $D = 11$ and reduced metrics are related by [8]

$$\begin{aligned} ds_{11}^2 &= e^{-\frac{1}{3}\vec{a}\cdot\vec{\phi}} ds_D^2 + \sum_i e^{(-\frac{4}{3}\vec{a}-\vec{a}_i)\cdot\vec{\phi}} (h^i)^2 &\quad (20a)\\ h^i &= dz^i + \mathcal{A}_1^i + \mathcal{A}_{0j}^i\, dz^j\ , &\quad (20b)\end{aligned}$$

where the reduction coordinates are denoted by z^i and the dilaton-vector coefficients $\vec{a}$, $\vec{a}_i$ determine the Einstein-frame couplings of the dilatonic scalars $\vec{\phi}$ to the various field strengths occurring in the D-dimensional reduced action

$$\begin{aligned} I_D = \int d^Dx\sqrt{-g}\Big[&R - \tfrac{1}{2}(\partial\vec{\phi})^2 - \tfrac{1}{48}e^{\vec{a}\cdot\vec{\phi}}F_{[4]}^2 - \tfrac{1}{12}\sum_i e^{\vec{a}_i\cdot\vec{\phi}}(F_{[3]}^i)^2 \\ &-\tfrac{1}{4}\sum_{i<j} e^{\vec{a}_{ij}\cdot\vec{\phi}}(F_{[2]}^{ij})^2 - \tfrac{1}{4}\sum_i e^{\vec{b}_i\cdot\vec{\phi}}(\mathcal{F}_{[2]}^i)^2 \\ &-\tfrac{1}{2}\sum_{i<j<k} e^{\vec{a}_{ijk}\cdot\vec{\phi}}(F_{[1]}^{ijk})^2 - \tfrac{1}{2}\sum_{ij} e^{\vec{b}_{ij}\cdot\vec{\phi}}(\mathcal{F}_{[1]}^{ij})^2\Big] + \mathcal{L}_{FFA}\ , \end{aligned} \qquad (21)$$

where the other dilaton vectors in (21) are determined in terms of $\vec{a}$, $\vec{a}_i$ by

$$\begin{aligned} \vec{a}_{ij} &= \vec{a}_i + \vec{a}_j - \vec{a} & \qquad \vec{b}_i &= -\vec{a}_i + \vec{a} \\ \vec{a}_{ijk} &= \vec{a}_i + \vec{a}_j + \vec{a}_k - 2\vec{a} & \qquad \vec{b}_{ij} &= -\vec{a}_i + \vec{a}_j\ . \end{aligned} \qquad (22)$$

The straight-backed $F_{[n]}$ field strengths in (21) arise from the reduction of $F_{[4]}$ in $D = 11$; the calligraphic $\mathcal{F}_{[n]}$ arise from the field strengths of Kaluza-Klein vectors.

In order to relate the electric and magnetic charges in the various dimensions, we need to use the reductions of the field strengths and their duals:

$$\begin{aligned} \hat{F}_{[4]} &= F_{[4]} + F_{[3]}^i \wedge h^i + \tfrac{1}{2}F_{[2]}^{ij}\wedge h^i \wedge h^j + \tfrac{1}{6}F_{[1]}^{ijk}\wedge h^i\wedge h^j\wedge h^k &\quad (23a)\\ {}^*\hat{F}_{[4]} &= e^{\vec{a}\cdot\vec{\phi}}\, {}^*F_{[4]}\wedge v + e^{\vec{a}_i\cdot\vec{\phi}}\, {}^*F_{[3]}^i \wedge v^i + \tfrac{1}{2}e^{\vec{a}_{ij}\cdot\vec{\phi}}\, {}^*F_{[2]}^{ij}\wedge v^{ij} + \tfrac{1}{6}e^{\vec{a}_{ijk}\cdot\vec{\phi}}\, {}^*F_{[1]}^{ijk}\wedge v^{ijk}\ , &\quad (23b)\end{aligned}$$

where the forms v, v_i, v_{ij} and v_{ijk} appearing in (23b) are given by

$$\begin{aligned} v &= \tfrac{1}{(11-D)!}\epsilon_{i_1\cdots i_{11-D}}\, h^{i_1}\wedge\cdots\wedge h^{i_{11-D}} & \qquad v_{ij} &= \tfrac{1}{(9-D)!}\epsilon_{iji_3\cdots i_{11-D}}\, h^{i_3}\wedge\cdots\wedge h^{i_{11-D}} \\ v_i &= \tfrac{1}{(10-D)!}\epsilon_{ii_2\cdots i_{11-D}}\, h^{i_2}\wedge\cdots\wedge h^{i_{11-D}} & \qquad v_{ijk} &= \tfrac{1}{(8-D)!}\epsilon_{ijki_4\cdots i_{11-D}}\, h^{i_4}\wedge\cdots\wedge h^{i_{11-D}} \end{aligned} \qquad (24)$$

The electric and magnetic charges in D dimensions take the forms

$$\begin{aligned} Q_{\rm e} &= \int \big(e^{\vec{c}\cdot\vec{\phi}}\,{}^*F + \kappa(A)\big) &\quad (25a)\\ Q_{\rm m} &= \int \tilde{F}\ , &\quad (25b)\end{aligned}$$

where $\tilde{F} = dA$, $F = \tilde{F}$ + (Kaluza-Klein modifications) (*i.e.* modifications involving lower-order forms arising in the dimensional reduction) and $\vec{c}$ is the dilaton vector corresponding to F in the dimensionally-reduced action (21). The term $\kappa(A)$ in (25a) is the analogue of the term $\frac{1}{2}A_{[3]}\wedge F_{[4]}$ in (3). From the expressions (23) for the reduced

Table 1. Relations between Q^{11} and Q^D

	$F_{[4]}$	$F^i_{[3]}$	$F^{ij}_{[2]}$	$F^{ijk}_{[1]}$
Electric $Q^{11}_{\rm e} =$	$Q^D_{\rm e} V$	$Q^D_{\rm e} \frac{V}{L_i}$	$Q^D_{\rm e} \frac{V}{L_i L_j}$	$Q^D_{\rm e} \frac{V}{L_i L_j L_k}$
Magnetic $Q^{11}_{\rm m} =$	$Q^D_{\rm m}$	$Q^D_{\rm m} L_i$	$Q^D_{\rm m} L_i L_j$	$Q^D_{\rm m} L_i L_j L_k$

field strengths and their duals, one obtains the relations between the original charges in $D = 11$ and those in the reduced theory given in Table 1, where $L_i = \int dz^i$ is the compactification period of the reduction coordinate z^i and $V = \int d^{11-D}z = \prod_{i=1}^{11-D} L_i$ is the total compactification volume. Note that the factors of L_i cancel out in the various products of electric and magnetic charges only for charges belonging to the same field strength in the reduced dimension D.

5. – CHARGE QUANTISATION CONDITIONS AND DIMENSIONAL REDUCTION

Now we shall tie together the various threads that have run through our story of the p-form charges. There are two main schemes for applying dimensional reduction in the context of supergravity p-brane solutions. The first of these is in fact just standard dimensional reduction, in which the reduction coordinate z is taken to be one of the worldvolume x^μ coordinates, on which p-brane solutions like (3,4) do not depend. Since this reduces simultaneously a spacetime dimension and a worldvolume dimension, one ends up with a $(p-1)$ brane in $(D-1)$ dimensions, hence this scheme is termed "diagonal" dimensional reduction.[9] Strictly speaking, the reduction procedure is not quite as simple as just declaring that solutions should be independent of z and checking that the p-brane solutions properly satisfy this condition: there is also a conventional requirement of ensuring that the reduced theory remains in the Einstein frame, as in (21). Arranging this, however, amounts simply to making a field-redefinition Weyl transformation.

The second dimensional reduction scheme uses the transverse space of a p-brane solution.[9] Since the isotropic p-brane solution given in (8) is not translationally invariant in its transverse directions, dimensional reduction cannot be directly effected using such an isotropic solution. However, any harmonic function $H(y)$ serves to determine a p-brane solution (6,7), and so one may generalise the $SO(D-d)$ isotropic solutions to solutions with $SO(D-d-1) \times \mathbb{R}$ symmetry, analogous to lines of charge in ordinary Maxwell electrodynamics. It is the translation invariance in the $\mathbb{R}$ direction that enables one to make a dimensional reduction. Another way to view this procedure is first to note that the single-charge-center solution (8) may be generalised to two-charge-center solutions (noting that the transverse Laplace equation has multiple-charge-center solutions). Continuing this procedure by adding further charge centers, one may construct a "deck" of charges, generating an $\mathbb{R}$ translational symmetry in a transverse direction as is needed for dimensional reduction.[9] Since this deck-stacking procedure leaves unchanged the worldvolume directions of the solution, it prepares for a reduction from an $SO(10-d) \times \mathbb{R}$ invariant p-brane deck in D dimensions down to an (isotropic) p-brane solution in $D-1$ dimensions. Accordingly, this second process is called "vertical" dimensional reduction.[9]

Finally, let us consider the various possible schemes for dimensional reduction in the presence of a dual electric/magnetic pair of branes. Depending on whether the reduction coordinate z belongs to the worldvolume or the transverse space of each brane, we have four reduction possibilities for the electric/magnetic pair: diagonal/diagonal, diagonal/vertical, vertical/diagonal and vertical/vertical. Only the mixed cases preserve dirac sensitivity in the lower dimension.

This is most easily illustrated by considering the diagonal/diagonal case, for which z belongs to the worldvolumes of both branes. With such a shared worldvolume direction, one has clearly fallen into the measure-zero set of Dirac-insensitive configurations with $Q^{\rm el}_{[p]} \wedge Q^{\rm mag}_{[\hat{p}]} = 0$ in the higher dimension D. Correspondingly, in $(D-1)$ dimensions one finds that the diagonally reduced electric $(p-1)$ brane is supported by an $n = p+1$ form field strength, but the diagonally reduced magnetic $(\hat{p}-1)$ brane is supported by an $n = p+2$ form; since only branes supported by the same field strength can have a Dirac quantisation condition, this diagonal/diagonal reduction properly corresponds to a Dirac-insensitive configuration.

Now consider the mixed reductions, *e.g.* diagonal/vertical. In performing a vertical reduction of a magnetic $\hat{p}$-brane by stacking up an infinite deck of single-center branes in order to create the $\mathbb{R}$ translational invariance necessary for the reduction, the total magnetic charge clearly will diverge. Thus, in a vertical reduction it is necessary to reinterpret the magnetic charge $Q_{\rm m}$ as a charge density per unit z compactification length. Before obtaining the Dirac quantisation condition in the lower dimension, it is necessary to restore a gravitational-constant factor of κ^2 that should have appeared in the quantisation conditions (18,19). As one may verify, the electric and magnetic charges as defined in (3,4) are not dimensionless. Thus, (18) in $D = 11$ should properly have been written $Q_{\rm e}Q_{\rm m} = 2\pi\kappa_{11}^2 n$. Accordingly, letting the compactification length be denoted by L, one obtains a Dirac phase $\exp({\rm i}\kappa_{D-1}^{-2} Q_{\rm e}Q_{\rm m}L)$ in the D-dimensional theory prior to reduction. However, this fits precisely with another aspect of dimensional reduction: the gravitational constants in dimensions D and $D-1$ are related by $\kappa_D^2 = L\kappa_{D-1}^2$. Thus, one obtains in dimension $D-1$ the expected quantisation condition $Q_{\rm e}Q_{\rm m} = 2\pi\kappa_{D-1}^2 n$. Note, correspondingly, that upon making a mixed diagonal/vertical reduction the electric and magnetic branes remain dual to each other, supported by the same $n = p-1+2 = p+1$ form field strength. The opposite mixed vertical/diagonal reduction case goes similarly, except that the dual branes are then supported by the same $n = p+2$ form field strength.

Finally, under vertical/vertical reduction, Dirac sensitivity is lost upon reduction, not because of the orientation of the branes, but because in this case both the electric and the magnetic charges need to be interpreted as densities per unit compactification length, and so one obtains a phase $\exp({\rm i}\kappa_D^{-2} Q_{\rm e}Q_{\rm m}L^2)$. Only one factor of L is absorbed into κ_{D-1}^2, and so one has $\lim_{L\to 0} L^2/\kappa_D^2 = 0$. Correspondingly, the two dimensionally reduced branes are supported by different field strengths: an $n = p+2$ form for the electric brane and an $n = p+1$ form for the magnetic brane.

REFERENCES

1. M. Bremer, H. Lü, C.N. Pope and K.S. Stelle, "Dirac quantisation conditions and Kaluza-Klein reduction," `hep-th/9710244`.
2. D.N. Page, *Phys. Rev.* D **28**, 2976 (1983).
3. G.W. Gibbons, G.T. Horowitz and P.K. Townsend, *Class. Quantum Grav.* **12**, 297 (1995).
4. E. Bergshoeff, E. Sezgin and P.K. Townsend, *Phys. Lett.* B **189**, 75 (1987).
5. P.K. Townsend, in *Proc. European Res. Conf. on Advanced Quantum Field Theory, La Londe-les-Maures, Sept. 1996*, `hep-th/9609217`.

6. R. Nepomechie, "Magnetic monopoles from antisymmetric tensor gauge fields," *Phys. Rev.* D **31**, 1921 (1985).
7. C. Teitelboim, *Phys. Lett.* B **67**, 63, 69 (1986).
8. H. Lü and C.N. Pope, *p-brane solitons in maximal supergravities*, *Nucl. Phys.* B **465**, 127 (1996), hep-th/9512012.
9. H. Lü, C.N. Pope and K.S. Stelle, "Vertical versus diagonal dimensional reduction for *p*-branes," *Nucl. Phys.* B **481**, 313 (1996), hep-th/9605082.

SECTION II
Neutrino Masses

KARMEN-UPGRADE: IMPROVEMENT IN THE SEARCH FOR NEUTRINO OSCILLATIONS AND FIRST RESULTS

J. Kleinfeller,[1] B. Armbruster,[1] I.M. Blair,[2] B.A. Bodmann,[3] N. E. Booth,[4] G. Drexlin,[1] V. Eberhard,[1] J.A. Edgington,[2] C. Eichner,[5] K Eitel,[1] E. Finckh,[3] H. Gemmeke,[1] J. Hößl,[3] T. Jannakos,[1] P. Jünger,[3] M. Kleifges,[1] W. Kretschmer,[3] R. Maschuw,[5] C. Oehler,[1] P. Plischke,[1] J. Rapp,[1] C. Ruf,[5] M. Steidl,[1] J. Wolf,[1] and B. Zeitnitz,[1]

[1]Institut für Kernphysik I, Forschungszentrum Karlsruhe, Institut für experimentelle Kernphysik, Universität Karlsruhe, D-76021 Karlsruhe, Postfach 3640, Germany
[2]Physics Department, Queen Mary and Westfield College Mile End Road, London E1 4NS, United Kingdom
[3]Physikalisches Institut, Universität Erlangen-Nürnberg Erwin Rommel Straße 1, D-91058 Erlangen, Germany
[4]Department of Physics, University of Oxford Keble Road, Oxford OX1 3RH, United Kingdom
[5]Institut für Strahlen- und Kernphysik, Universität Bonn Nußallee 14-16, D-53115 Bonn, Germany

INTRODUCTION

KARMEN, the **Ka**rlsruhe–**R**utherford **M**edium **E**nergy **N**eutrinoexperiment at the pulsed spallation neutron facility ISIS uses the beam stop neutrinos ν_μ, ν_e and $\bar{\nu}_\mu$ from π^+ and μ^+ decay at rest to search for neutrino oscillations in the appearance channels $\nu_\mu \rightarrow \nu_e$ and $\bar{\nu}_\mu \rightarrow \bar{\nu}_e$. The signature for both oscillations is based on charged current neutrino nuclear interaction spectroscopy in a high resolution 56 t liquid scintillator calorimeter. The data acquired from 1990 to 1995 are equivalent to 9122 Coulomb of protons on target. The KARMEN experiment has found no evidence for ν–oscillations in either of the investigated channels. The limits in $\sin^2 2\theta$ derived from the analysis of these data are $\sin^2 2\theta < 4 \times 10^{-2}$ for the $\nu_\mu \rightarrow \nu_e$ channel and $\sin^2 2\theta < 8.6 \times 10^{-3}$ for the $\bar{\nu}_\mu \rightarrow \bar{\nu}_e$ channel. Details of the neutrino source ISIS, the detector and experimental results obtained by KARMEN from 1990–1995 have been published elsewhere [1–2].

This report will focus on a description of the upgrade of KARMEN1 to KARMEN2 which has been carried out in 1996 and first results with the upgrade in operation. (Data acquisition February – September 1997, 1414 Coulomb of protons on target).

Physics of Mass
Edited by Kursunoglu *et al.*, Kluwer Academic / Plenum Publishers, New York, 1999

KARMEN UPGRADE

The sensitivity of the KARMEN1 experiment to neutrino oscillations in the $\nu_\mu \rightarrow \nu_e$ channel is limited by statistics only because the signature eliminates almost any cosmogenic background [1–2]. On the other hand, the sensitivity in the $\bar{\nu}_\mu \rightarrow \bar{\nu}_e$ channel is mainly limited by background consisting of high energy neutrons created by cosmogenic muons in the massive steel blockhouse. This background can only be eliminated through tagging of the muons which generate the neutrons in deep inelastic scattering reactions in the steel. The detector system has therefore been upgraded with an additional veto layer embedded in the walls and the roof of the shielding blockhouse (fig. 1). There is at least 1 m of steel between the new veto layer and the central detector. Neutrons created outside of the new veto by un–tagged muons are sufficiently suppressed (1 % of primary intensity), neutrons created by tagged muons within the space enclosed by the veto can be identified and discarded.
The new veto counter consists of 136 slabs of BC–412 (Bicron USA), each 50 mm thick and 650 mm wide; the lengths vary from 3 m to 4 m, the total area is 301 m^2. The veto counter system covers 84.3 % of the solid angle. Each module is read out by a set of 4 gain–matched XP2262 2"–phototubes (Philips) on either end. The scintillator modules are sandwiched between boron loaded polythene slabs. Design and position of the veto modules was determined by extensive Monte Carlo simulations to achieve a maximum in neutron suppression.

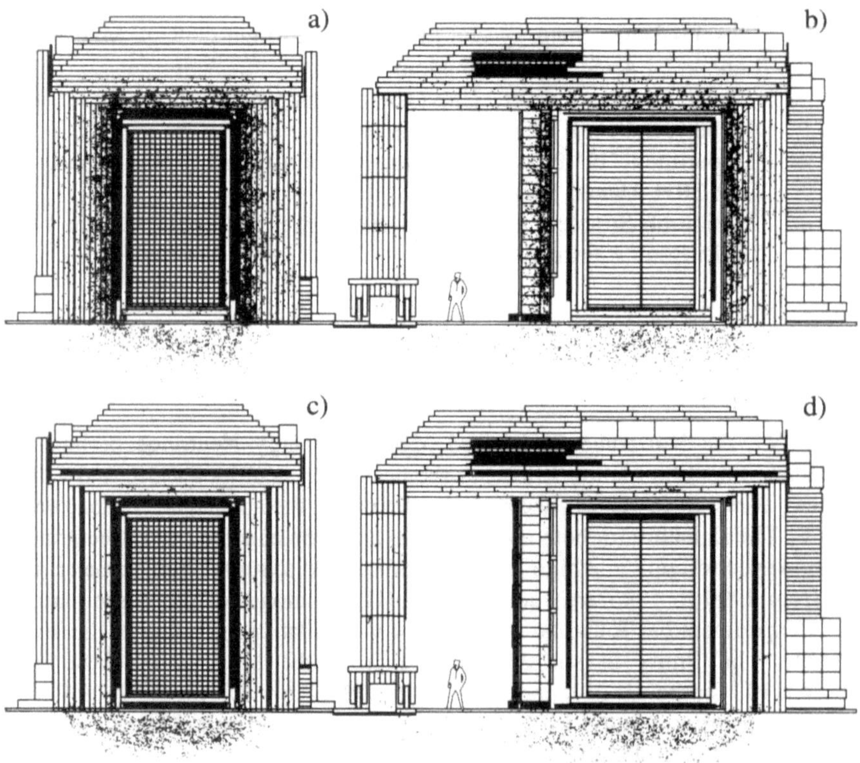

Figure 1. Front and side views of the KARMEN detector system inside the shielding blockhouse with Monte Carlo simulation of the points of creation of high energy neutrons due to cosmogenic μ, (a+b) before the upgrade installation, only 10 % of the simulated events are displayed, (c+d) after the upgrade installation with veto rejection enabled, full simulation sample. Neutron production in the roof area above the veto counters is not displayed in this figure.

FIRST RESULTS

KARMEN2 is taking data since February 1997. The efficiency of the veto counter system for cosmic ray muons has been measured to be 99.6 % in very good agreement with the Monte Carlo expectation of 99.4 %. The analysis of the first data (1414 Coulomb) shows that the desired background reduction has been achieved. For energies above 20 MeV the cosmogenic background in the spectra of single prong events (fig. 2a) is now only 3 % of its previous level. The sequential background (fig. 2b) limiting the sensitivity in the $\bar{\nu}_\mu \rightarrow \bar{\nu}_e$ oscillation channel is reduced to 2 % of its previous level. The irreducible background below 20 MeV is due to nuclear capture reactions of stopped μ^- in the central detector, it does not contribute to the background in the $\bar{\nu}_\mu \rightarrow \bar{\nu}_e$ oscillation search which is focused on the energy range from 20–50 MeV.

The successful suppression of cosmogenic background allows neutrino reactions to be analysed on an event to event basis. KARMEN2 is now capable of background free neutrino spectroscopy. The remaining number of cosmic ray induced background with a signature similar to $\bar{\nu}_\mu \rightarrow \bar{\nu}_e$ oscillations is only one event per year of data acquisition if the duty cycle factor of ISIS is taken into account.

A scan of the data for $\bar{\nu}_\mu \rightarrow \bar{\nu}_e$ signatures with the veto rejection enabled ends in a null result. The expected cosmic ray induced background amounts to only 0.26 events, whereas the total expected background rate including random coincidences and signatures from neutrino nucleus interactions is 2.9 events. This means more than 2.31 events due to $\bar{\nu}_\mu \rightarrow \bar{\nu}_e$ transitions can be excluded at the 90 % CL. Oscillations with probability P=1 would have produced 375 event signatures, we therefore deduce a limit in the 2-ν mixing amplitude of $\sin^2 2\theta < 6.2 \times 10^{-3}$ (90 % CL.). The corresponding exclusion plot is displayed in figure 3 in comparison with the results of KARMEN1 and the 90 % and 95 % likelihood regions favoured by the LSND [3] experiment for positive evidence of neutrino oscillations. This preliminary result from KARMEN2 is consistent with the result obtained from the KARMEN1 measurements of 1990–1995, it is also not yet in conflict with the positive result of LSND.

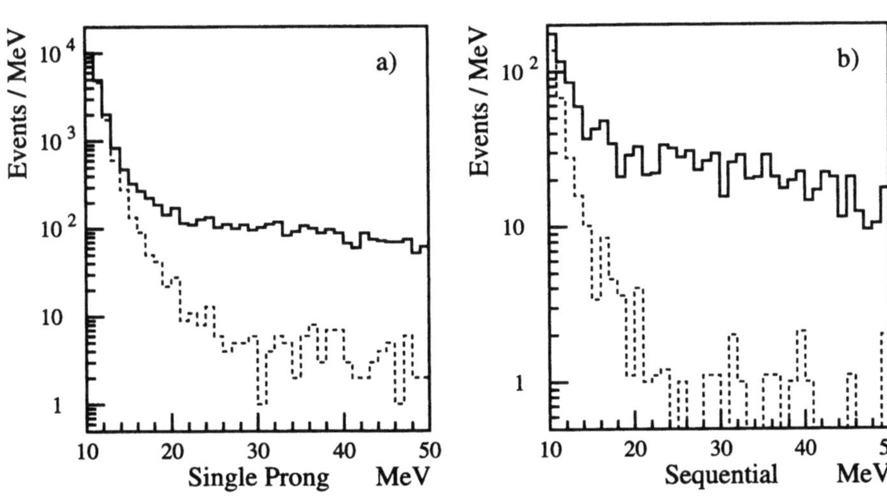

Figure 2. Background suppression for single prong events (a) and sequential signatures (b). Solid lines veto rejection disabled, dashed lines veto rejection enabled.

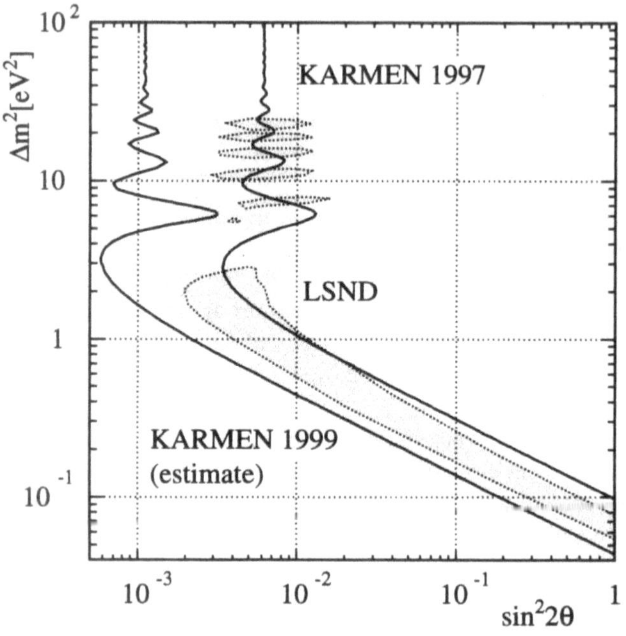

Figure 3. KARMEN2 90 % CL. exclusion for $\bar{\nu}_\mu \rightarrow \bar{\nu}_e$ oscillations (1414 C, February-September 1997) and expected 90 % CL. exclusion after 3 years of measurement, if no signal is seen.

ESTIMATE OF FUTURE KARMEN2 SENSITIVITY

The KARMEN experiment is scheduled to run for another 2 years. Due to the extremely low level of cosmic ray induced background the sensitivity of KARMEN2 will increase linearly with measuring time. The expected sensitivity of KARMEN2 (1997-99) if no oscillation is observed is also shown in figure 3. This estimate is based on an accumulated proton charge of 7500 C (corresponding to 3 ISIS years with 200 μA proton beam intensity and 90 % reliability) and has been tested in an analysis of thousands of Monte Carlo generated data samples either by maximum likelihood analysis (20-50) MeV or by a basic window method (36-50 MeV). The improvement in sensitivity with respect to the oscillation limits extracted from the data taken by KARMEN1 from 1990 to 1995 is almost one order of magnitude. KARMEN2 will then probe the entire parameter space favoured by LSND.

ACKNOWLEDGMENTS

We gratefully acknowledge the financial support of the German Bundesministerium für Bildung, Wissenschaft, Forschung und Technologie (BMBF)

REFERENCES

1. B. Zeitnitz et al. Neutrino-nuclear interactions with KARMEN, Prog. in Part. and Nucl. Phys. 32:351 (1994).
2 G. Drexlin et al. KARMEN: Precision tests of the standard model with neutrinos from muon and pion decay, Prog. in Part. and Nucl. Phys. 32:375 (1994).
3. C. Athanassopoulos et al. (LSND), Evidence for neutrino oscillations from muon decay at rest, Phys. Rev. C 54:2685 (1996).

ATMOSPHERIC NEUTRINO FLUX STUDIES WITH THE SUPER-KAMIOKANDE DETECTOR

Lawrence Wai

Physics Department
University of Washington
Seattle, WA 98195

Super-Kamiokande Collaboration

INTRODUCTION

Super-Kamiokande (see figure 1) is a cylindrical 50 kiloton ring imaging water Cherenkov detector. An optical barrier separates the inner volume of water, used for the basic measurements, from the outer layer, which is used for anti-coincidence of cosmic ray muons. The inner cylindrical volume is lined with 11,146 50-cm diameter photomultiplier tubes (PMT) which view inwards upon a fiducial volume of 22.5 kilotons of purified water. The fiducial volume is defined as the region at least 2 meters away from the surface of the inward facing PMTs. The outer detector is 2.6 meters thick, instrumented with 1,885 20-cm PMTs, and completely surrounds the inner volume.

In addition to observing solar neutrinos, the detector is used to measure atmospheric neutrinos originating in the decays of pions produced in hadronic showers from primary cosmic rays high in the atmosphere. The flux of atmospheric neutrinos has been studied in some detail [2]. The basic expectation is that there should be roughly 2 muon neutrinos for every electron neutrino. The distribution of light in the Cherenkov ring allows μ/e separation at the 99% level. See figure 2 for an example of a μ-like event in the detector.

PRELIMINARY RESULTS ON ν_μ DISAPPEARANCE

At the time of the writing of this report, the Super-Kamiokande detector has collected 25.5 kiloton-years of analyzed atmospheric neutrino data. The basic results are:

- ratio of μ-like to e-like events significantly smaller than expected (see table 1)
- an upward/downward ratio for μ-like events significantly smaller than expected (see figure 3)

Physics of Mass
Edited by Kursunoglu *et al.*, Kluwer Academic / Plenum Publishers, New York, 1999

detector	sub-GeV observed	sub-GeV expected	multi-GeV observed	multi-GeV expected
Kamiokande	0.60	1.00 ± 0.12	0.57	1.00 ± 0.24
Super-Kamiokande	0.61	1.00 ± 0.09	0.66	1.00 ± 0.15

Table 1. $(\mu/e)_{DATA}/(\mu/e)_{MC}$ ratios in Kamiokande and Super-Kamiokande for the Honda flux. The errors are shown for the null hypothesis.

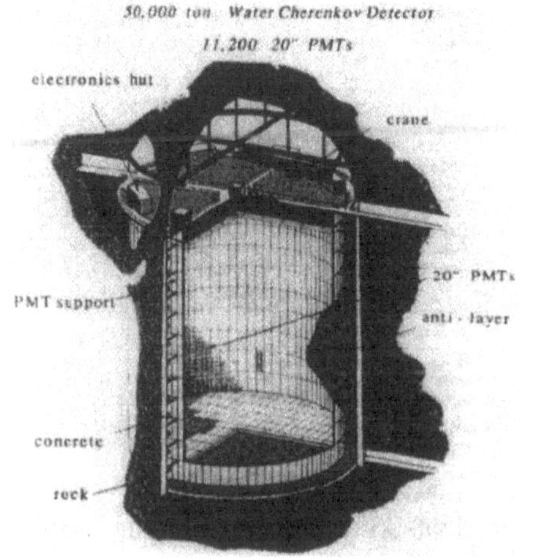

Figure 1: A sketch of the Super-Kamiokande detector.

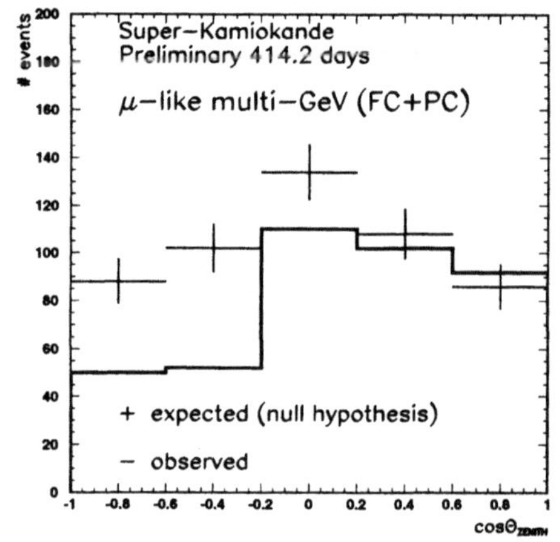

Figure 3: Zenith angle distribution for 414.2 days of Super-Kamiokande data (preliminary result).

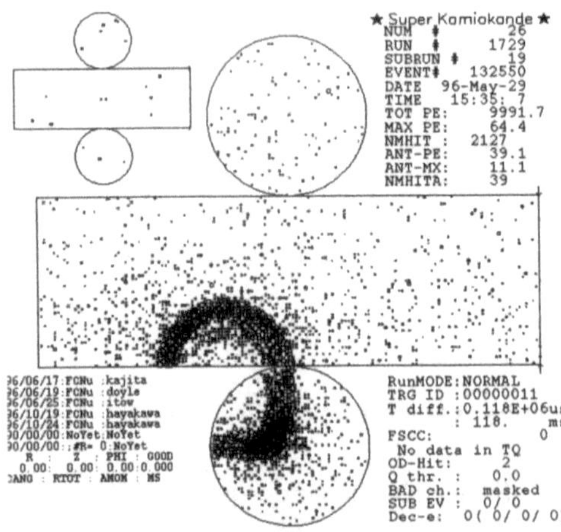

Figure 4: Preliminary confidence region obtained from the Super-Kamiokande data using the same analysis program used to obtain the Kamiokande result.

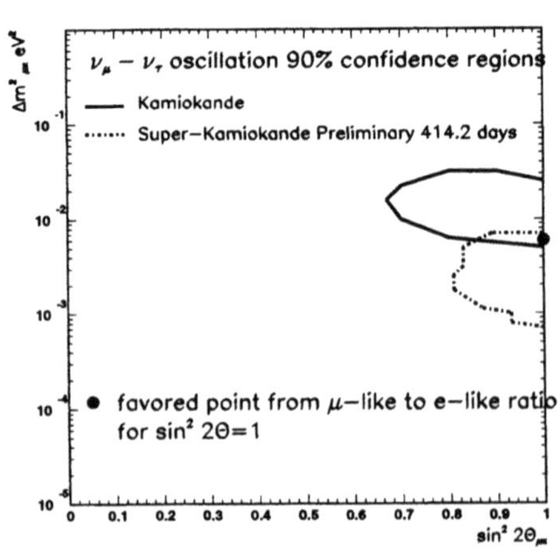

Figure 2: An event display of a μ-like event. The cylindrical surfaces of PMTs have been peeled back in this display.

A preliminary result using the same analysis program applied to the Kamiokande data [1] has been obtained for the $\nu_\mu \rightarrow \nu_\tau$ oscillation case,[1] shown in figure 4. The Super-Kamiokande 90% confidence region has a rather flat χ^2 dependence, and does not strongly favor any particular point. However, the average μ-like to e-like ratio favors a well defined Δm^2 for $sin^2 2\Theta = 1$, and this point agrees with the overlap of the Super-Kamiokande and Kamiokande confidence regions.

References

[1] Kamiokande Collaboration. *Atmospheric ν_μ/ν_e ratio in the multi-GeV energy range*, Physics Letters B 335 (1994), pp.237-245.

[2] M.Honda et al., Phys. Rev. **D52**(1995)4985

[1] Courtesy of Kenji Kaneyuki, Tokyo Institute of Technology.

NUMERICAL FIELD THEORY ON THE CONTINUUM

Stephen Hahn and G. S. Guralnik
Department of Physics, Brown University
Providence , RI 02912-1843

Abstract

An approach to calculating approximate solutions to the continuum Schwinger–Dyson equations is outlined, with examples for ϕ^4 in $D = 1$. This approach is based on the source Galerkin methods developed by García, Guralnik and Lawson. Numerical issues and opportunities for future calculations are also discussed briefly.

1 Introduction

The now conventional technique of numerical calculation of quantum field theory involves evaluating the path integral

$$\mathcal{Z}(J) = \int_{\Gamma} d\phi \exp\left[-\mathcal{S}(\phi) + J\phi\right] \tag{1}$$

on a spacetime lattice using Monte Carlo integration methods. Monte Carlo methods have been successful for an interesting class of problems; however, these techniques have had little success in evaluating theories with actions that are not manifestly positive definite or which have important effects from the details of fermionic interactions beyond the quenched approximation. In these cases, we have respectably an algorithmic failure or a massive inadequacy of compute power.

We are constructing an alternative computational method which works both on the lattice and the continuum and which handles fermions as easily as bosons. Furthermore, our "source Galerkin" method is less restrictive as to the class of allowed actions. Source Galerkin tends to use significantly less compute time than Monte Carlo methods but can consume significantly larger amounts of memory.[1] This talk is confined to continuum applications; examples of lattice calculations have been given elsewhere [2, 7, 6]. While it is not clear that Source Galerkin can replace Monte Carlo techniques, it appears that it will be able to solve some problems which are currently inaccessible.

Our approach begins with the differential equations satisfied by the vacuum functional $\mathcal{Z}$ for a quantum field theory with external sources. For the sake of simplicity

[1]That is, the iterative process for improving a source Galerkin calculation involves successively higher-point Green functions, whereas calculation of these correlations in a Monte Carlo scenario is optional.

in this talk we will mostly confine our attention to ϕ^4 interactions although we have studied non-linear sigma models and four fermion interactions and have gauge theory calculations in progress.

The vacuum persistence function $\mathcal{Z}$ for a scalar field ϕ with interaction $g\phi^4/4$ coupled to a scalar source $J(x)$ satisfies the equation:

$$(-\partial_D^2 + M^2)\frac{\delta \mathcal{Z}}{\delta J(x)} + g\frac{\delta^3 \mathcal{Z}}{\delta J^3(x)} - J(x)\mathcal{Z} = 0 \tag{2}$$

The source Galerkin technique is designed to directly solve functional differential equations of this type. Before we proceed to outline a solution technique, it is essential to point out that this equation by itself does not uniquely specify a theory [3].

This is dramatically illustrated by considering the special case of the above equation limited to one degree of freedom (zero dimensions).

$$M^2\frac{d\mathcal{Z}}{dJ} + g\frac{d^3\mathcal{Z}}{dJ^3} - J\mathcal{Z} = 0 \tag{3}$$

This is a third order differential equation, and therefore possesses three independent solutions. It is easy to see that one of the solutions for small g asymptotically approaches the perturbation theory solution, while the second asymptotically approaches the loop expansion "symmetry breaking" solution for small for small g and the third solution has an essential singularity as g becomes small. The situation becomes much more interesting for finite dimensions where the infinite class of solutions coalesce or become irrelevant in a way which builds the phase structure of the field theory. Any numerical study must be cognizant of the particular boundary conditions and hence solution of the class of solutions that is being studied. Care must be taken to stay on the same solution as any iterative technique is applied. The solutions discussed in this talk will be of the conventional nonperturbative sort which correspond to the solutions obtained from evaluating a path integral with the usual definitions for the region of integration. These are the solutions that are regular in the coupling, g.

We can for convenience write (2) in the form:

$$\hat{E}_j \mathcal{Z}(j) = 0 \tag{4}$$

The source Galerkin method is defined by picking an approximation $\mathcal{Z}^*(j)$ to the solution $\mathcal{Z}(j)$ such that

$$\hat{E}_j \mathcal{Z}^*(j) = R \tag{5}$$

where R is a residual dependent on j and the further requirement that this residual as small as possible on the average. To give this statement a meaning, we must define an inner product over the domain of j: *i.e.* $(A, B) \equiv \int d\mu(j)\, A(j)B(j)$. In addition we assume we have a collection of test functions which are members of a complete set) $\{\varphi_i(j)\}$ The source Galerkin minimization of the residual R is implemented by setting the parameters of our test function $\mathcal{Z}^*(j)$ so that projections of test functions against the residual vanish so that $||\mathcal{Z}^* - \mathcal{Z}||_2 \to 0$ as the number of test functions $\to \infty$.

The equations defining the quantum field theory are differential equations in the field sources and spacetime. While it is straightforward to deal with the spacetime problem by resorting to a lattice, we can deal with the continuum by taking advantage if our knowledge of functional integration. We know how to evaluate Gaussian functional integrals on the continuum:

$$\int [dj] \exp\left[\int_{xy} j(x)A(x,y)j(y) + \int_x j(x)\beta(x)\right]$$

$$= \frac{1}{\sqrt{\det A}} \exp\left[\int_{xy} \beta^*(x) A^{-1}(x,y)\beta(y)\right] \tag{6}$$

Consequently, we can evaluate integrals of the form:

$$I = \int [dj] \exp\left[-j^2(x)/\epsilon^2\right] P(j) \tag{7}$$

Using this we can define an inner product of sources on the continuum as follows:

$$(j(x_1)\cdots j(x_n), j(y_1)\cdots j(y_m))_j = \begin{cases} \epsilon^{n+m}\delta_+\{x_1\cdots x_n y_1\cdots y_m\} & n+m \text{ even} \\ 0 & \text{otherwise} \end{cases} \tag{8}$$

where we have absorbed a factor of 2 by redefining ϵ. δ_+ is defined by

$$\delta_+\{x\alpha\beta\cdots\} = \delta(x-\alpha)\delta_+\{\beta\cdots\} + \delta(x-\beta)\delta_+\{\alpha\cdots\} + \cdots, \tag{9}$$

$$\delta_+\{x\alpha\} = \delta(x-\alpha). \tag{10}$$

In addition to this inner product definition, we need good guesses for approximate form for $\mathcal{Z}^*$ and numerical tools to calculate, symbolically or numerically, various functions and their integrals, derivatives, and so on. For most of our calculations we have found it very useful to choose a lesser known class of functions, with very suitable properties for numerical calculation, known as Sinc functions. We take our notation for the Sinc functions from Stenger [8]:

$$S(k,h)(x) = \frac{\sin(\pi(x-kh)/h)}{\pi(x-kh)/h} \tag{11}$$

Some of the identities that Sinc approximations satisfy are given below:

$$S(k,h)(lh) = \delta_{kl} \tag{12}$$

$$\int_x S(k,h)(x)S(l,h)(x) = \delta_{kl} \tag{13}$$

$$\int_x F(x) \approx h\sum_{k=-N}^{N} F(kh) \tag{14}$$

$$F(x) \approx \sum_{k=-N}^{N} F(kh)S(k,h)(x) \tag{15}$$

$$F'(x) \approx \sum_{k=-N}^{N} F(kh)S'(k,h)(x) \tag{16}$$

$$F^{(n)}(lh) \approx \sum_{k=-N}^{N} F(kh)\delta^{(n)}_{k-l} \tag{17}$$

These properties, whice are proven and expanded upon greatly in [8], make these functions very easy to use for Galerkin methods, collocation, integration by parts, and integral equations.

With the definition of a norm and set of expansion functions, we can postulate an ansätz for $\mathcal{Z}$

$$\mathcal{Z}^* = \exp\left[\sum \int_{xy} j(x)G_2(x-y)j(y) + \cdots\right] \tag{18}$$

where each Green function, G_n, is represented by a d-dimensional Sinc expansion. For $d = 4$:

$$G_2(x-y) = \sum G_2^{ijkl}$$
$$S(i,h)(x^0-y^0)S(j,h)(x^1-y^1)S(k,h)(x^2-y^2)S(l,h)(x^3-y^3) \quad (19)$$

It is easy to examine this expansion in the case of a free field where we limit the approximation to the terms quadratic in $j(x)$. While the example is trivial, it shows that, as always, numerical approximations must be handled with care. Results of this calculation are shown in Figure 1.

This straightforward expansion works fine for interacting theories with more than G_2, the computational costs and storage costs become overwhelming: G_{2n} requires $N^{(2n-1)d}$ storage units. There are many ways that storage costs can be reduced, but in general these approaches are difficult, not particularly elegant, and eventually reaches a limit due to the exponential growth in the number of coefficients of the representation.

We can use our knowledge of the spectral representations of field theory and graphical approaches to introduce a much more beautiful and intuitive approach to producing candidates for $\mathcal{Z}^*$. We introduce regulated Lehmann representations. These build in the appropriate spacetime Lorentz structure into our approximations and remove the growth of operational cost with spacetime dimension shown by our previous naïve decomposition into complete sets of functions. Any exact two-point function can be represented as a sum over free two-point functions. We choose as the basis of our numerical solutions, a regulated Euclidean propagator structure:

$$\Delta(m;x) \equiv \int (dp) \frac{e^{ip\cdot x - p^2/\Lambda^2}}{p^2+m^2} = \int (dp) \int_0^\infty ds\, e^{ip\cdot x - p^2/\Lambda^2 - s(p^2+m^2)} \quad (20)$$

$$= \frac{1}{(2\pi)^d} \int_0^\infty ds \left[\frac{\pi}{s+1/\Lambda^2}\right]^{d/2} e^{-sm^2 - \frac{x^2}{4(s+1/\Lambda^2)}} \quad (21)$$

This regulation assures that we never have to deal with infinities in any calculated amplitude as long as we keep the cutoff finite.

This integral can be approximated using Sinc methods

$$\Delta(m;x) \approx \frac{h}{(2\pi)^d} \sum_{k=-N}^{N} \frac{1}{e^{kh}} \left[\frac{\pi}{z_k+1/\Lambda^2}\right]^{d/2} \exp\left[-z_k m^2 - \frac{x^2}{4(z_k+1/\Lambda^2)}\right], \quad (22)$$

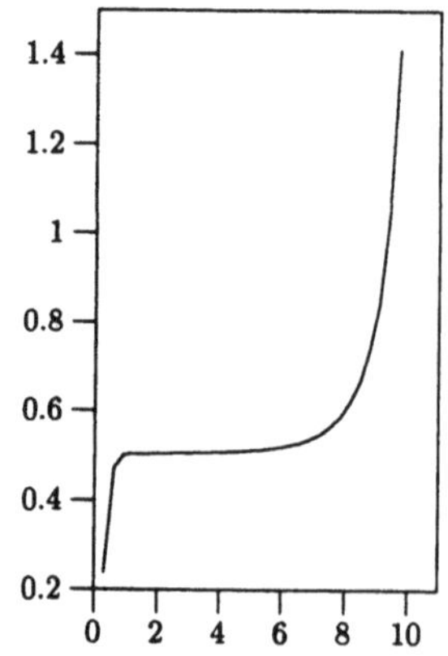

Figure 1. Mass, m^*, versus distance, x, for two-dimensional free scalar field. Note breakdown near origin (approximation of δ-function) and at large distance (spatial truncation). $m_0 = 0.5$.

Exact	0.654038297612956387655447
$N = 10$	0.686283404027993439164852
20	0.655465233659481763033090
40	0.654037798068595478871392
80	0.654038297613089305988864
100	0.654038297612956949544002

Table 1. Convergence of Sinc approximation to integral. (Exact integral calculated using Maple V, with 40 digits of precision.) $x^2 = 10$, $m = 1$, $\Lambda^2 = 10$.

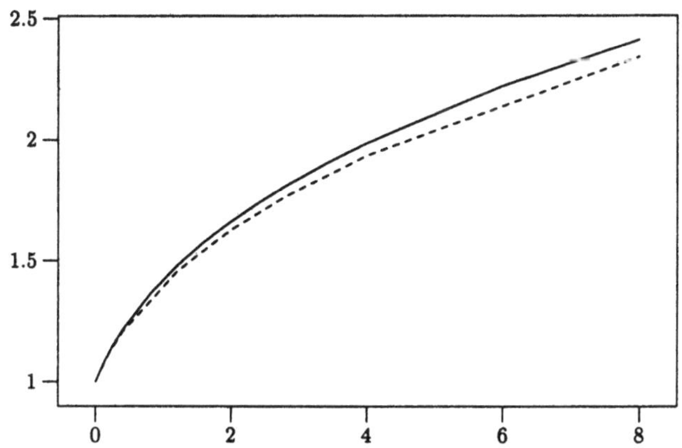

Figure 2. One dimensional ϕ^4 mass gap versus coupling (dashed line gives exact from Hioe and Montroll, 1975)

The example in Table 1 demonstrates that we can have as many digits as are necessary for the calculation, with the associated increase in compute time. For practical purposes, 80 terms is appropriate for most hardware floating-point representations. Thus we have a form for a two-point scalar Green function, regulated by the scale Λ^2 with constant computational cost regardless of spacetime dimension. We can take derivatives explicitly or by construction:

$$\partial^2 \Delta(m;x) = m^2 \Delta(m;x) - \bar{\delta}(x) \tag{23}$$

where $\bar{\delta}(x) = e^{-x^2\Lambda^2/4}$

From this representation, we can directly construct a fermion two-point function:

$$S(m;x) = (\gamma \cdot \partial - m)\Delta(m;x) \tag{24}$$

These representations mean that free scalar and free fermion results are exact and immediate in any Galerkin evaluation of these trivial cases. Furthermore, because of this simplicity, we have the basis for a complete numerical approach to conventional perturbation theory.

2 Results with Lehmann representation: ϕ^4

We itemize some results obtained using a regulated single propagator with parameters set by the Source Galerkin method. At lowest order, our ansätz for the generating functional is

$$\mathcal{Z}^* = \exp \int \frac{1}{2} j_x G_{xy} j_y. \tag{25}$$

Results for this ansätz are given in Figure 2. These results are strikingly accurate and can matched up essentially exactly with results of Monte Carlo calculations in two and higher dimensions.

We can enhance these results by including additional 4 source terms in $\mathcal{Z}^*$. Some simple additional terms that we include with weights and masses to be calculated using the Source Galerkin technique are the terms of the forms given in Figure 3. The effect of adding a fourth order term is shown in Figure 4.

In addition to the illustrations given here, we have examined $(\bar{\psi}\psi)^2$ in the mean field in two dimensions and have found rapid convergence to the known results from large-N expansions. It therefore appears that, at least for the simple cases studied, we have produced a numerical method which draws on the structural information already known in general through symmetry and spectral representations which when combined with Galerkin averaging to set parameters converges with very simple guesses for the vacuum amplitude to known correct answers produced through other methods of solution including Monte Carlo methods. More complicated gauge problems are under study.

3 Numerical issues

In this very brief presentation we have avoided discussion of many of the difficult numerical issues involved in constructing this approach. We note some of these issues here without discussion to have on record.

- both interpolative and spectral problems result in medium- to large-scale nonlinear systems; systems solvable using many variable Newton's method
- finite storage is the key constraint for interpolative representations, which must be constrained to two-point 'connectors', particularly in high dimensions

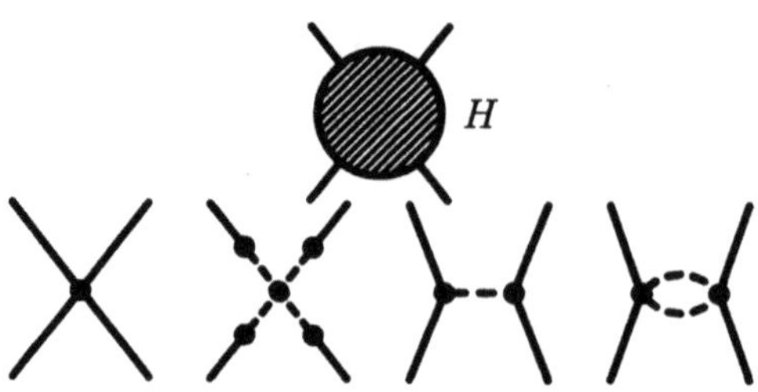

Figure 3. Additional connector-based ansätzen for the four-point function, H, in $\lambda\phi^4$. In the bottom row, we have two contact ansätzen on the left, followed by two mediated ansätzen.

- storage is a non-issue for spectral representations (memory use entirely for caching calculated quantities)
- both methods are also time-bound to 'connector'-based representations for higher point functions
- time cost from internal loops; however, algorithm can be made parallel via partitioning of sums
- resolution of elementary pole structures (*i.e.* differentiating between δ and $(p^2+m^2)^{-1}$) may be addressable with arbitrary precision libraries
- arbitrary precision may also be useful for regulated perturbation calculations

These numerical issues, and related general numerical techniques for source Galerkin are discussed in greater detail in [1, 5, 4].

4 Conclusions

We have discussed a method for numerical calculations for field theories on the continuum; this method being based on the source Galerkin technique introduced in [2]. The direct approach, using Sinc functions for interpolation, is effective but will ulti-

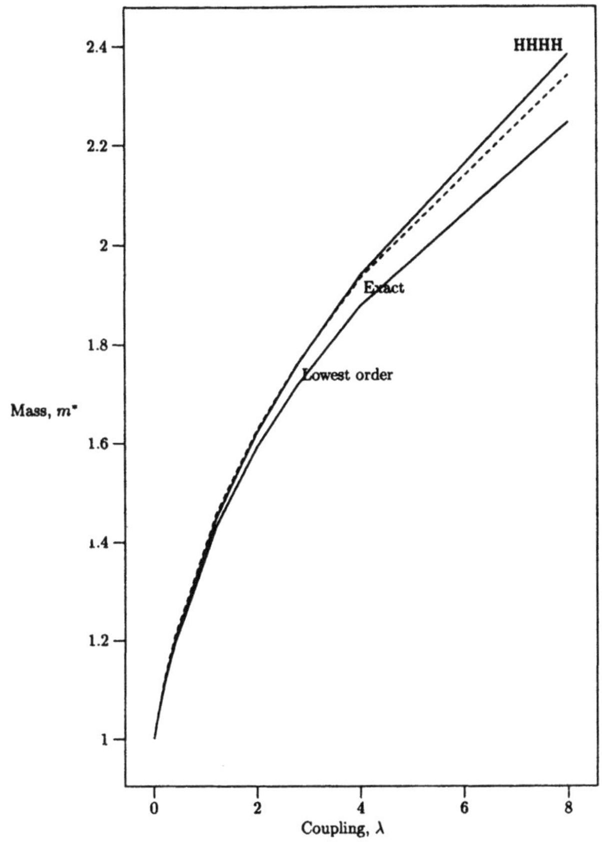

Figure 4. Comparison of four-H approximation with lowest order and exact answer, in one-dimension ($\Lambda^2 = 70$).

mately be limited by the finite nature of current computate resources. The Lorentz-invariant regulated representation derived from Lehmann representation does not suffer from these limitations, and is applicable to both perturbation- and mean-field-based ansätzen. This approach has the computational advantages of minimal memory utilization and parallelizable algorithms and also allows direct representation of fermionic Green functions. Finally, a number of useful peripheral calculations can made using this appoximate representation: one can calculate diagrams in a regulated perturbation theory, as well as calculating dimensionally regularized loops numerically. In general, this technique of evaluating field theories takes advantage of the symmetries of the Lorentz group; future work includes the extension of the method to more general internal groups, such as gauge groups or supersymmetry.

Acknowledgements

This work was supported in part by U. S. Department of Energy grant DE FC00-01-ER-40588—Task D. The authors have been the beneficiaries of many valuable conversations with S. García, Z. Guralnik, J. Lawson, K. Platt, and P. Emirdağ. Certain results in this work were previously published in Hahn [4]. Computational work in support of this research was performed at the Theoretical Physics Computing Facility at Brown University.

References

[1] S. García. *A new numerical method for quantum field theory.* PhD thesis, Brown University, 1993.

[2] S. García, G. S. Guralnik, and J. W. Lawson. A new approach to numerical quantum field theory. *Physics Letters*, B333:119, 1994.

[3] S. García, Z. Guralnik, and G. S. Guralnik. Theta vacua and boundary conditions of the Schwinger–Dyson equations. `hep-th/9612079`, 1996.

[4] S. C. Hahn. *Functional methods of weighted residuals and quantum field theory.* PhD thesis, Brown University, 1998.

[5] J. W. Lawson. *Numerical method for quantum field theory.* PhD thesis, Brown University, 1994.

[6] J. W. Lawson and G. S. Guralnik. New numerical method for fermion field theory. *Nuclear Physics*, B459:612, 1996. `hep-th/9507131`.

[7] J. W. Lawson and G. S. Guralnik. Source Galerkin calculations in scalar field theory. *Nuclear Physics*, B459:589, 1996. `hep-th/9507130`.

[8] F. Stenger. *Numerical Methods Based on Sinc and Analytic Functions.* Springer Series in Computational Mathematics. Springer–Verlag, 1993.

SECTION III
Progress on New and Old Ideas

LEPTOQUARKS REVISITED

Paul H. Frampton

Department of Physics
University of North Carolina
Chapel Hill, NC 27599-3255

DATA FROM HERA

In the second half of February 1997, the two collaborations H1[1] and ZEUS[2] working on e^+p collisions at HERA: e^+(27GeV) + p(820GeV) simultaneously submitted to Z.Phys. announcements of small-statistics discrepancies from the Standard Model(SM). The two papers can also be downloaded from the World Wide Web at the URL: http://info.desy.de/

From the Web we may learn much interesting information. For example H1 has 400 members from 12 countries while ZEUS has 430 members from 12 countries.

As for physics, H1 finds 12 events with $Q^2 > 15,000\text{GeV}^2$ where the SM predicts 4.71 ± 0.76. Coincidentally, ZEUS finds 2 events with $Q^2 > 35,000\text{GeV}^2$ for which the SM prediction is 0.145 ± 0.013. The probability that these data result from a statistical fluctuation is 0.5%. This is the same as the probability that four dice thrown together will roll to a common number.

Another provocative fact, although not as clearcut when H1 and ZEUS are compared, is that these excess high-Q^2 events *may* cluster around a common x value. The mass $M = \sqrt{xs}$ of a direct-channel resonance would be $\sim$ 200GeV for H1 and perhaps somewhat higher for ZEUS.

There are 20 times more data in e^+p collisions compared to e^-p. For the latter there has been an ion trapping problem at HERA. The integrated luminosities are $34.3pb^{-1}$ of e^+p data and $1.53pb^{-1}$ of e^-p data.

The conservative reaction is that (i) we have been chasing deviations from the SM for two decades and experience has shown that more precise data always remove the discrepancy. (ii) instead of extrapolating the QCD to higher Q^2 we should make an overall fit including the new events. Actually (ii) seems unlikely to work because here we are dealing with valence quarks, not gluons, with structure functions that are better known - it is not like the situation for high transverse energy jets at FermiLab where the gluon structure functions could be modified to explain the data.

For the present purposes, we take the experimental result seriously at face value.

Physics of Mass
Edited by Kursunoglu *et al.*, Kluwer Academic / Plenum Publishers, New York, 1999

THEORY RESPONSE

There were 30 theory papers involving 63 theorists in the 10 weeks following the HERA announcement which for 14 events corresponds to over 2 papers and 4 theorists per event!

The 30 papers, with author(s) and hep-ph/97mmnnn numbers, are listed in chronological order of appearance in Refs.[3-32].

These 30 papers break down as follows:

- 1 on consistency of the data.[15]
- 2 on compositeness.[3, 32]
- 5 on contact interactions.[13, 18, 21, 22, 31]
- 9 on R-symmetry breaking squark. [4, 6, 7, 11, 19, 25, 26, 28, 29]
- 13 on light leptoquark. [5, 8, 9, 10, 12, 14, 16, 17, 20, 23, 24, 27, 30]

We see that a new light ($\sim$ 200GeV) particle is the most popular explanation with 22 papers out of the 30.

First, I shall review these 30 papers written by theorists in 1997.

Then I will review my own 1992 paper [33] predicting weak-scale leptoquarks.

Finally I shall comment on some other experimental constraints up to December 1997.

Consistency of Data

This is from Drees, hep-ph/9703332. He compares H1 with ZEUS. First look at the distribution of $M(LQ) = \sqrt{xs}$. One finds 200.3 ± 1.2 GeV (7 events) for H1 and 219.3 ± 5.5 GeV (4 events) for ZEUS. These disagree by 3.4σ but overall uncertainty in the energy reduces this to 1.8σ, or an 8% chance occurrence.

Next compare the absolute event numbers 7 for H1 and 4 for ZEUS. The integrated luminosities are $14.2pb^{-1}$ for H1 and $20.1pb^{-1}$ for ZEUS. Zeus has "looser" cuts and the likelihood of this outcome is not higher than 5%.

Combining the two effects (x and # events) gives a 0.5% compatibility between H1 and ZEUS. This is about the same as the compatibility of the combined HERA data with the Standard Model. So the size of statistical fluctuation is comparable!

Compositeness

Adler (hep-ph/9702378) is the first 1997 theory of the HERA effect. In his "frustrated SU(4)" for preons, the positron interacts with a gluon and makes a transition to a E^+ state, a kind of leptogluon which decays into a e^+ and a jet.

Akama, Katsuura and Terazawa (hep-ph/9704327) revive a 1977 model and calculate cross-sections for several different composite states. The conclusion is that certain composites are possible:

- leptoquark, e^{+*}

while excluded are:

- Z^*,q^* (by pre-existing mass bounds)

Contact Interactions

A general approach is due to Buchmuller and Wyler (hep-ph/9704317) who parametrize physics beyond the SM by four-fermion contact interactions; quark-lepton universality, absence of FCNC color-independence, flavor-independence and consistency with atomic parity violation results are imposed. There is then a unique current-current form of contact interaction:

$$\frac{\epsilon}{\Lambda^2} J^\mu J_\mu \tag{1}$$

where:

$$J^\mu = \bar{u}\gamma^5\gamma^\mu u + \bar{d}\gamma^5\gamma^\mu d + \bar{e}\gamma^5\gamma^\mu e + \bar{\nu}\gamma^5\gamma^\mu \nu \tag{2}$$

This contact term shares the U(45) invariance of the standard model when the gauge and Yukawa couplings vanish.

Rare processes like $K_L \to e^-\mu^+$ and LEP2 data constrain $\Lambda > 5$TeV. The HERA data require, on the other hand, $\Lambda < 3$TeV. So Buchmuller and Wyler conclude that *if* there is new physics there must be sizeable breaking of quark-lepton symmetry and/or new particles.

Nelson(hep-ph/9703379) looks at the atomic parity violation constraints.

Di Bartolomeno and Fabbrichesi(hep-ph/9703375) *and* Barger, Cheung, Hagiwara and Zeppenfeld (hep-ph/9703311) concur that $\Lambda < 3$TeV is needed to fit the HERA data.

Gonzales-Garcia and Novaes (hep-ph/9703346) examine constraints on contact interactions from the one-loop contribution to $Z \to e^+e^-$.

Leptoquark [R Breaking Squark]

As mentioned above, 22 of the 30 papers in 1997 are in this direction. From here on we assume a direct-channel resonance in e^+q.

Note that a *valence* quark is more likely than a *sea* antiquark because the latter is going to give e^-p effects conflicting with the data, despite the twenty times smaller integrated luminosity.

Also: a scalar is more likely than a vector. A vector leptoquark has a coupling $\phi_\mu^\dagger G^{\mu\nu}\phi_\nu$ to gluons and would be produced by $\bar{q}q \to g \to \phi^\dagger\phi$ at the Tevatron. We expect therefore the leptoquark to be a scalar corresponding to $(e^+u)^{5/3}$ and/or $(e^+d)^{2/3}$.

There are seven scalar leptoquark couplings to $e^\pm q$, three of which involve e^+q. All demand that the LQ is an SU(2) doublet:

$$O = \lambda_{ij} L_i \Phi d_j^C \tag{3}$$

$$O' = \lambda'_{ij} L_i \Phi' u_j^C \tag{4}$$

$$O'' = \lambda''_{ij} Q_i e_j^C \Phi'' \tag{5}$$

For $\mathcal{O}$ to explain HERA data, $\lambda_{11} \sim 0.05$.

The branching ratio $B(K_L \to e^+e^-\pi^0) \leq 4.3 \times 10^{-9}$ implies that $\lambda_{11}\lambda_{12} \leq 1.5 \times 10^{-4}$. This means $\lambda_{12} \leq 3 \times 10^{-3}$, an unusual flavor hierarchy.

The rare decays $K^+ \to \pi^+\nu\bar{\nu}$ and $\mu \to e\gamma$ also constrain the off-diagonal λ_{ij}. The conclusion about the flavor couplings of the leptoquark is:

- The leptoquark scalar needs to couple nearly diagonally to mass-eigenstate quarks.

If one believes in weak-scale SUSY, a special case of scalar leptoquark is an R-symmetry breaking squark. The quantity $R = (-1)^{3B+L+2S}$ clearly must be broken because e^+q has $R = +1$ and $\tilde{q}$ has $R = -1$.

In the superpotential $\lambda_{ijk}L_iL_j\bar{E}_k + \lambda'_{ijk}L_iQ_kD_k + \lambda''_{ijk}U_iD_jD_k$, we put $\lambda''_{ijk} = 0$ for B conservation. Then to explain the data we need $\lambda'_{1j1} \sim 0.04/\sqrt{B}$ where $B(\tilde{q} \to e^+d)$ is the branching ratio.

Neutrinoless double beta decay $(\beta\beta)_{0\nu}$ requires $\lambda'_{111} < 7 \times 10^{-3}$ which excludes $\tilde{q} = \tilde{u}$.

Can the squark be $\tilde{q} = \tilde{c}$? If so, it implies that the rare decay, $K^+ \to \pi^+\nu\bar{\nu}$, being measured at Brookhaven is close to its current bound, which is interesting.

Finally, if $\tilde{q} = \tilde{t}$, atomic parity violation implies that $|\lambda'_{131}| < 0.5$, so the branching ratio B can be much less than 1 and still be consistent.

THE SU(15) POSSIBILITY

The paper[33], published in 1992, *predicts* light leptoquarks in e^-p, the mode in which HERA was then running.

Such scalar leptoquarks predicted by SU(15) lie at the *weak* scale. The situation can be contrasted with SU(5) grand unification where "leptoquark" gauge bosons couple to $\bar{5}$ and **10**, are simultaneously "biquarks", and hence must have mass $\sim 10^{16}$GeV.

At the Warsaw Rochester Conference in July 1996, the speakers from HERA and Tevatron report failure to find leptoquarks so I thought my idea might be wrong.

Now it is worth spending a few minutes to review SU(15).

The SU(15) GUT was inspired by the desire to *remove proton decay*. With:

$$15 = (u_1, u_2, u_3, d_1, d_2, d_3, \bar{u}_1, \bar{u}_2, \bar{u}_3, \bar{d}_1, \bar{d}_2, \bar{d}_3, e^+, \nu, e^-) \tag{6}$$

three times, every gauge boson has definite B and L which are therefore conserved in the gauge sector.

The symmetry breaking may be assumed to follow the steps:

$$SU(15) \overset{M_{GUT}}{\to} SU(12)_q \times SU(3)_l \tag{7}$$

$$\overset{M_B}{\to} SU(6)_L \times SU(6)_R \times U(1) \tag{8}$$

$$\overset{M_A}{\to} SU(3) \times SU(2) \times U(1) \tag{9}$$

M_{GUT} can be as low as 6×10^6GeV, for $M_A = M_W$. Once we select M_A, the other two scales M_B and M_{GUT} are calculable from the renormalization group equations.

In SU(15), the family-diagonal scalar leptoquark is in the **120** representation. The content of **120** is shown in [33]. It contains $(3, 2)_{Y=7/6}$ under $SU(3) \times SU(2) \times U(1)$. This SU(2) doublet has the quantum numbers of (e^+u) and (e^+d), required to explain the HERA data.

The point is that this irreducible representaion contains the standard Higgs doublet so there is every reason to expect the scalar leptoquarks to lie near the weak scale $\sim$ 250GeV. SU(15) predicts further leptoquark and "bifermion" states at or near the weak scale.

The discovery of light leptoquarks would be evidence as compelling as proton decay for grand unification.

BOUNDS ON LEPTOQUARKS FROM OTHER EXPERIMENTS

Both CDF[34] and D0[35] have found severe constraints for leptoquarks coupling to $e^\pm$ jet, in the first generation. If the branching ratio to this channel is 100% then D0 excludes masses $M < 225$GeV and CDF excludes $M < 213$GeV. If the branching ratio to the first generation channel is lower, then the lower limit on the mass is reduced *e.g.* for $B_{qe} = 0.5$ only $M < 176$GeV is excluded.

There are other limits from LEP2 but these are less restrictive *e.g.* excluding $M < 130$GeV.

For consistency with the Tevatron data, the first-generation branching ratio must satisfy $B_{eq} < 0.7$.

CONSTRAINTS FROM S AND T

Extra states beyond the Standard Model tend to *increase* the values of the parameters S and T and take us out of agreement with the precision electroweak data. However, if there are *two* doublets of leptoquarks with charges (5/3, 2/3) and (2/3, -1/3) and if the charge 2/3 states mix then it is possible to obtain negative contributions to S and T.

A second mechanism that gives negative contributions is on the basis of bileptonic gauge bosons[36].

Both of these mechanisms were examined in detail in [37]. In particular, it was shown that the SU(15) theory, discussed above, is consistent. This is non-trivial because the three families of mirror fermions necessary to cancel chiral anomalies give $\Delta S = 3/\pi$, a too-large value. It turns out, however, that because SU(15) contains *both* leptoquarks and bileptons, that there is a significant region of parameter space where S and T are acceptable[37].

SUMMARY

The HERA data may suggest a leptoquark resonance in e^+q at about 200GeV, although the case is not stronger now in December 1997 than it was in early 1997! There are at least 50 theory papers analysing this data, many of whch have been discussed above. Other exeriments, especially D0 and CDF at Fermilab, restrict the possibility of a first-generation leptoquark with 100% branching ratio to (e^+d); lower branching ratios below 70% may be admissable.

ACKNOWLEDGEMENT

This work was supported in part by the US Department of Energy under Grant No. DE-FG02-97ER41036.

REFERENCES

1. H1 COLLABORATION: Z. Phys. **C74,** 191 (1997).
2. ZEUS COLLABORATION, Z. Phys. **C74,** 207 (1997).
3. S.L. Adler, hep-ph/9702378.
4. D. Choudhury and S. Raychaudhuri, hep-ph/9702392. Phys. Lett. **B401,** 54 (1997)
5. T.K. Kuo and T. Lee, hep-ph/9703255. Mod. Phys. Lett. **A13,** 2367 (1997).
6. G. Altarelli, J. Ellis, S. Lola, G.F. Guidice and M.L. Mangano, hep-ph/9703276. Nucl. Phys. **B506,** 3 (1997)
7. H. Dreiner and P. Morawitz, hep-ph/9703279. Nucl. Phys. **B503,** 55 (1997).
8. R. Fiore, L.L. Jenkovsky, F. Paccanoni and E. Predazzi, hep-ph/9703283. Nucl. Phys. **B503,** 55 (1997)
9. M.A. Doncheski and S. Godfrey, hep-ph/9703285. Mod. Phys. Lett. **A12,** 1719 (1997).
10. J. Bluemlein, hep-ph/9703287. Z. Phys. **C74,** 605 (1997).
11. J. Kalinowski, R. Rueckl, H. Spiesberger and P.M. Zerwas, hep-ph/9703288. Z. Phys. **C74,** 595 (1997).
12. K.S. Babu, C. Kolda, J. March-Russell and F. Wilczek, hep-ph/9703299. Phys. Lett. **402,** 367 (1997).
13. V. Barger, K. Cheung, K. Hagiwara and D. Zeppenfeld, hep-ph/9703331. Phys. Lett. **404,** 147 (1997); and hep-ph/9707412. Phys. Rev. **D57,** 391 (1998).
14. M. Suzuki, hep-ph/9703316.
15. M. Drees, hep-ph/9703332. Phys. Lett. **403,** 353 (1997).
16. J.L. Hewett and T. Rizzo, hep-ph/9703337. Phys. Rev. **D56,** 1778 (1997).
17. G.K. Leontaris and J.D. Vergados, hep-ph/9703338. Phys. Lett. **B409,** 283 (1997).
18. M.C. Gonzalez-Garcia and S.F. Novaes, hep-ph/9703346. Phys. Lett. **B407,** 255 (1997).
19. D. Choudhury and R. Raychaudhury, hep-ph/9703369. Phys. Rev. **D56,** 1778 (1997).
20. C.G. Papadopoulos, hep-ph/9703372.
21. N. Di Bartolomeo and M. Fabbrichesi, hep-ph/9703375. Phys. Lett. **B406,** 237 (1997).
22. A.E. Nelson, hep-ph/9703379. Phys. Rev. Lett. bf 78, 4159 (1997).
23. Z. Kunszt and W.J. Stirling, hep-ph/9703427. Z. Phys. **C75,** 453 (1997).
24. T. Plehn, H. Spiesberger, M. Spira and P.M. Zerwas, hep-ph/9703433. Z. Phys. **C74,** 611 (1997).
25. J. Kalinowski, R. Rueckl, H. Spiesberger and P.M. Zerwas, hep-ph/9703436. Phys. Lett. **B406,** 314 (1997).
26. H. Dreiner, E. Perez and Y. Sirois, hep-ph/9703444. In Proceedings of the Workshop, Future Physics at HERA, DESY, Hamburg, September 1997.
27. C. Friberg, E. Norrbin and T. Sjostrand, hep-ph/9704214. Phys. Lett. **B403,** 329 (1997).
28. T. Kon and T. Kobayashi, hep-ph/9704221. Phys. Lett. **B409,** 265 (1997).
29. R. Barbieri, Z. Berezhiani and A. Strumia, hep-ph/9704275. Phys. Lett. **B407,** 250 (1997).
30. I. Montvay, hep-ph/9704280. Phys. Lett. **B407,** 250 (1997).
31. W. Buchmuller and D. Wyler, hep-ph/9704317. Phys. Lett. **B407,** 147 (1997).
32. K. Akama, K. Katsuura and H. Terazawa, hep-ph/9704327. Phys. Rev. **D56,** 2490 (1997).
33. P.H. Frampton, Mod. Phys. Lett. **A7,** 559 (1992).
34. F. Abe *et al.* (CDF Collaboration) Phys. Rev. Lett. **79,** 4327 (1997).
35. B. Abbott *et al.* (D0 Collaboration) Phys. Rev. Lett. **79,** 4321 (1997).
36. K. Sasaki, Phys. Lett. **B308,** 297 (1993).
37. P.H. Frampton and M. Harada, UNC-Chapel Hill Reports IFP-746-UNC(1997) and IFP-748-UNC(1997).

THE RELATION OF SPIN, STATISTICS, LOCALITY AND TCP

O.W. Greenberg[1]

Center for Theoretical Physics, Department of Physics,
University of Maryland, College Park, Maryland 20742-4111

Abstract

The spin-statistics theorem has been confused with another theorem, the "spin-locality theorem." The spin-statistics theorem properly depends on the properties of asymptotic fields which are free fields. Ghosts evade both theorems, because they are fields with an indefinite metric. Either the canonical (anticanonical) commutation relations or the locality of commutators of observables, such as currents, can serve as the basis of the spin-statistics theorem for fields without asymptotic limits, such as quark and gluon fields. The requirements for the TCP theorem are extremely weak.

INTRODUCTION

I have two purposes in this talk[1]. The first is to make clear the difference between the spin-statistics theorem: *particles* that obey Bose statistics must have integer spin and *particles* that obey Fermi statistics must have odd half-integer spin[2, 3], and what I suggest should be called the spin-locality theorem: *fields* that commute at spacelike separation must have integer spin and *fields* that anticommute at spacelike separation must have odd half-integer spin[4, 5, 6, 7]. My second purpose is to emphasize the weakness of the conditions under which the TCP theorem[10] holds and in that way to distinguish it from the spin-statistics theorem and the spin-locality theorem. In so doing I amplify Res Jost's example[11] of a field that has the wrong spin-statistics connection, but obeys the TCP theorem. The thrust of this talk is to separate these three theorems that are sometimes lumped together.

SPIN-STATISTICS AND SPIN-LOCALITY

Since the "right" cases of both the spin-statistics theorem and the spin-locality theorem agree, I emphasize what fails in each of the theorems for the "wrong" cases. Spacelike commutativity (locality) of observables fails for the wrong cases of the spin-statistics theorem. For example, as I

[1]Supported in part by a Semester Research Grant from the General Research Board of the University of Maryland and by a grant from the National Science Foundation.

Physics of Mass
Edited by Kursunoglu *et al.*, Kluwer Academic / Plenum Publishers, New York, 1999

will discuss in detail in amplifying Jost's example, for a neutral spin-0 scalar field that obeys Fermi statistics, observables, such as currents, fail to be local. By contrast, a neutral spin-0 scalar field whose anticommutator is local does not exist–it is identically zero. The obvious corresponding wrong cases for spin-1/2 and higher spin have the corresponding failures.

Spin-statistics: Because the spin-statistics theorem refers to the statistics of particles, its formulation in field theory should involve the operators that create and annihilate particles. These operators are the asymptotic fields, the in- and out-fields. Since the asymptotic fields (at least for massive particles) are free fields, the proof of the spin-statistics theorem only requires using the properties of free fields. The assumptions necessary for the proof are (1) that the space of states is a Hilbert space, i.e., the metric is positive-definite, (2) the fields smeared with test functions in the Schwartz space $\mathcal{S}$ have a common dense domain in the Hilbert space, (3) the fields transform under a unitary representation of the restricted inhomogeneous Lorentz group, (4) the spectrum of states contains a unique vacuum and all other states have positive energy and positive mass, and (5) the bilinear observables constructed from the (free) asymptotic fields commute at spacelike separation (local commutativity of observables). Using these assumptions for free fields of any spin, Fierz[2] and Pauli[3] proved that integer-spin particles must be bosons and odd half-integer spin particles must be fermions. They used locality of observables as the crucial condition for integer-spin particles and positivity of the energy as the crucial condition for the odd half-integer case. Weinberg[12] showed that one can use the locality of observables for both cases if one requires positive-frequency modes to be associated with annihilation operators and negative-frequency modes to be associated with creation operators.

I assume that for non-gauge theories with no massless particles the asymptotic fields are an irreducible set of operators. I will show that the conserved observables such as the energy-momentum operators and, for theories with conserved currents, the current operators, must be a sum of the free field functionals of the asymptotic fields, where the sum runs over the independent asymptotic fields, including those for bound states if there are bound states in the theory. To see this, require–say for the in-fields–

$$i[P^\mu, \phi^{in}(x)]_- = \partial^\mu \phi^{in}(x) \tag{1}$$

for the case of the energy-momentum operator and a neutral scalar field. The general expansion in the in-fields for P^μ is

$$P^\mu = \sum_{n=0}^{\infty} \frac{1}{n!} \int f^{(n)\,\mu}(x_1, \cdots, x_n) \overleftrightarrow{\partial}_{x_1}^{\mu_1} \cdots \overleftrightarrow{\partial}_{x_n}^{\mu_n} : \phi^{in}(x_1) \cdots \phi^{in}(x_n) : d\Sigma_{\mu_1}(x_1) \cdots d\Sigma_{\mu_n}(x_n) \tag{2}$$

and inserting it in Eq.(1) shows that only the constant term and the bilinear free functional of the in-fields can enter P^μ. The requirement that the vacuum have zero energy eliminates the constant term. The equation for the bilinear term is

$$\int f^{(2)}(x_1, x) \overleftrightarrow{\partial}_{x_1}^{\mu_1} : \phi^{in}(x_1) : d\Sigma_{\mu_1}(x_1) = \partial^\mu \phi^{in}(x). \tag{3}$$

The integrals over the spacelike surfaces $\Sigma(x_i)$ are independent of the time because of the time-translation invariance of the Klein-Gordon scalar product. Thus the solution of Eq.(3),

$$f^{(2)}(x_1, x) = -\partial_x^\mu \Delta(x - x_1), \tag{4}$$

leads to the usual result for P^μ using $\Delta(0, \mathbf{x}) = 0$, $\partial_0 \Delta(0, \mathbf{x}) = -\delta(\mathbf{x})$, and the Klein-Gordon equation for $\Delta(x)$. Thus the arguments of Fierz, Pauli, and Weinberg for free fields hold in the case of interacting theories that have an irreducible set of in- (or out-) fields. For example the charge density for a charged spin-zero field is

$$j^\mu(x) = i : \phi^{as\,\dagger}(x) \overleftrightarrow{\partial}^\mu \phi^{as}(x) : \tag{5}$$

and for a charged spin-one-half field it is

$$j^{\mu}(x) =: \bar{\psi}^{as}(x)\gamma^{\mu}\psi^{as}(x) : \tag{6}$$

For a spin-zero field, the commutator of the currents $[j^{\mu}(x), j^{\nu}(y)]_{-}$ will contain the local distribution $i\Delta(x-y)$ if the annihilation and creation operators obey Bose commutation relations and the nonlocal distribution $\Delta^{(1)}(x-y)$ if the annihilation and creation operators obey Fermi commutation relations. For a spin-one-half field, the commutator $[j^{\mu}(x), j^{\nu}(y)]_{-}$ will contain the local distribution $iS(x-y)$ if the particle operators obey Fermi rules and the nonlocal distribution $S^{(1)}(x-y)$ if the particles obey Bose rules. (Here I assume that the fields are expanded in annihilation operators for the positive frequency modes and in creation operators for the negative frequency modes. If a Dirac field is expanded in annihilation operators for both types of modes, then the commutator of the field and its Pauli adjoint will be the local distribution $iS(x-y)$, but the energy operator will be unbounded below.[13])

Spin-locality: For the spin-locality theorem, Lüders and Zumino[4] and Burgoyne[5] replaced assumption (5) of the spin-statistics theorem by (5′) that the fields either commute

$$[A_{\mu}(x), A_{\nu}^{\dagger}(y)]_{-} = 0, (x-y)^2 < 0, \tag{7}$$

or anticommute

$$[\psi_{\alpha}(x), \bar{\psi}_{\beta}(y)]_{+} = 0, (x-y)^2 < 0. \tag{8}$$

at spacelike separation. Here $[A, B]_{\pm} = AB \pm BA$. Since in general the fields are not observables, this is not an assumption about physical quantities. I will call such fields local or antilocal and, as mentioned above, I will call the theorem the spin-locality theorem. The Lüders-Zumino and Burgoyne proof shows that if the fields have the wrong commutation relations, i.e. integer-spin fields are antilocal and odd-half-integer-spin fields are local, the fields vanish. This assumption does not relate directly to particle statistics and for that reason this theorem should not be called the spin-statistics theorem. Thus the assumptions of the spin-statistics theorem and of the spin-locality theorem differ; further, the conclusions of the two theorems differ for the case of the wrong association between spin and either statistics or type of locality.

Ghosts: There is a case in practical calculations in which both the spin-statistics theorem and the spin-locality theorem seem to be violated: namely, the ghosts of gauge theory. These are scalar fields that (1) anticommute at spacelike separation and thus seem to violate the spin-locality theorem and (2) whose asymptotic limits are quantized obeying Fermi particle statistics and seem to violate the spin-statistics theorem. Most discussions of gauge theory rely on path integrals and don't explicitly consider the commutation or anticommutation relations of ghost fields. N. Nakanishi and I. Ojima[14] give

$$[C_i^{as}(x), \bar{C}_j^{as}(y)]_{+} = -\delta_{ij} D(x-y), \tag{9}$$

where i and j run over the adjoint representation of the gauge group, as the anticommutator between the ghost and antighost fields. The anticommutators of C^{as} with itself and of $\bar{C}^{as}$ with itself vanish. The arguments used in the proof of the spin-locality theorem show that the two-point functions $\langle 0|C(x)C(y)|0\rangle$ and $\langle 0|\bar{C}(x)\bar{C}(y)|0\rangle$ both vanish. Since C and $\bar{C}$ are hermitian, if the metric of the space were positive-definite the fields C and $\bar{C}$ would annihilate the vacuum and the fields would vanish. Because the space of states is indefinite, this conclusion does not follow. The off-diagonal form of these anticommutators is connected with the fact that the ghost and antighost fields create zero norm states. Nakanishi and Ojima take the ghost and antighost fields to be independent hermitian fields, so the assumptions of neither the spin-statistics nor the spin-locality theorem hold and there is no violation of either theorem.[14] I give the anticommutation relations of Nakanishi and Ojima for the asymptotic fields for the annihilation and creation operators of the ghosts and antighosts to illustrate from the particle point of view how these fields evade the two theorems,

$$[C^{(as)}(k), \bar{C}^{(as)\,\dagger}(l)]_+ = i2E_k\delta(\mathbf{k}-\mathbf{l}), \quad [\bar{C}^{(as)}(k), C^{(as)\,\dagger}(l)]_+ = -i2E_k\delta(\mathbf{k}-\mathbf{l}), \tag{10}$$

where other anticommutators vanish, and I used relativistic normalization for the annihilation and creation operators,

$$C^{(as)}(x) = \frac{1}{(2\pi)^{3/2}}\int \frac{d^3k}{2E_k}[C^{(as)}(k)exp(-ik\cdot x) + C^{(as)\,\dagger}(k)exp(ik\cdot x)] \tag{11}$$

and a similar formula for $\bar{C}^{(as)}$. These two anticommutators go into each other under hermitian conjugation. The i factors are what allow the anticommutator $[C^{(as)}(x), \bar{C}^{(as)}(y)]_+$ to be $-D(x-y)$ (for the massless case), rather than a multiple of $D^{(1)}(x-y)$.

Fields without an asymptotic limit: For fields that do not have asymptotic fields, such as quark or gluon fields, one needs a condition on the fields that can replace the condition on the asymptotic fields in deriving the spin-statistics theorem. I have argued[17] that the c-number equal-time canonical commutation (anticommutation) rules for the fields lead to the commutation (anticommutation) relations for the asymptotic fields using the LSZ weak asymptotic limit. This suggests that the requirement of local commutativity of observables that is satisfied by asymptotic fields by having either Bose or Fermi statistics can be satisfied for fields that do not have asymptotic fields by having either the canonical equal-time commutators or the canonical equal-time anticommutators to be c-numbers. This alternative replaces the alternative of either locality or antilocality of the fields of the Lüders-Zumino and Burgoyne theorem. The commutator of currents at equal times will involve a sum of terms with either equal-time commutators or equal-time anticommutators of the fields. Since these are c-numbers, they will vanish, except at coincident points, only if they have the correct choice of integer or odd half-integer spin. Thus the requirement that the observable densities commute at equal times, except at coincident points, again leads to the correct association of spin with either c-number canonical equal-time commutators or anticommutators.

Gauge theories in covariant gauges have a space of states with an indefinite metric. Since both the spin-statistics theorem and the spin-locality theorem assume a positive-definite metric, we have to understand how these theorems can apply to the particles and fields in gauge theories. The qualitative answer is that such gauge theories have a physical space of states (called H_{phys} by Nakanishi and Ojima) that has a positive-definite metric. The space H_{phys} is the quotient of a subspace (called $\mathcal{V}_{phys}$ by Nakanishi and Ojima) of an indefinite metric space (called $\mathcal{V}$) and the space of zero norm states (called $\mathcal{V}_0$ by Nakanishi and Ojima) and the theorems presumably hold in this physical space.

The local observable point of view allows a very general discussion of the spin-statistics connection based on the principles of locality, relativistic invariance, and spectrum without reference to fields. The literature on this point of view can be traced from the book by R. Haag[15] and the talk by S. Doplicher[16].

TCP

Now I turn to the TCP theorem[10]. The TCP theorem in Jost's formulation (given for simplicity for a single charged field) states that the necessary and sufficient condition for TCP to be a symmetry of the theory in the sense that there is an antiunitary operator θ such that

$$\theta\phi(x)\theta^{-1} = \phi^{\dagger}(-x), \quad \theta\phi^{\dagger}(x)\theta^{-1} = \phi(-x), \quad \theta|0\rangle = |0\rangle, \tag{12}$$

is that the field ϕ obey weak local commutativity in a real neighborhood of a Jost point. Weak local commutativity is equality of the vacuum matrix elements of a product of fields at points that are totally spacelike when the order of the fields is inverted, i.e.

$$\rangle|\phi(x_1)\phi(x_2)\cdots\phi(x_n)|\rangle = \rangle|\phi(x_n)\cdots\phi(x_2)\phi(x_1)|\rangle, \tag{13}$$

if all $(x_i - x_j)^2 < 0$. Jost points are the points where all convex sums of the successive difference vectors of the points in a vacuum matrix element are purely spacelike. Local commutativity implies weak local commutativity, but weak local commutativity is much weaker than local commutativity. I amplify Jost's example[11] of a free relativistic neutral scalar field quantized with Fermi statistics, repeat Jost's proof that this field obeys the TCP theorem, and find the Hamiltonian density for this field.

Expand the field in terms of annihilation and creation operators that obey relativistic normalization,

$$\phi(x) = \frac{1}{(2\pi)^3}\int \frac{d^3k}{2E_k}(A(k)exp(-ik\cdot x) + A^{\dagger}(k)exp(ik\cdot x)). \tag{14}$$

The annihilation and creation operators obey

$$[A(k), A^{\dagger}(l)]_+ = 2E_k\delta(\mathbf{k}-\mathbf{l}), \tag{15}$$

$$[A(k), Al)]_+ = 0, \quad [A^{\dagger}(k), A^{\dagger}(l)]_+ = 0. \tag{16}$$

The anticommutator of the field is

$$[\phi(x), \phi(y)]_+ = \Delta^{(1)}(x-y), \tag{17}$$

which is not local. With a vacuum that is annihilated by the annihilation operators, $A(k)|0\rangle = 0$, this is a theory of free, neutral scalar fermions. This example is nonlocal, but the field does not vanish. It obeys the TCP theorem, because its vacuum matrix elements are sums of products of two-point vacuum matrix elements and its two-point vacuum matrix elements obey local commutativity from the properties of spectrum and Lorentz invariance. (To further emphasize how weak a condition TCP invariance is, note that even free quon fields obey TCP.[18])

The Hamiltonian for this free theory is

$$H = \int \frac{d^3k}{2E_k} E_k A^{\dagger}(k)A(k). \tag{18}$$

Translate this into position space using

$$A(k) = i\int e^{ik\cdot x}\overleftrightarrow{\partial}^0 \phi(x)\frac{d^3x}{(2\pi)^{3/2}}, \tag{19}$$

$$A^{\dagger}(k) = -i\int e^{-ik\cdot x}\overleftrightarrow{\partial}^0 \phi(x)\frac{d^3x}{(2\pi)^{3/2}}. \tag{20}$$

The result is

$$H = \int d^3x d^3y\ [i\partial^0_x\Delta^{(1)}(x-y)]\ \overleftrightarrow{\partial}^0_x\overleftrightarrow{\partial}^0_y\!: \phi^{\dagger}(x)\phi(y) :], \tag{21}$$

which is the integral of a (nonlocal) energy density,

$$H = \int d^3x\mathcal{H}(x), \tag{22}$$

$$
\begin{aligned}
\mathcal{H}(x) &= i\int d^3\rho[\dot{\Delta}^{(1)}(\rho) :\dot{\phi}^{\dagger}(x+\rho/2)\,\dot{\phi}\,(x-\rho/2) : -\ddot{\Delta}^{(1)}(\rho) : \phi^{\dagger}(x+\rho/2)\,\dot{\phi}\,(x-\rho/2) : + \\
&\quad \ddot{\Delta}^{(1)}(\rho) :\dot{\phi}^{\dagger}(x+\rho/2)\phi(x-\rho/2) : -\dddot{\Delta}^{(1)}(\rho) : \phi^{\dagger}(x+\rho/2)\phi(x-\rho/2) :]. \qquad (23)
\end{aligned}
$$

This energy density is nonlocal in both senses: it is not a pointlike functional of the fields and it does not commute with itself at spacelike separation. This result for $\mathcal{H}(x)$ also follows from Eq.(1). The difference between the spin-0 field quantized with Fermi statistics (the wrong case) and with Bose statistics (the right case) is that for the wrong case the $\Delta^{(1)}(x)$ distribution enters rather than $\Delta(x)$, and the zero-time values of $\Delta^{(1)}(x)$ are not local, in contrast to the vanishing of the zero-time value of $\Delta(x)$ and the locality of its time derivative at zero time.

SUMMARY

One should distinguish three theorems: *The spin-statistics theorem*: Given the choice between Bose and Fermi statistics, particles with integer spin must obey Bose statistics and particles with odd half-integer spin must obey Fermi statistics. *The spin-locality theorem*: Given the choice between commutators that vanish at spacelike separation and anticommutators that vanish at spacelike separation, fields with integer spin must have local commutators and fields with odd half-integer spin must have local anticommutators. *The TCP theorem*: The necessary and sufficient condition for the existence of an antiunitary operator θ such that $\theta\phi(x)\theta^{-1} = \phi^{\dagger}(-x)$, $\theta\phi^{\dagger}(x)\theta^{-1} = \phi(-x), \theta|0\rangle = |0\rangle$, is weak local commutativity at Jost points.

For the spin-statistics theorem, the basis of the theorem is the requirement that observables commute at spacelike separation. If the wrong choice is made, observable densities fail to commute at spacelike separation. For fields that don't have asymptotic fields, the choice is between fields whose canonical variables have c-number equal-time commutators and fields whose canonical variables have c-number equal-time anticommutators. For the spin-locality theorem, if the wrong choice is made, the field vanishes. The TCP theorem can hold even if the field and its particles have the wrong connection of spin and statistics; clearly it can hold under very general conditions.

ACKNOWLEDGEMENTS

It is a pleasure to thank Professor Behram Kursunoglu for his hospitality during the Orbis. I am happy to thank Xiangdong Ji, Jim Swank, and Ching-Hung Woo for stimulating questions and helpful suggestions. I especially thank Joe Sucher for many suggestions to improve this talk. I thank Sergio Doplicher and Kurt Haller for bringing relevant references to my attention.

References

[1] O.W. Greenberg, Phys. Lett. B 416 (1998) 144.

[2] M. Fierz, Helv. Phys. Acta 12 (1939) 3.

[3] W. Pauli, Phys. Rev. 58 (1940) 716.

[4] G. Lüders and B. Zumino, Phys. Rev. 110 (1958) 1450.

[5] N. Burgoyne, Il Nuovo Cimento VIII (4) (1958) 607.

[6] It is also true that these two theorems are not exclusive theorems; rather the spin-statistics theorem should include the cases in which Bose is replaced by parabose and Fermi is replaced by parafermi, in each case of any integer order p, with $p = 1$ being the Bose or Fermi case and the spin-locality theorem should include the corresponding cases of paralocality and para-antilocality[7, 9].

[7] H.S. Green, Phys. Rev. 90 (1953) 270, introduced parastatistics. Other references include O.W. Greenberg and A.M.L. Messiah, Phys. Rev. 136 (1964) B248; O.W. Greenberg, in *Mathematical Theory of Elementary Particles*, edited by R. Goodman and I. Segal (MIT Press, Cambridge, MA, 1966), p. 29; O.W. Greenberg and C.A. Nelson, Phys. Rep. 32C (1977) 69; Y. Ohnuki and S. Kamefuchi, *Quantum Field Theory and Parastatistics* (Springer, Berlin, 1982); O.W. Greenberg and K.I. Macrae, Nucl. Phys. B219 (1983) 358; and O.W. Greenberg, D.M. Greenberger and T.V. Greenbergest, in *Quantum Coherence and Reality*, eds. J.S. Anandan and J.L. Safko, (World Scientific, Singapore, 1994), p. 301.

[8] I. Duck and E.C.G. Sudarshan, Am. J. Phys. 66 (1998) 284.

[9] G.F. Dell'Antonio, O.W. Greenberg and E.C.G. Sudarshan, in *Group Theoretical Concepts and Methods in Elementary Particle Physics*, (Gordon and Breach, New York, 1964), ed. F. Gürsey, p. 403 give a proof of the spin-locality theorem for parastatistics fields using the Green ansatz for a general field theory. I don't know of a published proof of the spin-statistics theorem for parastatistics particles. Such a proof can be made using the point of view of this talk. The requirement is that observable densities bilinear in asymptotic parastatistics fields must commute with each other at spacelike separation. This will be possible for the correct connection of spin and parastatistics, but will fail for the wrong connection.

[10] R. Jost, Helv. Phys. Acta 30 (1957) 409.

[11] R. Jost, *The General Theory of Quantized Fields* (Am. Math Soc., Providence, 1965), pp. 103-104.

[12] S. Weinberg, Phys. Rev. 133 (1964) B1318 and 134 (1964) B882.

[13] W. Pauli, Prog. Theor. Physics 5 (1950) 526.

[14] N. Nakanishi and I. Ojima, *Covariant Operator Formalism of Gauge Theories and Quantum Gravity*, (World Scientific, Singapore, 1990), Chapter 4; K. Haller and E. Lim-Lombridas, Found. of Physics 24 (1994) 217, and cited references, discuss the algebra of ghosts in quantum electrodynamics.

[15] R. Haag, *Local Quantum Physics*, (Springer-Verlag, Berlin, 1992).

[16] S. Doplicher, in *Proc. Int. Congress of Mathematicians, Kyoto 1990*, Vol 2, (Springer-Verlag, Tokyo, 1991), ed. I. Satake, p1319.

[17] O.W. Greenberg, Princeton Thesis, (1956).

[18] O.W. Greenberg, Phys. Rev. D 43 (1991) 4111.

TESTING A CPT- AND LORENTZ-VIOLATING EXTENSION OF THE STANDARD MODEL

V. Alan Kostelecký

Physics Department
Indiana University
Bloomington, IN 47405
U.S.A.

INTRODUCTION AND BACKGROUND

The standard model of particle physics is invariant under a variety of continuous symmetry operations, including translations, Lorentz transformations, and gauge transformations. The model is also invariant under the action of the product CPT of charge conjugation C, parity reflection P, and time reversal T. Indeed, CPT symmetry is known to be a characteristic of all local relativistic field theories of point particles [1]. It has been experimentally tested to high accuracy in a variety of situations [2]. The general validity of CPT symmetry for particle theories and the existence of high-precision tests means CPT breaking is an interesting candidate experimental signal for new physics beyond the standard model, such as might emerge in the context of string theory [3, 4, 5].

In a talk [6] delivered at the previous meeting in this series (Orbis Scientiae 1997-I), I discussed the possibility that CPT and Lorentz symmetry might be broken in nature by effects emerging from a fundamental theory beyond the standard model. String theory, which currently represents the most promising framework for a consistent quantum theory of gravity incorporating the known particles and interactions, is a candidate theory in which effects of this type might occur. The point is that strings are extended objects, so the standard axioms underlying proofs of CPT invariance are inappropriate. In fact, it is known that spontaneous CPT and Lorentz violation can occur in the context of string theory [3, 7].

If the fundamental theory has Lorentz and CPT symmetry and is naturally formulated in more than four spacetime dimensions, then some kind of spontaneous breaking of the higher-dimensional Lorentz group presumably must occur to produce an effective low-energy theory with only four macroscopic dimensions. This situation exists

Physics of Mass
Edited by Kursunoglu *et al.*, Kluwer Academic / Plenum Publishers, New York, 1999

for some string theories, for example. An interesting issue is whether the spontaneous breaking generates apparent Lorentz and CPT violation in our four spacetime dimensions. It might seem natural for this to happen, since there is no evident reason why four dimensions would be preferred in the higher-dimensional theory. However, no experimental evidence exists for Lorentz or CPT breaking, so if it occurs it must be highly suppressed at the level of the standard model. If the standard model is regarded as an effective low-energy theory emerging from a realistic string theory, then the natural dimensionless suppression factor for observable Lorentz or CPT violation would be the ratio r of the low-energy scale to the Planck scale, $r \sim 10^{-17}$. Relatively few experiments would be sensitive to such effects.

In the previous talk [6], I outlined the low-energy description of effects from spontaneous Lorentz and CPT breaking in an underlying theory. At this level, the potentially observable Lorentz and CPT violations appear merely as consequences of the vacuum structure, so many desirable properties of Lorentz-invariant models are maintained. The low-energy theory acquires additional terms with a generic form [4, 5]. More specifically, at the level of the standard model, refs. [8, 9] have identified the most general terms that can arise from spontantaneous Lorentz violation (both with and without CPT breaking) while maintaining SU(3) × SU(2) × U(1) gauge invariance and power-counting renormalizability. The existence of this explicit extension of the standard model offers the possibility of quantitative investigations of a variety of experimental signals for apparent Lorentz and CPT breaking.

Some possible consequences of the additional terms in the standard-model extension were presented at the previous meeting [6]. Among the most interesting quantitative tests of CPT are experiments with neutral-meson oscillations in the K system [3, 4, 5, 10], the two B systems [5, 11, 12], and the D system [5, 13]. Implications of CPT violation in other contexts, such as baryogenesis [14], were also described.

In the present talk, I provide an update of some developments that have occurred in the months since the previous meeting. Possible experimental tests of the QED limit of the standard-model extension have been examined [9, 15, 16]. The sensitivity of tests of CPT violation in neutral-meson systems has been investigated [10], and the first experimental results have been obtained constraining CPT violation in the neutral-B system [17, 18].

EXTENDED QUANTUM ELECTRODYNAMICS

The general Lorentz-violating extension of the standard model (including terms with and without CPT violation), explicitly given in refs. [8, 9], follows from imposing two requirements. One is that the form of the additional terms must be compatible with an origin from spontaneous Lorentz breaking in an underlying theory. The other is that the usual properties of SU(3) × SU(2) × U(1) gauge invariance and power-counting renormalizability must be maintained. These criteria suffice to keep relatively small the number of new terms in the action. A framework for treating the implications of apparent Lorentz and CPT violation has also been presented in the above works.

One limit of this extended standard model is an extension of quantum electrodynamics (QED) [9]. This is of particular interest because QED is a well-established theory for which numerous experimental tests exist. Here, I give only the lagrangian for the extended theory of photons, electrons, and positrons, which has a relatively simple form.

The usual lagrangian is

$$\mathcal{L}^{\text{QED}} = \overline{\psi}\gamma^\mu(\tfrac{1}{2}i \overleftrightarrow{\partial_\mu} e - qA_\mu)\psi - m\overline{\psi}\psi - \tfrac{1}{4}F_{\mu\nu}F^{\mu\nu} \quad . \tag{1}$$

The CPT-violating terms are

$$\mathcal{L}_e^{\text{CPT}} = -a_\mu\overline{\psi}\gamma^\mu\psi - b_\mu\overline{\psi}\gamma_5\gamma^\mu\psi \quad ,$$

$$\mathcal{L}_\gamma^{\text{CPT}} = \tfrac{1}{2}(k_{AF})^\kappa\epsilon_{\kappa\lambda\mu\nu}A^\lambda F^{\mu\nu} \quad . \tag{2}$$

The Lorentz-violating but CPT-preserving terms are

$$\mathcal{L}_e^{\text{Lorentz}} = c_{\mu\nu}\overline{\psi}\gamma^\mu(\tfrac{1}{2}i \overleftrightarrow{\partial^\nu} - qA^\nu)\psi + d_{\mu\nu}\overline{\psi}\gamma_5\gamma^\mu(\tfrac{1}{2}i \overleftrightarrow{\partial^\nu} - qA^\nu)\psi - \tfrac{1}{2}H_{\mu\nu}\overline{\psi}\sigma^{\mu\nu}\psi$$

$$\mathcal{L}_\gamma^{\text{Lorentz}} = -\tfrac{1}{4}(k_F)_{\kappa\lambda\mu\nu}F^{\kappa\lambda}F^{\mu\nu} \quad . \tag{3}$$

The coefficients of the various terms can be regarded as Lorentz- and CPT-violating couplings. The reader is directed to refs. [8, 9] for details of notations and conventions as well as for more information about the various terms, including issues such as the effect of field redefinitions and the possibility of other couplings.

As mentioned above, many conventional tests of Lorentz and CPT symmetry are expected to be insensitive to effects from the additional terms in the extension of QED because of the expected small size of the couplings. Nonetheless, certain kinds of experiment can provide constraints.

First, consider the fermion sector. One important class of tests consists of Penning-trap experiments measuring anomaly and cyclotron frequencies with exceptional precision [19, 20, 21, 22]. These have been investigated in the present context in refs. [15], where possible signals are identified, appropriate figures of merit are introduced, and estimates are given of limits on Lorentz and CPT violation that would be attainable in present and future experiments. A summary of the results of these works can be found in a separate contribution to the present volume [16]. As one example, the spacelike components of the coefficient b_μ can be bounded by experiments comparing the anomalous magnetic moments of the electron and positron. The associated figure of merit for CPT violation could be constrained to about one part in 10^{20}. This is comparable to the ratio of the electron mass to the Planck scale at which suppressed but observable effects from an underlying theory might be expected. Some interesting constraints on a subset of couplings in the fermion sector of extended QED might also arise from high-precision experiments of various other kinds, including clock-comparison tests [23].

Next, consider the photon sector of the QED extension [8, 9]. The CPT-breaking term with coefficient $(k_{AF})_\mu$ has theoretical difficulties in that the associated canonical energy can be negative and arbitrarily large. This suggests that the coefficient should vanish, which in turn provides an interesting theoretical consistency check of the model. The point is that, even if this coefficient vanishes at tree level, it would typically be expected to acquire radiative corrections involving CPT-breaking couplings from the fermion sector, which in the present context could cause difficulty with the positivity of the theory. However, it has been shown [9] that no such radiative corrections arise in the context of the standard-model extension described above. At the experimental level, limits from cosmological birefringence restrict the components of $(k_{AF})_\mu$ to $\lesssim 10^{-42}$ GeV [24], although there exist disputed claims [25] for a nonzero effect corresponding to $|\vec{k}_{AF}| \sim 10^{-41}$ GeV.

In contrast, a nonzero contribution from the CPT-preserving, Lorentz-breaking term in the photon sector of the QED extension would maintain the positivity of the total canonical energy density and appears to be theoretically allowed [9]. Moreover, even if the coefficients $(k_F)_{\kappa\lambda\mu\nu}$ vanish at tree level, one-loop corrections from the fermion sector are induced. It is therefore of interest to examine possible experimental constraints on this type of term. One irreducible component of $(k_F)_{\kappa\lambda\mu\nu}$ is rotation invariant and can be bounded to $\lesssim 10^{-23}$ by the existence of cosmic rays [26] or by other tests. The remaining components violate rotation invariance and might in principle be bounded by cosmological birefringence. The attainable bounds are substantially weaker than those discussed above for the CPT-breaking term because, unlike $(k_{AF})_\mu$, the coefficients $(k_F)_{\kappa\lambda\mu\nu}$ are dimensionless and so are suppressed by the energy scale of the radiation involved. Further details about the photon sector of the QED extension can be found in ref. [9].

NEUTRAL-MESON OSCILLATIONS

Since the last meeting in this series, there have been several developments concerning the possibility of testing the standard-model extension using neutral-meson oscillations. In what follows, a generic neutral meson is denoted by P, where $P \equiv K$, D, B_d, or B_s.

Interferometry with P mesons can involve two types of (indirect) CP violation: T violation with CPT invariance, or CPT violation with T invariance. These are phenomenologically described by complex parameters ϵ_P and δ_P, respectively, that are introduced in the effective hamiltonian for the time evolution of a neutral-meson state. Within the context of the standard-model extension, it can be shown that the CPT-violating parameter δ_P depends only on one of the types of additional coupling [10]. Only CPT-violating terms in the lagrangian of the form $-a_\mu^q \overline{q} \gamma^\mu q$ are relevant, where q is a quark field and the coupling a_μ^q is constant in spacetime but depends on the quark flavor q. It is also noteworthy that the parameters δ_P are the only quantities known to be sensitive to the couplings a_μ.

To define δ_P, one must work in a frame comoving with the P meson. It can be shown that the CPT and Lorentz breaking introduces a dependence of δ_P on the boost and orientation of the meson. Let the P-meson four-velocity be $\beta^\mu \equiv \gamma(1, \vec{\beta})$. Then, at leading order in all Lorentz-breaking couplings in the standard-model extension, δ_P is given by [10]

$$\delta_P \approx i \sin\hat{\phi} \exp(i\hat{\phi}) \gamma(\Delta a_0 - \vec{\beta} \cdot \Delta\vec{a})/\Delta m \quad . \tag{4}$$

In this expression, $\Delta a_\mu \equiv a_\mu^{q_2} - a_\mu^{q_1}$, where q_1 and q_2 denote the valence-quark flavors in the P meson. The quantity $\hat{\phi}$ is given by $\hat{\phi} \equiv \tan^{-1}(2\Delta m/\Delta\gamma)$, where Δm and $\Delta\gamma$ are the mass and decay-rate differences between the P-meson eigenstates, respectively. Note that a subscript P is suppressed on all variables on the right-hand side of Eq. (4).

One implication of the above results for experiment is a proportionality between the real and imaginary components of δ_P [4, 5]. A second is the possibility of a variation of the magnitude of δ_P with P, arising from the flavor dependence of the couplings a_μ^q

[5]. Other implications arise from the momentum and orientation dependences in Eq. (4), which offer the possibility of striking signals for Lorentz and CPT breaking [10]. The momentum and orientation dependences also imply an enhanced signal for boosted mesons and suggest that published bounds on δ_P from distinct experiments could represent different CPT sensitivities. Experiments involving highly boosted mesons, such as the K-system experiment E773 at Fermilab [27], would be particularly sensitive to Planck-scale effects.

The tightest neutral-meson bounds on CPT violation at present are from experiments with the neutral-K system. The possibility exists that relatively large CPT violation might occur in the behavior of heavier neutral mesons. At the time of the previous meeting in this series, no bounds existed on CPT violation in the D or B systems. My talk at that meeting [6] emphasized that sufficient data already existed to place bounds on CPT violation in the B_d system [12]. Since then, two experimental groups at CERN have performed the suggested measurement. The OPAL collaboration has published the result [17] $\mathrm{Im}\,\delta_{B_d} = 0.020 \pm 0.016 \pm 0.006$, while the DELPHI collaboration has released a preliminary measurement [18] $\mathrm{Im}\,\delta_{B_d} = -0.011 \pm 0.017 \pm 0.005$. Other analyses of CPT violation in heavy-meson systems are presently underway.

ACKNOWLEDGMENTS

I thank Orfeu Bertolami, Robert Bluhm, Don Colladay, Rob Potting, Neil Russell, Stuart Samuel, and Rick Van Kooten for collaborations. This work is supported in part by the United States Department of Energy under grant number DE-FG02-91ER40661.

REFERENCES

1. See, for example, R.G. Sachs, *The Physics of Time Reversal* (University of Chicago Press, Chicago, 1987).
2. See, for example, R.M. Barnett *et al.*, Review of Particle Properties, Phys. Rev. D **54** (1996) 1.
3. V.A. Kostelecký and R. Potting, Nucl. Phys. B **359** (1991) 545; Phys. Lett. B **381** (1996) 89.
4. V.A. Kostelecký, R. Potting, and S. Samuel, in S. Hegarty *et al.*, eds., *Proceedings of the 1991 Joint International Lepton-Photon Symposium and Europhysics Conference on High Energy Physics,* World Scientific, Singapore, 1992; V.A. Kostelecký and R. Potting, in D.B. Cline, ed., *Gamma Ray–Neutrino Cosmology and Planck Scale Physics* (World Scientific, Singapore, 1993) (hep-th/9211116).
5. V.A. Kostelecký and R. Potting, Phys. Rev. D **51** (1995) 3923.
6. V.A. Kostelecký, in B.N. Kursunoglu, ed., *Twenty-Five Coral Gables Conferences and Their Impact on High Energy Physics and Cosmology* (Plenum, New York, 1997) (hep-ph/9704264).
7. V.A. Kostelecký and S. Samuel, Phys. Rev. Lett. **63** (1989) 224; *ibid.*, **66** (1991) 1811; Phys. Rev. D **39** (1989) 683; *ibid.*, **40** (1989) 1886.
8. D. Colladay and V.A. Kostelecký, Phys. Rev. D **55** (1997) 6760.
9. D. Colladay and V.A. Kostelecký, preprint IUHET 359 (1997).

10. V.A. Kostelecký, Phys. Rev. Lett. **80** (1998) 1818.
11. D. Colladay and V.A. Kostelecký, Phys. Lett. B **344** (1995) 259.
12. V.A. Kostelecký and R. Van Kooten, Phys. Rev. D **54** (1996) 5585.
13. D. Colladay and V.A. Kostelecký, Phys. Rev. D **52** (1995) 6224.
14. O. Bertolami, D. Colladay, V.A. Kostelecký, and R. Potting, Phys. Lett. B **395** (1997) 178.
15. R. Bluhm, V.A. Kostelecký, and N. Russell, Phys. Rev. Lett. **79** (1997) 1432; Phys. Rev. D **57** (1998), in press (issue of April 1, 1998).
16. R. Bluhm, V.A. Kostelecký, and N. Russell, these proceedings.
17. OPAL Collaboration, R. Ackerstaff *et al.*, Z. Phys. C **76** (1997) 401.
18. DELPHI Collaboration, M. Feindt *et al.*, preprint DELPHI 97-98 CONF 80 (July 1997).
19. R.S. Van Dyck, Jr., P.B. Schwinberg, and H.G. Dehmelt, Phys. Rev. Lett. **59** (1987) 26; Phys. Rev. D **34** (1986) 722.
20. P.B. Schwinberg, R.S. Van Dyck, Jr., and H.G. Dehmelt, Phys. Lett. A **81** (1981) 119.
21. D.J. Heinzen and D.J. Wineland, Phys. Rev. A **42** (1990) 2977; W. Quint and G. Gabrielse, Hyperfine Int. **76**, 379 (1993).
22. G. Gabrielse *et al.*, Phys. Rev. Lett. **74** (1995) 3544.
23. V.W. Hughes, H.G. Robinson, and V. Beltran-Lopez, Phys. Rev. Lett. **4** (1960) 342; R.W.P. Drever, Philos. Mag. **6** (1961) 683; J.D. Prestage *et al.*, Phys. Rev. Lett. **54** (1985) 2387; S.K. Lamoreaux *et al.*, Phys. Rev. A **39** (1989) 1082; T.E. Chupp *et al.*, Phys. Rev. Lett. **63** (1989) 1541.
24. S.M. Carroll, G.B. Field, and R. Jackiw, Phys. Rev. D **41** (1990) 1231.
25. B. Nodland and J.P. Ralston, Phys. Rev. Lett. **78** (1997) 3043; J.F.C. Wardle, R.A. Perley, and M.H. Cohen, *ibid.*, **79** (1997) 1801; D.J. Eisenstein and E.F. Bunn, *ibid.*, **79** (1997) 1957; S.M. Carroll and G.B. Field, *ibid.*, **79** (1997) 2394.
26. S. Coleman and S.L. Glashow, Phys. Lett. B **405** (1997) 249; S.L. Glashow, these proceedings.
27. B. Schwingenheuer *et al.*, Phys. Rev. Lett. **74** (1995) 4376; R.A. Briere, Ph.D. thesis, University of Chicago, June, 1995; B. Schwingenheuer, Ph.D. thesis, University of Chicago, June, 1995.

TESTS OF CPT AND LORENTZ SYMMETRY IN PENNING-TRAP EXPERIMENTS

Robert Bluhm[1], V. Alan Kostelecký[2], and Neil Russell[2]

[1]Physics Department
Colby College
Waterville, ME 04901 USA

[2]Physics Department
Indiana University
Bloomington, IN 47405 USA

INTRODUCTION

The CPT theorem [1] states that local relativistic quantum field theories of point particles in flat spacetime must be invariant under the combined operations of charge conjugation C, parity reversal P, and time reversal T. As a result of this invariance, particles and antiparticles have equal masses, lifetimes, charge-to-mass ratios, and gyromagnetic ratios. The CPT theorem has been tested to great accuracy in a variety of experiments [2]. The best bound is obtained in experiments with neutral mesons, where the figure of merit is

$$r_K \equiv |(m_K - m_{\overline{K}})/m_K| \lesssim 2 \times 10^{-18} \quad . \tag{1}$$

Experiments in Penning traps have also yielded sharp bounds on CPT violation, including the best bounds on lepton and baryon systems. Two types of experimental tests are possible in Penning traps. Both involve making accurate measurements of cyclotron frequencies ω_c and anomaly frequencies ω_a of single isolated particles confined in the trap. The first compares the ratio $2\omega_a/\omega_c$ for particles and antiparticles. In the context of conventional QED, this ratio equals $g-2$ for the particle or antiparticle. A second experiment compares values of $\omega_c \sim q/m$, where $q > 0$ is the magnitude of the charge and m is the mass, and is therefore a comparison of charge-to-mass ratios.

Experiments comparing $g-2$ for electrons and positrons yield the figure of merit [3, 4]

$$r_g \equiv |(g_{e^-} - g_{e^+})/g_{\rm avg}| \lesssim 2 \times 10^{-12} \quad , \tag{2}$$

while the charge-to-mass-ratio experiments yield the bound [5]

$$r^e_{q/m} \equiv |\left[(q_{e^-}/m_{e^-}) - (q_{e^+}/m_{e^+})\right]/(q/m)_{\rm avg}| \lesssim 1.3 \times 10^{-7} \quad . \tag{3}$$

To date, no experiments measuring $g-2$ for protons or antiprotons have been performed in Penning traps because of the difficulty in obtaining sufficient cooling and an adequate signal for detection of the weaker magnetic moments. However, proposals have been put forward that might make these types of experiments feasible in the future [6]. The best current tests of CPT in proton and antiproton systems come from comparisons of the charge-to-mass ratios [7], which yield the bound

$$r^p_{q/m} \equiv |\left[(q_p/m_p) - (q_{\overline{p}}/m_{\overline{p}})\right]/(q/m)_{\rm avg}| \lesssim 1.5 \times 10^{-9} \quad . \tag{4}$$

It is interesting to note that in the neutral meson experiments which yield the bound on r_K in (1), measurements are made with an experimental uncertainty of approximately one part in 10^4. In contrast, measurements of frequencies in Penning traps have experimental uncertainties of about one part in 10^9. This raises some intriguing questions about the Penning-trap experiments as to why they do not provide better tests of CPT when they have better experimental precision. In the context of conventional QED, which does not permit CPT breaking, it is not possible to pursue these types of questions. Instead, one would need to work in the context of a theoretical framework that allows CPT breaking, making possible an investigation of possible experimental signatures. Only recently has such a framework been developed [8].

In this paper, we describe the application of this theoretical framework to experiments on electron-positron and proton-antiproton systems in Penning traps. Our results have been published in Refs. [9, 10].

THEORETICAL FRAMEWORK

The framework we use [8] is an extension of the SU(3)×SU(2)×U(1) standard model originating from the idea of spontaneous CPT and Lorentz breaking in a more fundamental model such as string theory [11, 12]. This framework preserves various desirable features of quantum field theory such as gauge invariance and power-counting renormalizability. It has two sectors, one that breaks CPT and one that preserves CPT, while both break Lorentz symmetry. The possible CPT and Lorentz violations are parametrized by quantities that can be bounded by experiments. Within this framework, the modified Dirac equation describing a fermion with charge q and mass m is given by

$$\left(i\gamma^\mu D_\mu - m - a_\mu\gamma^\mu - b_\mu\gamma_5\gamma^\mu - \tfrac{1}{2}H_{\mu\nu}\sigma^{\mu\nu} + ic_{\mu\nu}\gamma^\mu D^\nu + id_{\mu\nu}\gamma_5\gamma^\mu D^\nu\right)\psi = 0 \quad . \tag{5}$$

Here, ψ is a four-component spinor, $iD_\mu \equiv i\partial_\mu - qA_\mu$, A^μ is the electromagnetic potential in the trap, and a_μ, b_μ, $H_{\mu\nu}$, $c_{\mu\nu}$, $d_{\mu\nu}$ are the parameters describing possible violations of CPT and Lorentz symmetry. The transformation properties of ψ imply that the terms involving a_μ, b_μ break CPT while those involving $H_{\mu\nu}$, $c_{\mu\nu}$, $d_{\mu\nu}$ preserve it, and that Lorentz symmetry is broken by all five terms. Since no CPT or Lorentz breaking has been observed in experiments to date, the quantities a_μ, b_μ, $H_{\mu\nu}$, $c_{\mu\nu}$, $d_{\mu\nu}$ must all be small.

PENNING-TRAP EXPERIMENTS

We use this theoretical framework to analyze tests of CPT and Lorentz symmetry in Penning-trap experiments. To begin, we note that the time-derivative couplings in (5) alter the standard procedure for obtaining a hermitian quantum-mechanical hamiltonian operator. To overcome this, we first perform a field redefinition at the lagrangian level that eliminates the additional time derivatives. We also use charge conjugation to obtain a Dirac equation and hamiltonian for the antiparticle.

To test CPT, experiments compare the cyclotron and anomaly frequencies of particles and antiparticles. According to the CPT theorem, particles and antiparticles of opposite spin in a Penning trap with the same magnetic fields but opposite electric fields should have equal energies. The experimental relations $g-2 = 2\omega_a/\omega_c$ and $\omega_c = qB/m$ provide connections to the quantities g and q/m used in defining the figures of merit r_g, $r^e_{q/m}$, and $r^p_{q/m}$. We perform calculations using Eq. (5) to obtain possible shifts in the energy levels due to either CPT-breaking or CPT-preserving Lorentz violation. In this way, we examine the effectiveness of Penning-trap experiments as tests of both CPT-breaking and CPT-preserving Lorentz violation. From the computed energy shifts we determine how the frequencies ω_c and ω_a are affected and if the conventional figures of merit are appropriate.

For experiments in Penning traps, the dominant contributions to the energy come from interactions of the particle or antiparticle with the constant magnetic field of the trap. The quadrupole electric fields generate smaller effects. In a perturbative calculation, the dominant CPT- and Lorentz-violating effects can therefore be obtained by working with relativistic Landau levels as unperturbed states. Conventional perturbations, such as the anomaly, will lead to corrections that are the same for particles and antiparticles. CPT- and Lorentz-breaking effects will result in either differences between particles and antiparticles or in unconventional effects such as diurnal variations in the measured frequencies.

RESULTS

Our calculations for electrons and positrons in Penning traps [9] show that the leading-order effects due to CPT and Lorentz breaking cause corrections to the cyclotron and anomaly frequencies:

$$\omega_c^{e^-} \approx \omega_c^{e^+} \approx (1 - c_{00}^e - c_{11}^e - c_{22}^e)\omega_c \quad , \tag{6}$$

$$\omega_a^{e^\mp} \approx \omega_a \mp 2b_3^e + 2d_{30}^e m_e + 2H_{12}^e \quad . \tag{7}$$

Here, ω_c and ω_a represent the unperturbed frequencies, while $\omega_c^{e^\mp}$ and $\omega_a^{e^\mp}$ denote the frequencies including the corrections. Superscripts have also been added on the coefficients b_μ, etc. to denote that these are parameters of the electron-positron system. From these relations we find the electron-positron differences for the cyclotron and anomaly frequencies to be

$$\Delta\omega_c^e \equiv \omega_c^{e^-} - \omega_c^{e^+} \approx 0 \quad , \qquad \Delta\omega_a^e \equiv \omega_a^{e^-} - \omega_a^{e^+} \approx -4b_3^e \quad . \tag{8}$$

Evidently, in the context of this framework comparisons of cyclotron frequencies to leading order do not provide a signal for CPT or Lorentz breaking, since the corrections to ω_c for electrons and positrons are equal. On the other hand, comparisons of ω_a provide unambiguous tests of CPT since only the CPT-violating term with b_3 results in a nonzero value for the difference $\Delta\omega_a^e$.

We have also found that to leading order there are no corrections to the g factors for either electrons or positrons. This leads to some interesting and unexpected results concerning the figure of merit r_g in Eq. (2). With $g_{e^-} \approx g_{e^+}$ to leading order, we find that r_g vanishes, which would seem to indicate the absence of CPT violation. However, this cannot be true since the model contains explicit CPT violation. Furthermore, our calculations show that with $\vec{b} \neq 0$ the experimental ratio $2\omega_a/\omega_c$ is field dependent and is undefined in the limit of vanishing magnetic field. Thus, the usual relation $g - 2 = 2\omega_a/\omega_c$ does not hold in the presence of CPT breaking. For these reasons, the figure of merit r_g in Eq. (2) is misleading, and an alternative is suggested. Since the CPT theorem predicts that states of opposite spin in the same magnetic field have equal energies, we propose as a model-independent figure of merit,

$$r_{\omega_a}^e \equiv \frac{|E_{n,s}^{e^-} - E_{n,-s}^{e^+}|}{E_{n,s}^{e^-}} \quad , \tag{9}$$

where $E_{n,s}^{e^\mp}$ are the energies of the relativistic states labeled by their Landau-level numbers n and spin s. Our calculations show $r_{\omega_a}^e \approx |\Delta\omega_a^e|/2m_e \approx |2b_3^e|/m_e$, and we estimate as a bound on this figure of merit,

$$r_{\omega_a}^e \lesssim 10^{-20} \quad . \tag{10}$$

In Ref. [10], we describe additional possible signatures of CPT and Lorentz breaking. These include possible diurnal variations in the anomaly and cyclotron frequencies. Tests for these effects would provide bounds on various components of the parameters $c_{\mu\nu}^e$, $d_{\mu\nu}^e$, and $H_{\mu\nu}^e$ at a level of about one part in 10^{18}.

A similar analysis can also be performed on proton-antiproton experiments in Penning traps. In this context, it suffices to work at the level of an effective theory in which the protons and antiprotons are regarded as basic objects described by a Dirac equation. The coefficients a_μ^p, b_μ^p, $H_{\mu\nu}^p$, $c_{\mu\nu}^p$, $d_{\mu\nu}^p$ represent effective parameters, which at a more fundamental level depend on the underlying quark interactions. Comparisons of protons and antiprotons in the context of this model yield the results for the proton-antiproton frequency differences,

$$\Delta\omega_c^p \equiv \omega_c^p - \omega_c^{\bar{p}} = 0 \quad , \qquad \Delta\omega_a^p \equiv \omega_a^p - \omega_a^{\bar{p}} = 4b_3^p \quad . \tag{11}$$

Assuming an experiment could be made sensitive enough to measure ω_a^p and $\omega_a^{\bar{p}}$ with a precision similar to that of electron $g-2$ experiments, then the appropriate figure of merit would be

$$r_{\omega_a}^p \equiv \frac{|E_{n,s}^p - E_{n,-s}^{\bar{p}}|}{E_{n,s}^p} \quad . \tag{12}$$

A bound on this can be estimated as

$$r_{\omega_a}^p \lesssim 10^{-23} \quad . \tag{13}$$

It is apparent that an experiment comparing anomaly frequencies of protons and antiprotons in a Penning trap has the potential to provide a particularly tight CPT bound. Other signatures of CPT and Lorentz breaking involving diurnal variations in ω_a and ω_c are described in Ref. [10]. These additional signatures provide bounds on various components of $c^p_{\mu\nu}$, $d^p_{\mu\nu}$, and $H^p_{\mu\nu}$ estimated at about one part in 10^{21}.

CONCLUSIONS

We find that the use of a general theoretical framework incorporating CPT and Lorentz breaking permits a detailed investigation of possible experimental signatures in Penning-trap experiments. Our results indicate that the sharpest tests of CPT symmetry emerge from comparisons of anomaly frequencies in $g-2$ experiments. Our estimates of appropriate figures of merit provide bounds of approximately 10^{-20} in electron-positron experiments and of 10^{-23} for a plausible proton-antiproton experiment. Other signals involving possible diurnal variations provide additional bounds at the level of 10^{-18} in the electron-positron system and 10^{-21} in the proton-antiproton system. A table showing all our estimated bounds is presented in Ref. [10].

ACKNOWLEDGMENTS

This work was supported in part by the National Science Foundation under grant number PHY-9503756.

REFERENCES

1. See, for example, R.F. Streater and A.S. Wightman, *PCT, Spin, Statistics, and All That* (Benjamin Cummings, London, 1964).
2. See, for example, R.M. Barnett et al., Review of Particle Properties, Phys. Rev. D **54** (1996) 1.
3. R.S. Van Dyck, Jr., P.B. Schwinberg, and H.G. Dehmelt, Phys. Rev. Lett. **59** (1987) 26; Phys. Rev. D **34** (1986) 722.
4. L.S. Brown and G. Gabrielse, Rev. Mod. Phys. **58** (1986) 233.
5. P.B. Schwinberg, R.S. Van Dyck, Jr., and H.G. Dehmelt, Phys. Lett. A **81** (1981) 119.
6. D.J. Heinzen and D.J. Wineland, Phys. Rev. A **42** (1990) 2977; W. Quint and G. Gabrielse, Hyperfine Int. **76**, 379 (1993).
7. G. Gabrielse et al., Phys. Rev. Lett. **74** (1995) 3544.
8. D. Colladay and V.A. Kostelecký, Phys. Rev. D **55** (1997) 6760; Indiana University preprint IUHET 359 (1997).
9. R. Bluhm, V.A. Kostelecký and N. Russell, Phys. Rev. Lett. **79** (1997) 1432.
10. R. Bluhm, V.A. Kostelecký and N. Russell, Phys. Rev. D, in press.
11. V.A. Kostelecký and R. Potting, Nucl. Phys. B **359** (1991) 545; Phys. Lett. B **381** (1996) 389.
12. V.A. Kostelecký and S. Samuel, Phys. Rev. Lett. **63** (1989) 224; ibid. **66** (1991) 1811; Phys. Rev. D **39** (1989) 683; ibid. **40** (1989) 1886.

Mass Hierarchy and Flat Directions in String Models

Gerald B. Cleaver

Department of Physics and Astronomy
The University Of Pennsylvania
Philadelphia, Pennsylvania 19104-6396

ABSTRACT

I discuss a method for generating an effective μ parameter and quasi-realistic generational mass hierarchy based on intermediate scales between M_Z and M_{string}. Application to string models is briefly discussed. Additionally, a method based on the singular value decomposition (SVD) of a matrix is presented for determining, en masse, a complete basis of D-flat directions in supersymmetric models.

I. Mass Hierarchy of Standard Model Matter

The standard model (SM) and its supersymmetric counterpart, the minimal supersymmetric standard model (MSSM) have resolved many issues regarding the elementary particles and their interactions. Formation of these models is of the leading scientific accomplishments of this century. Several Nobel awards have been earned for work contributing to development of the SM. Yet the SM/MSSM still leaves many issues unresolved. In line with the theme of this conference, I would like to address one such issue related to the *Physics of Mass.* The topic I will address is the mass hierarchy between the three generations of elementary particles.

With the mass of the top quark now experimentally determined to within about 5% [1], the physical masses of all three generations of up-, down-, and electron-like particles are known to very good accuracy. These particle masses are displayed in Table 1. Therein, I have expressed all masses in units of the top mass. This normalization makes more apparent the approximate mass hierarchy of $10^{-5} : 10^{-3} : 1$ between the three generations of MSSM particles. (Notably, m_b and m_τ are exceptions to this pattern.)

In MSSM physics, it is generally assumed that quarks and electrons (and their supersymmetric partners) gain mass through superpotential couplings to Higgs bosons H_1 and H_2,[1]

$$W_{u_i} \sim \lambda_{u_i} \hat{H}_2 \hat{Q}_i \hat{U}_i^c \,; \quad W_{d_i} \sim \lambda_{d_i} \hat{H}_1 \hat{Q}_i \hat{D}_i^c \,; \quad W_{e_i} \sim \lambda_{e_i} \hat{H}_1 \hat{L}_i \hat{E}_i^c , \tag{1}$$

TABLE I. Fermion mass ratios with the top quark mass normalized to 1. The values of $u-$, $d-$, and s-quark masses used in the ratios (with the t-quark mass normalized to 1 from an assumed mass of 170 GeV) are estimates of the $\overline{\text{MS}}$ scheme current-quark masses at a scale $\mu \approx 1$ GeV. The c- and b-quark masses are pole masses. An additional mass constraint for stable light neutrinos is $\sum_i m_{\nu_i} \leq 6 \times 10^{-11}$ (i.e., 10 eV), based on the neutrino contributions to the mass density of the universe and the growth of structure.

m_u	:	m_c	:	m_t	=	3×10^{-5}	:	7×10^{-3}	:	1
m_d	:	m_s	:	m_b	=	6×10^{-5}	:	1×10^{-3}	:	3×10^{-2}
m_e	:	m_μ	:	m_τ	=	0.3×10^{-5}	:	0.6×10^{-3}	:	1×10^{-2}
m_{ν_e}	:	m_{ν_μ}	:	m_{ν_τ}	=	$< 6 \times 10^{-11}$	:	$< 1 \times 10^{-6}$	:	$< 1 \times 10^{-4}$

where i is the generation number. Effective mass terms appear when the Higgs acquire a typical soft supersymmetry breaking scale vacuum expectation value (VEV) $\langle H_{1,2}\rangle \sim m_{soft} = \mathcal{O}(M_Z)$.

Intergenerational mass ratios can be induced when the associated first and second generation superpotential terms contain effective couplings λ that include non-renormalizable suppression factors,

$$\lambda_{(u,d,e)_i} \sim (\frac{\langle S\rangle}{M_{\rm Pl}})^{P'_i}, \quad \text{for } i = 1, 2, \tag{2}$$

where S is a non-Abelian singlet, $M_{\rm Pl}$ is the Planck scale (which is replaced by the $M_{\rm string}$ for string models), and P' is a positive integer. VEVs only slightly below the Planck/string scale (as might occur through $U(1)$ anomaly cancellation – see Section III) imply large values of $P'_{1,2}$ for 10^{-5} and 10^{-3} suppression factors. On the other hand, intermediate scale VEVs (between M_Z and $M_{\rm string}$) require far lower values for P'_i. In a series of recent papers [2, 3, 4, 5] intermediate scales have been explored and occurrence of intermediate scales in actual models has been under investigation.

In the following section, I show how intermediate scales can occur and how, in theory, they could produce an intergenerational $10^{-5} : 10^{-3} : 1$ mass ratio. In Section III string realizations of intermediate scales are discussed The often first step towards intermediate scales in string models, removal of a $U(1)$ anomaly, is reviewed. D- and F-flat directions must cancel the dangerous Fayet-Iliopoulos (FI) term generated by the standard anomaly cancellation mechanism when models initially contain an anomalous $U(1)$ [7, 8]. Section IV introduces, as a tangential topic, a method for generating en masse a complete basis of D-flat directions.

II. Intermediate Scales from String Models

One method for inducing an intermediate scale in a phenomenologically viable manner involves extending the SM gauge group by an additional $U(1)'$. Then two SM singlets carrying respective $U(1)'$ charges Q_1 and Q_2 are minimally required. If the two $U(1)'$ charges are of opposite sign ($Q_1 Q_2 < 0$), then there is a $D-$flat scalar field direction S defined by,[2]

$$\langle S_1\rangle = \cos\alpha_Q \langle S\rangle, \quad \langle S_2\rangle = \sin\alpha_Q\langle S\rangle, \quad \text{where} \quad \tan^2\alpha_Q \equiv \frac{|Q_1|}{|Q_2|}. \tag{3}$$

[1] $\hat{X}$ denotes a generic superfield and X its bosonic component.

[2] See Section III for a discussion of D and F-flatness. In the absence of an extra $U(1)'$ only a single uncharged scalar field S_1 would be necessary.

$S = S_1 \cos\alpha_Q + S_2 \sin\alpha_Q$ will also be a renormalizable F-flat direction if (as I assume) $\hat{S}_1$ and $\hat{S}_2$ do not couple among themselves in the renormalizable superpotential. Consider the real component of this flat direction, $s = \sqrt{2}\mathrm{Re}S = s_1 \cos\alpha_Q + s_2 \sin\alpha_Q$. This scalar's running mass is,

$$m^2 = m_1^2(\mu)\cos^2\alpha_Q + m_2^2(\mu)\sin^2\alpha_Q = \left(\frac{m_1^2}{|Q_1|} + \frac{m_2^2}{|Q_2|}\right)\frac{|Q_1 Q_2|}{|Q_1| + |Q_2|}, \tag{4}$$

which generates a potential

$$V(s) = \frac{1}{2}m(\mu = s)^2 s^2. \tag{5}$$

I will assume that m^2 is positive at the string scale and of order $m^2_{\text{soft}} \sim \mathcal{O}(M_Z{}^2)$ (m_o^2 if universality is assumed). However, for various chooses of the supersymmetry breaking parameters A^0 (the universal Planck scale soft trilinear coupling) and $M_{1/2}$ (the universal Planck scale gaugino mass) normalized by the universal scalar electroweak (EW) scale soft mass-squared parameter m_0, Ref. [2] demonstrated that m^2 can be driven to negative values (of EW scale magnitude) by large Yukawa couplings (i) of S_1 to exotic triplets, $W = h\hat{D}_1\hat{D}_2\hat{S}_1$; (ii) of S_1 to exotic doublets and of S_2 to exotic triplets, $W = h_D\hat{D}_1\hat{D}_2\hat{S}_1 + h_L\hat{L}_1\hat{L}_2\hat{S}_2$; and (iii) of S_1 to varying numbers of additional SM singlets $W = h\sum_{i=1}^{N_p}\hat{S}_{ai}\hat{S}_{bi}\hat{S}_1$. $m(\mu = s)^2$ can turn negative anywhere between a scale of $\mu_{rad} = 10^4$ GeV and $\mu_{rad} = 10^{17}$ GeV (near the string scale). This leads to a minimum of the potential developing along the flat direction and S gains a non-zero VEV. In the case of only a mass term and no Yukawa contribution to $V(s)$, minimizing the potential

$$\frac{dV}{ds} = \left(m^2 + \frac{1}{2}\beta_{m^2}\right)\bigg|_{\mu=s} s = 0, \tag{6}$$

(where $\beta_{m^2} = \mu\frac{dm^2}{d\mu}$) shows that the VEV $\langle s\rangle$ is determined by

$$m^2(\mu = \langle s\rangle) = -\frac{1}{2}\beta_{m^2}. \tag{7}$$

Eqs. (6,7) are satisfied very close to the scale μ_{RAD} at which m^2 crosses zero. This scale is fixed by the renormalization group evolution of parameters from M_{string} down to the EW scale and will lie at some intermediate scale.

Location of the potential minimum can also be effected by non-renormalizable self-interaction terms,

$$W_{\text{NR}} = \left(\frac{\alpha_K}{M_{\text{Pl}}}\right)^K \hat{S}^{3+K}, \tag{8}$$

where $K = 1, 2...$ and α_K are coefficients. Such non-renormalizable operators (NROs) lift the flat direction for sufficiently large values of s. Generally, the running mass is the dominant factor when $\mu_{RAD} \ll 10^{12}$ GeV even if NROs exist. If they are present, the non-renormalizable operators NRO dominate when $\mu_{RAD} \gg 10^{12}$ GeV. NRO contributions to the potential transform (5) into

$$V(s) = \frac{1}{2}m^2 s^2 + \frac{1}{2(K+2)}\left(\frac{s^{2+K}}{\mathcal{M}^K}\right)^2, \tag{9}$$

where $\mathcal{M} = \mathcal{C}_K M_{\text{Pl}}/\alpha_K$, with $\mathcal{C}_K = [2^{K+1}/((K+2)(K+3)^2)]^{1/(2K)}$.

When the NRO factor dominates over the running mass effect in (9), the VEV of s is found to be

$$\langle s \rangle = \left[\sqrt{(-m^2)}\mathcal{M}^K\right]^{\frac{1}{K+1}} = \mu_K \sim (m_{soft}\mathcal{M}^K)^{\frac{1}{K+1}}, \tag{10}$$

where $m_{soft} = \mathcal{O}(|m|) = \mathcal{O}(M_Z)$ is a typical soft supersymmetry breaking scale.

While $\langle s \rangle$ is an intermediate scale VEV, the mass M_S of the physical field s is still on the order of the soft SUSY breaking scale: When the running mass dominates

$$M_S^2 \equiv \left.\frac{d^2V}{ds^2}\right|_{s=\langle s \rangle} = \left.\left(\beta_{m^2} + \frac{1}{2}\mu\frac{d}{d\mu}\beta_{m^2}\right)\right|_{\mu=\langle s \rangle} \simeq \beta_{m^2} \sim \frac{m_{soft}^2}{16\pi^2}, \tag{11}$$

whereas when the NRO term dominates,

$$M_S^2 = 2(K+1)(-m^2) \sim m_{soft}^2. \tag{12}$$

What powers P_i in (2) for first and second generation suppression factors in an NRO-dominated model could produce a mass ratio of order $10^{-5} : 10^{-3} : 1$? When $\langle s \rangle$ takes the form in (10), the suppression factors become

$$\left(\frac{m_{soft}}{M}\right)^{\frac{P_i'}{K+1}}, \tag{13}$$

where the coefficient α_K has been absorbed into the definition of the mass scale $M \equiv \mathcal{M}/\mathcal{C}_K$. The mass suppression factors corresponding to P' in the range 0 to 5 and K in the range of 1 to 7 are given in Table 2. This table shows that intermediate scale VEVs around 8×10^{14} GeV to 2×10^{15} GeV, resulting from $K = 5$ or $K = 6$ self-interactions terms of S, reproduce the required mass ratio for $P_1' = 2$ and $P_2' = 1$.

However, unless $tan\ \beta \equiv \frac{\langle H_2 \rangle}{\langle H_1 \rangle} \gg 1$, the intragenerational mass ratio of 10^{-2} : 10^{-2} : 1 for m_τ, m_b, and m_t is not realizable from a $K = 5$ or 6 NRO singlet term. For $tan\ \beta \sim 1$, m_τ, m_b are too small to be associated with a renormalizable coupling ($P = 0$) like is assumed for m_t, but are somewhat larger than predicted by $P' = 1$ for $K = 5$ or $K = 6$. Instead, m_b and m_τ might be associated with a different NRO involving the VEV of an entirely different singlet. In that event, Table 2 suggests another flat direction S', (formed from a second singlet pair S_1' and S_2') with a $K = 7$ self-interaction NRO and $P' = 1$ suppression factor for m_b and m_τ.

TABLE II. Non-Renormalizable MSSM mass terms via $\langle S \rangle$. For $m_{soft} \sim 100$ GeV, $M \sim 3 \times 10^{17}$ GeV.

	P or P'	$K=1$	$K=2$	$K=3$	$K=4$	$K=5$	$K=6$	$K=7$
$(\frac{m_{soft}}{M})^{\frac{1}{K+1}}$		2×10^{-8}	7×10^{-6}	1×10^{-4}	8×10^{-4}	3×10^{-3}	6×10^{-3}	1×10^{-2}
$\langle S \rangle$ (GeV)		5×10^{9}	2×10^{12}	4×10^{13}	2×10^{14}	8×10^{14}	2×10^{15}	3×10^{15}
	$K-1$	5×10^{7}	1×10^{5}	7×10^{3}	1×10^{3}	400	200	90
$\frac{\mu_{eff}}{m_{soft}}$	K	1	1	1	1	1	1	1
	$K+1$	2×10^{-8}	7×10^{-6}	1×10^{-4}	8×10^{-4}	3×10^{-3}	6×10^{-3}	1×10^{-2}
	0	1	1	1	1	1	1	1
	1	2×10^{-8}	7×10^{-6}	1×10^{-4}	8×10^{-4}	3×10^{-3}	6×10^{-3}	1×10^{-2}
	2	3×10^{-16}	5×10^{-11}	2×10^{-8}	6×10^{-7}	7×10^{-6}	4×10^{-5}	1×10^{-4}
$\frac{m_{Q,L}}{\langle H_i \rangle}$	3	6×10^{-24}	3×10^{-16}	2×10^{-12}	5×10^{-10}	2×10^{-8}	2×10^{-7}	2×10^{-6}
	4	1×10^{-31}	2×10^{-21}	3×10^{-16}	4×10^{-13}	5×10^{-11}	1×10^{-9}	2×10^{-8}
	5	2×10^{-39}	2×10^{-26}	5×10^{-20}	3×10^{-16}	1×10^{-13}	9×10^{-12}	2×10^{-10}

An intermediate scale VEV $\langle S\rangle$ can also solve the μ problem through a superpotential term

$$W_\mu \sim \hat{H}_1\hat{H}_2\hat{S}\left(\frac{\hat{S}}{M}\right)^{P}_{\mu}. \qquad (14)$$

With NRO-dominated $\langle S\rangle \sim (m_{soft}M^K)^{\frac{1}{K+1}}$, the effective Higgs μ-term takes the form,

$$\mu_{eff} \sim m_{soft}\left(\frac{m_{soft}}{M}\right)^{\frac{P-K}{K+1}}. \qquad (15)$$

The phenomenologically preferred choice among this class of terms is clearly $P = K$, since this yields a $K-$independent $\mu_{eff} \sim m_{soft}$.

Intermediate VEVs also provide various means by which neutrinos can acquire small masses. Some of these processes do not involve the traditional seesaw mechanism. Very light non-seesaw doublet neutrino Majorana masses

$$m_{L_iL_i} \sim \frac{\langle H_2\rangle^2}{M}\left(\frac{m_{soft}}{M}\right)^{\frac{P''_{L_iL_i}}{K+1}} \sim \langle H_2\rangle\left(\frac{m_{soft}}{M}\right)\times\left(\frac{m_{soft}}{M}\right)^{\frac{P''_{L_iL_i}}{K+1}} \ll 1 \text{ eV} \qquad (16)$$

are possible from Majorana doublet superpotential terms

$$W^{(\mathrm{Maj})}_{L_iL_i} \sim \frac{\left(\hat{H}_2\hat{L}_i\right)^2}{M}\left(\frac{\hat{S}}{M}\right)^{P''_{L_iL_i}}. \qquad (17)$$

The upper bound (corresponding to $P''_{L_iL_i} = 0$) on such neutrino masses is around 10^{-4} eV (using $\langle H_2\rangle \sim m_{soft} = 100$ GeV and $M = 3\times 10^{17}$ GeV), which is too small to be relevant to dark matter or MSW conversions in the sun [9].

If Majorana doublet terms are not present, then naturally heavier physical Dirac neutrino masses

$$m_{L_i\nu_i^c} \sim \langle H_2\rangle\left(\frac{m_{soft}}{M}\right)^{\frac{P'_{L_i\nu_i^c}}{K+1}}, \qquad (18)$$

result from Dirac superpotential terms like

$$W^{(\mathrm{Dir})}_{L_i\nu_i^c} \sim \hat{H}_2\hat{L}_i\hat{\nu}_i^c\left(\frac{\hat{S}}{M}\right)^{P'_{L_i\nu_i^c}} \qquad (19)$$

(with $\hat{\nu}\in\hat{L}$ the neutrino doublet component and $\hat{\nu}^c$ a neutrino singlet). For $K = 5$ the experimental neutrino upper mass limits given in Table 1 allow $P'_{L_1\nu_1^c} \geq 4$, $P'_{L_2\nu_2^c} \geq 3$, and $P'_{L_3\nu_3^c} \geq 2$. $P'_{L_i\nu_i^c} = 4$ (5) corresponds to $m_{L_i\nu_i^c} = 0.9$ eV (10^{-2} eV), which is in the interesting range for solar and atmospheric neutrinos, oscillation experiments, and dark matter.

Neutrino singlets can acquire Majorana masses

$$m_{\nu_i^c\nu_i^c} \sim m_{soft}\left(\frac{m_{soft}}{M}\right)^{\frac{\bar{P}_{\nu_i^c\nu_i^c}-K}{K+1}}, \qquad (20)$$

through superpotential terms

$$W^{(\mathrm{Maj})}_{\nu_i^c\nu_i^c} \sim \hat{\nu}_i^c\hat{\nu}_i^c\hat{S}\left(\frac{\hat{S}}{M}\right)^{\bar{P}_{\nu_i^c\nu_i^c}}. \qquad (21)$$

Such masses can be very large or small, depending on the sign of $\bar{P}_{\nu_i^c\nu_i^c} - K$.

When both $W^{(\mathrm{Dir})}_{L_i\nu^c_i}$ and $W^{(\mathrm{Maj})}_{\nu^c_i\nu^c_i}$ terms are present, the standard seesaw mechanism can produce light neutrinos via diagonalization of the mass matrix for eqs. (19,21). The light mass eigenstate is

$$m^{\mathrm{light}}_{\mathrm{seesaw}} \sim m^2_{L_i\nu^c_i}/m_{\nu^c_i\nu^c_i} \sim m_{soft}\left(\frac{m_{soft}}{M}\right)^{\frac{2P'_{L_i\nu^c_i}+K-\bar{P}_{\nu^c_i\nu^c_i}}{K+1}}, \tag{22}$$

while the heavy mass eigenstate is still to first order $m_{\nu^c_i\nu^c_i}$ as given by (20). Various combinations of K, $P'_{L_i\nu^c_i}$ and $\bar{P}_{\nu^c_i\nu^c_i}$ produce viable masses for three generations of light neutrinos. For example, with $K=5$ and $P'_{L_i\nu^c_i} = P'_i = 2$ (1) for $i=1$, (2), and with either $P'_{L_3\nu^c_3} = 1$ or $P'_{L_3\nu^c} = P'_{u_3} = 0$, the light eigenvalues of the three generations fall into the hierarchy of 3×10^{-5} eV, 1×10^{-2} eV, and either 1×10^{-2} eV or 5 eV for $\bar{P}_{\nu^c_i\nu^c_i} = P'_{L_i\nu^c_i}+1$. This is an interesting mass range for laboratory and non-accelerator experiments.

III. Anomalous $U(1)$ and Flat String Directions

In string models, one problem must generically be taken care of before possible flat directions for intermediate scale VEVs can be investigated. That is, most four dimensional string models contain an anomalous $U(1)_A$ (meaning $\mathrm{Tr}\, Q_A \neq 0$). Since the appearance of anomalous $U(1)$ in four-dimensional string models has been discussed in many prior papers *e.g.* [10, 3], I will only review the essentials here.

Often in a generic charge basis, a string model with an Abelian anomaly may actually contain not just one, but several anomalous $U(1)$ symmetries. However, all anomalies can all be transferred into a single $U(1)_A$ through the *unique* rotation

$$U(1)_{\mathrm{A}} \equiv c_A \sum_n \{\mathrm{Tr}\, Q_n\} U(1)_n, \tag{23}$$

with c_A a normalization factor. The remaining non-anomalous components of the original set of $\{U(1)_n\}$ may be rotated into a complete (non-unique) orthogonal basis $\{U(1)_a\}$

The Green-Schwarz (GS) relations,

$$\frac{1}{k_m k_A^{1/2}} \mathop{\mathrm{Tr}}_{G_m} T(R) Q_A = \frac{1}{3k_A^{3/2}} \mathrm{Tr}\, Q_A^3 = \frac{1}{k_a k_A^{1/2}} \mathrm{Tr}\, Q_a^2 Q_A = \frac{1}{24 k_A^{1/2}} \mathrm{Tr}\, Q_A \neq 0, \tag{24}$$

$$\frac{1}{k_m k_a^{1/2}} \mathop{\mathrm{Tr}}_{G_m} T(R) Q_a = \frac{1}{3k_a^{3/2}} \mathrm{Tr}\, Q_a^3 = \frac{1}{k_A k_a^{1/2}} \mathrm{Tr}\, Q_A^2 Q_a = \frac{1}{24 k_a^{1/2}} \mathrm{Tr}\, Q_a = 0, \tag{25}$$

and additional generalizations involving $Q_A Q_a Q_b$, $Q_a Q_b Q_c$, etc.. guarantee the elimination of all triangle anomalies except those involving one or three $U(1)_A$ gauge bosons. These relations result from stringy modular invariance constraints and have no parallels in standard field theory. Thus, in a generic field-theoretic model or in a strongly coupled string model, (23) would not necessarily place the entire anomaly into a single $U(1)_A$: $\mathrm{Tr}\, Q_a$ and $\mathrm{Tr}\, Q_a^3$ may be independent.

The standard anomaly cancellation mechanism [7, 8] breaks $U(1)_{\mathrm{A}}$, while simultaneously generating a FI D-term,

$$\xi = \frac{e^{\phi} M^2_{\mathrm{Pl}}}{192\pi^2 k_A^{1/2}} \mathrm{Tr}\, Q_A\,, \tag{26}$$

where ϕ is the dilaton and $g \equiv e^{\phi/2}$ is the physical four-dimensional gauge coupling. The FI D-term breaks spacetime supersymmetry unless it is cancelled by appropriate

VEVs of the scalar components φ_j of supermultiplets Φ_j that carry non-zero anomalous charge,

$$D_{\rm A} = \sum_j Q_j^{(A)} |\varphi_j|^2 + \frac{g^2 M_{\rm Pl}^2}{192\pi^2} {\rm Tr} Q_{\rm A} = 0 \; . \tag{27}$$

Constraints are imposed on possible VEV directions $\sum_j |\varphi_j|$ by D-flatness in the non-anomalous directions,

$$D_a = \sum_j Q_j^{(a)} |\varphi_j|^2 = 0, \tag{28}$$

along with F-flatness,

$$F_j = \frac{\partial W}{\partial \Phi_j} = 0; \; W = 0. \tag{29}$$

In [3] my colleagues and I at Penn developed methods for classifying D- and F-flat directions in anomalous string models. In [5] we then applied this process to all of the free fermionic three generation $SU(3) \times SU(2)$ models introduced in [6]. We determined (i) which models have anomaly-cancelling directions that involve only VEVs of non-Abelian singlet fields, and (ii) since anomaly-free $U(1)_a$ are also generally broken along with $U(1)_A$, which (if any) singlet field anomaly-cancelling flat directions retain a good hypercharge and at least one additional $U(1)'$ for generating an intermediate scale VEV.

IV. D-Flat Basis From Matrix Decomposition

For the remaining part of my talk, I would like to discuss a new approach I found for generating a complete basis of D-flat directions. Most often (including in [3, 5]) D-flat directions for the non-anomalous $U(1)_n$ are found via their the one-to-one correlation with holomorphic gauge invariant polynomials of fields [11]. Here a D-flat basis corresponds to a maximal set of independent holomorphic monomials. I have found an alternate approach involving singular value decomposition (SVD) of a matrix [12]. While the monomial and matrix methods are essentially different langauges for the same process, the matrix decomposition method generates, en masse, a complete basis of D-flat directions for non-Abelian singlet states.

The matrix method is based on the mathematical fact that any $(M \times N)$-dimensional matrix $\mathbf{D}$ whose number of rows M is greater than or equal to its number of columns N, can be written as the product of an $M \times N$ column-orthogonal matrix $\mathbf{U}$, an $N \times N$ diagonal matrix $\mathbf{W}$ containing only semi-positive-definite elements, and the transpose of an $N \times N$ orthogonal matrix $\mathbf{V}$ [12],

$$\mathbf{D}_{M\times N} = \mathbf{U}_{M\times N} \cdot \mathbf{W}^{\rm diag}_{N\times N} \cdot \mathbf{V}^T_{N\times N}, \quad \text{for } M \geq N \, . \tag{30}$$

This decomposition is always possible, no matter how singular the matrix is. The decomposition is nearly unique also, up to (i) making the same permutation of the columns of U, diagonal elements of W, and columns of V, or (ii) forming linear combinations of any columns of U and V whose corresponding elements of W are degenerate. If $M < N$, then a $(N - M) \times N$ zero-matrix can be appended onto D so this decomposition can be performed: $D_{(M<N),N} \rightarrow D_{(M=N),N}$.

This decomposition is extremely useful when the matrix D is associated with a set of M simultaneous linear equations expressed by,

$$\mathbf{D} \cdot \vec{x} = \vec{b}, \tag{31}$$

where x and b are vectors. (I assume hereon that D has been enhanced by a zero submatrix if necessary so that $M \geq N$.) Eq. (31) defines a linear mapping from N-dimensional vector space x to M-dimensional vector-space b. When $\mathbf{D}$ is singlar (corresponding to the M constraints not all being linearly independent) there is a subspace of $\vec{x}$ termed the *nullspace* that is mapped to $\vec{0}$ in b-space by $\mathbf{D}$. The dimension of this nullspace is referred to as the nullity.

The subspace of $\vec{b}$ that *can* be reached by the matrix $\mathbf{D}$ acting on $\vec{x}$ is called the *range* of $\mathbf{D}$ and the dimension of the range is denoted as the *rank* of $\mathbf{D}$. Clearly $\text{rank}\,\mathbf{D} + \text{nullity}\,\mathbf{D} = N$, with $\text{rank}\,\mathbf{D} \equiv \#$ of independent constraint equations $\equiv M' \leq M$

In the decomposition of $\mathbf{D}$, the set of the jth columns of $\mathbf{U}$ corresponding to the jth non-zero diagonal components of $\mathbf{W}$ form an orthonormal set of basis vectors that span the range of D. The columns of $\mathbf{V}$ whose corresponding diagonal components of W are zero form an orthonormal basis for the nullspace.

This method is directly applicable to constructing D-flat directions, especially when only non-Abelian singlet states are allowed VEVs, Let M (M_I) denote the number of (independent) D-flat constraints and N denote the number of fields allowed to take on VEVs. Then the $D_{i,j}$ component of the matrix $\mathbf{D}$ is the $Q_j^{(i)}$ charge of the state ϕ_j. (i takes on the value A for the anomalous $U(1)_A$ and values $\{a = 1 \text{ to } M-1\}$ for the set of non-anomalous $U(1)_a$.) The components of the vector x are the values of $|\langle\phi_j\rangle|^2$, and b has all zero-components except in its row corresponding to the anomalous $U(1)_A$. The value of b in the anomalous position is $-\xi$, as defined by eq. (26).

Let $\mathbf{D}'$ be the matrix that excludes the row of anomalous charges in $\mathbf{D}$. In this language, the dimension of the moduli space $\mathcal{M}_{null}$ of flat directions for the $M' \equiv M-1$ non-anomalous $U(1)_{a=1 \text{ to } M'}$ is the nullity of matrix $\mathbf{D}'$, formed from the M' non-anomalous D-flat constraints. In other words, the nullity of $\mathbf{D}'$ is all VEVs formed from combinations of states that have zero net charge in each non-anomalous direction. The dimension of $\mathcal{M}_{null}$ is in the range

$$N - M' \leq \dim \mathcal{M}_{null} = N - M_I' \leq N, \tag{32}$$

where N is the number of states allowed VEVs and M_I' is the number of independent non-anomalous constraints.

For the complete matrix $\mathbf{D}$ containing also anomalous constraint, the elements of the range corresponding to true $\mathbf{D}$-flat directions are those formed solely from linear combinations of elements of the nullity of $\mathbf{D}'$ that generate an anomalous component for $\vec{b}$ of opposite sign to that of ξ. The nullspace of $\mathbf{D}$ will likewise be formed from linear combinations of $\mathbf{D}$'s nullity elements that generate a zero anomalous component for $\vec{b}$. The dimension of the $\mathbf{D}'$ nullity subset that projects into the range of $\mathbf{D}$, denoted by $\dim \mathcal{M}_R$, is 1 (since the anomalous constraint must necessarily be independent of the non-anomalous constraints). An $(N - M_I' - 1)$-dimensional subset of $\mathcal{M}_{null}$ forms the nullity of $\mathbf{D}$.

Using this decomposition process to form a basis of D-flat directions requires one additional projection be applied to the the range and nullity of $\mathbf{D}$, That is, components of a vector x can only have positive real values if they are to truly represent the norms of VEVs, $|\langle\phi_j\rangle|^2$. So while generically the orthogonal basis elements of nullspace of $\mathbf{D}'$, obtained from $\mathbf{V}$ via decomposition of $\mathbf{D}'$, will have components with negative values, we must project onto the sub-space containing only vectors with positive components. Without having to impose this sign constraint, a $\mathbf{D}$-flat direction is automatically guaranteed whenever the projection of the nullspace of $\mathbf{D}'$ onto the range of $\mathbf{D}$ was not empty.

There is one important caveat to this. An effective negative $|\langle\phi_j\rangle|^2$ can result when there are vector-like pairs of states (i.e., that carry exactly opposite charges for all $U(1)_a$ and the $U(1)_A$). If the VEVs of these two states are $|\langle\phi_j\rangle|^2$ and $|\langle\phi_{\bar{j}}\rangle|^2$, respectively, then each row in $\mathbf{D} \cdot \vec{x}$ will contain pairs of terms

$$Q_j^{(i)}|\langle\phi_j\rangle|^2 + Q_{\bar{j}}^{(i)}|\langle\phi_{\bar{j}}\rangle|^2 \tag{33}$$

that can be rewritten as

$$Q_j^{(i)}|\langle\phi_j\rangle|^2 - Q_j^{(i)}|\langle\phi_{\bar{j}}\rangle|^2 = Q_j^{(i)}(|\langle\phi_j\rangle|^2 - |\langle\phi_{\bar{j}}\rangle|^2) \equiv Q_j^{(i)}|\langle\phi_{j\bar{j}}\rangle|^2, \tag{34}$$

where $|\langle\phi_{j\bar{j}}\rangle|^2$ may take on any real value. We can consider an effective $|\langle\phi_{j\bar{j}}\rangle|^2$ as originating from a single field and thus can reduce the number of columns of $\mathbf{D}$ and $\mathbf{D}'$ by one for each vector pair of fields.

As a general rule, the more vector pairs of non-Abelian singlets there are, the more likely a D-flat (i.e. anomaly cancelling) direction can be formed from the nullspace of $\mathbf{D}'$ composed solely of non-Abelian singlets. Having all states in vector-like pairs is equivalent to totally relaxing the "positivity" constraint after projecting the nullspace of $\mathbf{D}'$ onto the range and nullspace of $\mathbf{D}$. Then for flat directions to exists, the projection of the nullspace of $\mathbf{D}'$ onto the range of $\mathbf{D}$ need only be non-empty.

V. Comments

I have discussed how intermediate VEVs can generate quasi-realistic intergenerational mass ratios among MSSM quarks and leptons in both string and field-theoretic models. For string models, usually (near) string scale VEVs must first cancel the FI D-term contribution from an anomalous $U(1)_A$. Since some non-anomalous $U(1)_a$ are simultaneously broken by the string scale VEVs, which $U(1)_a$ might be associated with intermediate scale VEVs depends on the particular set of (near) string-scale VEVs chosen. Lastly, a matrix decomposition method for generating D-flat directions was introduced.

ACKNOWLEDGMENTS

G.C. thanks all of the coauthors for stimulating the group collaborations. G.C. wishes to thank the organizers of ORBIS SCIENTIAE '97 II, in particular Behram N. Kursunoglu, for producing (as always) a very stimulating and enjoyable conference.

References

[1] R.M. Barnett *et al.*, *Phys. Rev.* **D54** (1996) 1.

[2] G. Cleaver, M. Cvetič, L. Everett, J.R. Espinosa, and P. Langacker, "Intermediate Scales, μ Parameter, and Fermion Masses from String Models," UPR-0750-T; IEM-FT-155/97; [hep-ph/9705391].

[3] G. Cleaver, M. Cvetič, L. Everett, J.R. Espinosa, and P. Langacker, "Classification of Flat Directions in Perturbative Heterotic Superstring Vacua with Anomalous $U(1)$," UPR-0779-T; CERN-TH/97-338; IEM-FT-166/97 [hep-th/9711178].

[4] G. Cleaver, "Aspects of CHL Three Generation String Models," UPR-0772-T.

[5] G. Cleaver, M. Cvetič, L. Everett, J.R. Espinosa, and P. Langacker, "Flat Directions in Three Generation Free Fermionic String Models," UPR-0784-T.

[6] S. Chaudhuri, G. Hockney and J. Lykken, *Nucl. Phys.* **B469** (1996) 357.

[7] M. Dine, N. Sieberg, and E. Witten, *Nucl. Phys.* **B289** (1987) 585.

[8] J. Atick, L. Dixon and A. Sen, *Nucl. Phys.* **B292** (87) 109; M. Dine, I. Ichinose, and N. Seiberg, *Nucl. Phys.* **B293** (87) 253; M. Dine and C. Lee, *Nucl. Phys.* **B336** (90) 317; L. Dixon and V. Kaplunovsky, unpublished.

[9] For detailed reviews, see G. Gelmini and E. Roulet, Rept. Prog. Phys. **58**, 1207 (1995); P. Langacker in *Testing The Standard Model*, ed. M. Cvetič and P. Langacker (World, Singapore, 1991) p. 863.

[10] T. Kobayashi and H. Nakano, *"Anomalous U(1) Symmetry in Orbifold String Models,"* INS-REP-1179; NIIG-DP-96-3; hep-th/9612066; G. Cleaver and A. Faraggi, *"On the Anomalous U(1) in Free Fermionic Superstring Models,"* UPR-0773-T; UFIFT-HEP-97-28; [hep-ph/9711339] and references therein.

[11] M.A. Luty and W. Taylor IV, *Phys. Rev.* **D53** (1996) 37 and references therein.

[12] W. Press, B. Flannery, S. Teukolsky, and W. Vetterling, *"Numerical Recipes,"* (Cambridge University Press, Cambridge, 1989).

MEASUREMENT OF THE MASS OF THE INTERMEDIATE VECTOR BOSONS AT LEP

Paolo Privitera

Dipartimento di Fisica, Università di Roma II and INFN, Tor Vergata, I-00173, Rome, Italy

1 INTRODUCTION

The history of unification of electromagnetic and weak interactions [1] is intimately connected to the experimental study of Intermediate Vector Bosons. Thanks to the discovery and precision measurements of the W and Z bosons, what was considered 'the simplest model' [2] in the 70's is now called with deference 'the Standard Model' of electroweak interaction.

It is known since long time [3] that electron-positron colliding beam experiments are a privileged laboratory to study the exchange of a neutral weakly interacting vector boson: a strong resonant peak will appear in the cross section, and the measurement of its shape parameters allows an unambiguous theoretical interpretation. On the other hand, pairs of charged bosons are produced in e^+e^- collisions once the center of mass energy exceeds the threshold. From this point of view, the LEP experimental program represents a great success.

The Z mass is measured at LEP with a relative precision of $\simeq 2\cdot10^{-5}$ which is at the same level of the Fermi constant G_F. The achieved precision allows a stringent scrutiny of the Standard Model, by comparing predictions of the model with measurements of various observables. A recent study [4], reported in Table 1, confirms the internal consistency of the Standard Model, with a few notable exceptions. Details of the Z mass measurement are given in Section 2.

The measurement of the charged W boson mass is an important item of the physics program of LEP II, with the goal of $\simeq$ 30 MeV precision. Since 1996 LEP is running above the threshold of W^+W^- pair production, and first results on the W mass are already available. The progress achieved in this field is summarized in Section 3.

2 THE Z MASS MEASUREMENT

Large statistical samples and the detailed control of systematic uncertainties, together with a robust theoretical framework to interpret the data, are the ingredients of a precision measurement.

Physics of Mass
Edited by Kursunoglu *et al.*, Kluwer Academic / Plenum Publishers, New York, 1999

Table 1. Summary of measurements included in the combined analysis of Standard Model parameters. Section a) summarises LEP averages, Section b) SLD results ($\sin^2\theta_{\rm eff}^{\rm lept}$ includes $A_{\rm LR}$ and the polarised lepton asymmetries), Section c) the LEP and SLD heavy flavour results and Section d) electroweak measurements from $\rm p\bar{p}$ colliders and νN scattering. The total errors in column 2 include the systematic errors listed in column 3. The determination of the systematic part of each error is approximate. The Standard Model results in column 4 and the pulls (difference between measurement and fit in units of the total measurement error) in column 5 are derived from the Standard Model fit including all data with the Higgs mass treated as a free parameter.
[a] The systematic errors on $m_{\rm Z}$ and $\Gamma_{\rm Z}$ contain the errors arising from the uncertainties in the LEP energy only.

	Measurement with Total Error	Systematic Error	Standard Model	Pull
$\alpha(m_{\rm Z}^2)^{-1}$	128.896 ± 0.090	0.083	128.898	0.0
a) LEP				
line-shape and				
lepton asymmetries:				
$m_{\rm Z}$ [GeV]	91.1867 ± 0.0020	[a]0.0015	91.1866	0.0
$\Gamma_{\rm Z}$ [GeV]	2.4948 ± 0.0025	[a]0.0015	2.4966	−0.7
$\sigma_{\rm h}^0$ [nb]	41.486 ± 0.053	0.052	41.467	0.4
R_ℓ	20.775 ± 0.027	0.024	20.756	0.7
$A_{\rm FB}^{0,\ell}$	0.0171 ± 0.0010	0.0007	0.0162	0.9
τ polarisation:				
$\mathcal{A}_\tau$	0.1411 ± 0.0064	0.0040	0.1470	−0.9
$\mathcal{A}_{\rm e}$	0.1399 ± 0.0073	0.0020	0.1470	−1.0
$\rm q\bar{q}$ charge asymmetry:				
$\sin^2\theta_{\rm eff}^{\rm lept}$ ($\langle Q_{\rm FB}\rangle$)	0.2322 ± 0.0010	0.0008	0.23152	0.7
$m_{\rm W}$ [GeV]	80.48 ± 0.14	0.05	80.375	0.8
b) SLD				
$\sin^2\theta_{\rm eff}^{\rm lept}$ ($A_{\rm LR}$)	0.23055 ± 0.00041	0.00014	0.23152	−2.4
c) LEP and SLD Heavy Flavour				
$R_{\rm b}^0$	0.2170 ± 0.0009	0.0007	0.2158	1.3
$R_{\rm c}^0$	0.1734 ± 0.0048	0.0038	0.1723	0.2
$A_{\rm FB}^{0,\rm b}$	0.0984 ± 0.0024	0.0010	0.1031	−2.0
$A_{\rm FB}^{0,\rm c}$	0.0741 ± 0.0048	0.0025	0.0736	0.1
$\mathcal{A}_{\rm b}$	0.900 ± 0.050	0.031	0.935	−0.7
$\mathcal{A}_{\rm c}$	0.650 ± 0.058	0.029	0.668	−0.3
d) $\rm p\bar{p}$ and νN				
$m_{\rm W}$ [GeV] ($\rm p\bar{p}$)	80.41 ± 0.09	0.07	80.375	0.4
$1 - m_{\rm W}^2/m_{\rm Z}^2$ (νN)	0.2254 ± 0.0037	0.0023	0.2231	0.6
$m_{\rm t}$ [GeV] ($\rm p\bar{p}$)	175.6 ± 5.5	4.2	173.1	0.4

Table 2. Comparison of systematic uncertainties at the beginning and at the end of LEP I.

	La Thuile 91	Orbis Scientiae 97
relative precision		
selection	$\simeq 4 \cdot 10^{-3}$	$\simeq 0.8 \cdot 10^{-3}$
luminosity (exp.)	$\simeq 7 \cdot 10^{-3}$	$\simeq 0.7 \cdot 10^{-3}$
luminosity (theor.)	$\simeq 5 \cdot 10^{-3}$	$\simeq 1.1 \cdot 10^{-3}$
center of mass energy	20 MeV	$\simeq 1.5$ MeV

The successful operation of LEP from 1990 to 1995 around the Z peak has enabled the four LEP experiments, ALEPH, DELPHI, L3 and OPAL, to collect $\simeq 16 \cdot 10^6$ Z decays.

The Z mass is determined by the measurement of the cross section as a function of the center of mass energy, $\sqrt{s}$. In Fig. 1 the measured hadronic cross section as a function of $\sqrt{s}$ is shown, with the clear Z resonance. Thus, a detailed knowledge of detector performances is needed for the event selection efficiency calculation. The measurement of luminosity must be well understood, both from the experimental and theoretical point of view. Last but not least, it is essential to precisely determine the center of mass energy. Systematic uncertainties at the time of La Thuile winter conference in 1991 [5] are compared with current results [4] in Table 2. The improvement achieved in the understanding of systematic uncertainties is really impressive. Reduction of the errors by factors 5 to 10 have been reached.

In order to achieve these results, luminosity monitors were improved or rebuilt by the LEP Collaborations, while theoreticians provided a more precise calculation of the Bhabha cross section for the luminosity determination [6]. The routine implementation of the resonant depolarization calibration [7] boosted the knowledge of the LEP center

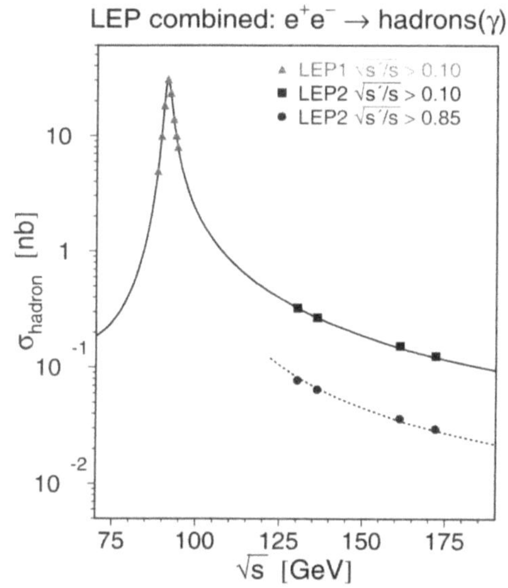

Figure 1. The hadronic cross section measured at LEP as a function of the center of mass energy.

of mass energy, allowing the identification of several subtle systematic effects. Tides, for example, produce a change of the LEP circumference of $\simeq$1 mm over 27 Km, corresponding to a systematic shift of the center of mass energy of $\simeq$8 MeV. Another interesting effect more recently discovered involved vagabund currents from electrical trains travelling above LEP. The return current of these trains, which should go back to the power supply along the iron railtracks, in part finds its way in the ground around the railroad. Then, it follows the LEP beampipe which acts as a good conductor. In Fig. 2 the perfect time correlation found between the voltage measured on rails, the one on the LEP beam pipe, and the magnetic field near the pipe is shown. Systematic shifts of $\simeq$4 MeV were attributed to this source and could be corrected. The preliminary estimate of the systematic uncertainty due to the LEP energy calibration is $\simeq$ 1.5 MeV on both the mass and the width of the Z boson [8].

A model independent parametrization [9], with radiative corrections properly taken into account avoids ambiguities in the theoretical interpretation. The Z line shape can be, in fact, expressed by:

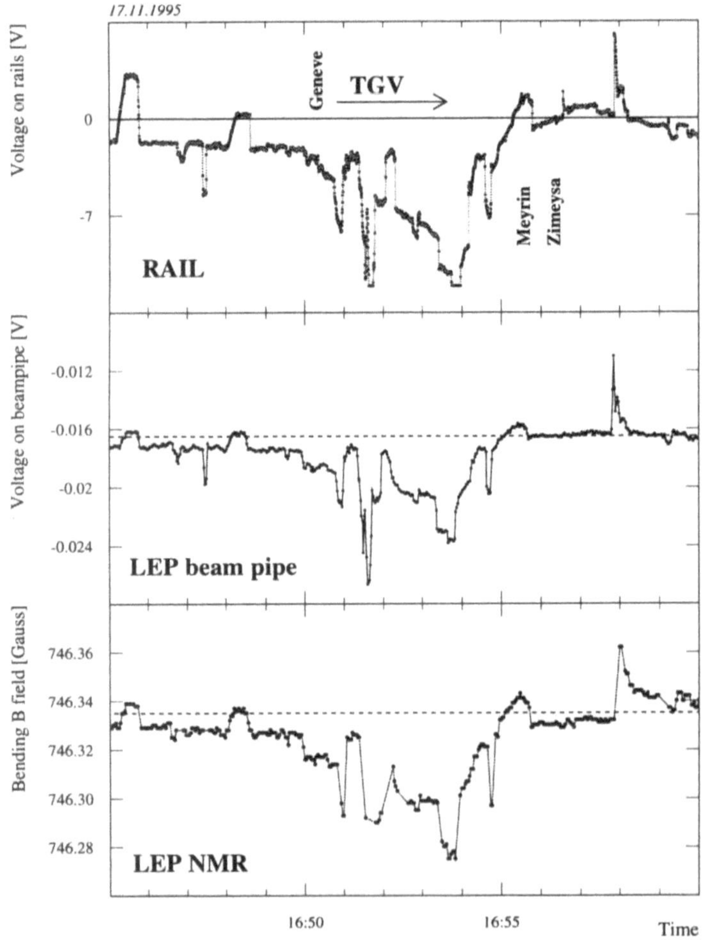

Figure 2. From top to bottom: railtrack potential, LEP beampipe potential, NMR reading.

where:

$$\sigma(e^+e^- \to f\bar{f}) = \frac{s}{s - m_Z^2 + \frac{s^2\Gamma_Z^2}{m_Z^2}} \left(12\pi\frac{\Gamma_e\Gamma_f}{m_Z^2} + \frac{I_f N_c\,(s - m_Z^2)}{s}\right) + \frac{4\pi}{3}N_c Q_f^2\frac{\bar{\alpha}^2}{s}, \quad (1)$$

- The photon exchange term is the known QED result, with a running α (the fine structure constant).

- The Z exchange term is modulated by a Breit-Wigner factor and at the peak (s=m_Z^2) reads as:

$$\sigma_f^0 = \frac{12\pi}{m_Z^2}\frac{\Gamma_e}{\Gamma_Z}\frac{\Gamma_f}{\Gamma_Z}, \quad (2)$$

that is the product of the unitary bound for J=1 channel and the branching ratios of initial and final state (Γ_e/Γ_Z, Γ_f/Γ_Z).

- The $\gamma - Z$ interference term cannot be written in terms of the shape parameters (m_Z, Γ_Z, Γ_l, etc.). However, the I_f contribution to the cross section is small (0 at the peak and $\leq 0.5\%$ in the range $m_Z \pm \Gamma_Z$).Thus, I_f is fixed to its Standard Model value and a small model dependence is introduced.

The described parametrization, derived just using Quantum Field Theory and the well established QED theory, gives a precise definition of the actual physical parameters m_Z,Γ_Z,Γ_e,Γ_f, with minimal dependence on the Standard Model interpretation.

Radiative corrections to the line shape can be included very precisely [10]. The electroweak corrections, like the self-energy corrections of vector bosons and virtual W and Z loops, are small but contain very interesting physics information (m_t, m_H, new physics). The bulk of these corrections can be naturally absorbed in the definition of the physical parameters. Photonic corrections, with the dominant contribution given by the initial state bremsstrahlung, are large, compared to the experimental precision (the peak cross section is reduced by about 25% and the peak position is shifted by about 110 MeV). However they are known at the level of a few 10^{-3} precision and can be properly taken into account convoluting the non radiative cross section with a suitable radiator function.

The measurements of the Z boson mass and width performed by the LEP experiments are summarized in Table 3. The total error on the LEP averages (2 MeV for the mass and 2.5 MeV for the width) include the contribution of 1.5 MeV from the LEP energy calibration.

The precision reached by the measurement has also driven new effort on the theoretical side in order to test different approaches in the definition of the mass of the Z boson. In the S-matrix model of the line shape the Z is associated to a complex pole in the S matrix [11]. The mass and the width are defined in terms of the pole in the energy plane via

Table 3. Measurements of the mass and width of the Z boson by the LEP Collaborations.

	Z mass (MeV)	Z width (MeV)
ALEPH	91188.3 ± 3.1	2495.1 ± 4.3
DELPHI	91186.6 ± 2.9	2489.3 ± 4.0
L3	91188.6 ± 2.9	2499.9 ± 4.3
OPAL	91184.1 ± 2.9	2495.8 ± 4.3
LEP Average	91186.7 ± 2.0	2494.8 ± 2.5
χ^2/DoF	2.0/3	3.9/3

$$\bar{s} = \overline{M_Z}^2 - i\overline{M_Z}\overline{\Gamma_Z}, \tag{3}$$

which is characterized by an s-independent Z width. The cross section is then:

$$\sigma(e^+e^- \to f\bar{f}) = \frac{4}{3}\pi\alpha^2 \left[\frac{g_f}{s} + \frac{j_f(s - \overline{M_Z}^2) + r_f s}{(s - \overline{M_Z}^2)^2 + \overline{M_Z}^2\overline{\Gamma_Z}^2} \right]. \tag{4}$$

The parameters g_f, j_f and r_f are fitted together with the mass and width. In particular, j_f is related to the $\gamma - Z$ interference. Including the new cross section data of LEP II significantly helps in constraining this term, thus reducing the correlation with the fitted mass. A fit to the data using the S-matrix approach yields [4]:

$$j_{had} = 0.14 \pm 0.12$$

which should be compared with the Standard Model expectation of 0.22, and

$$m_Z = 91188.2 \pm 2.9 \text{ MeV}.$$

The agreement between determinations using different theoretical approaches gives confidence in the current understanding of the mass as a fundamental property of the Z boson.

3 THE W MASS MEASUREMENT

A precise measurement of the W mass represents an important element in our understanding of the Standard Model. A global fit of electroweak observables using the Standard Model (see Table 1) allows an indirect determination of the W mass. Its consistency with a precise measurement would further strengthen the confidence in the Standard Model or open the door to physics beyond it. The current situation is shown in Fig. 3, in which the 68% CL probability contours corresponding to indirect determinations and direct measurements are plotted on the W - top mass plane. Also, specific models can be tested in the same way. In Fig. 4 the band corresponding to the predictions of the Minimal Supersymmetric Standard Model (with the hypothesis of no new particle found at LEP II) is clearly separated from the one corresponding to the Standard Model.

Two experimental tecniques have been exploited so far at LEP for the M_W measurement.

The first one involves the measurement of the W^+W^- production cross section close to the threshold. The optimal sensitivity to the mass is found at $\sqrt{s}$ =161 GeV, where LEP provided 10 pb^{-1} per experiment in 1996. The measured cross section of W^+W^- pairs production is shown in Fig. 5. From the cross section measured at threshold the W mass was determined to be [12]:

$$m_W = 80.40 \pm 0.22 \text{ GeV} \quad \text{(Threshold)}.$$

The threshold measurement fitted well in the schedule of increasing LEP energy by steps. In principle, the method is able to reach the goal precision of 30 MeV. However it has the disadvantage of running LEP in a region of small cross section without exploiting the maximum energy reachable by the accelerator, to the detriment of other aspects of the experimental program (e.g. searches of new particles).

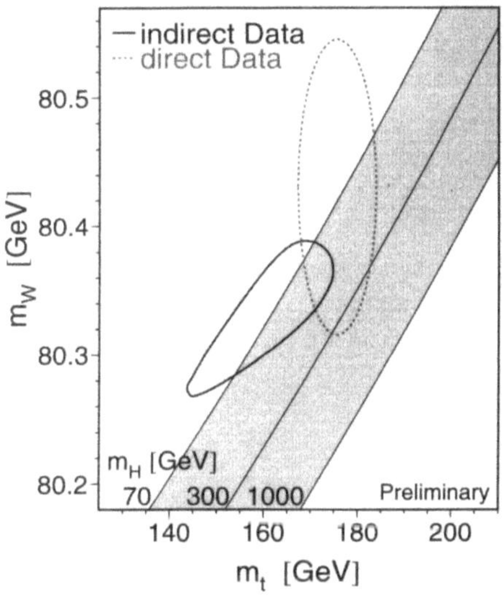

Figure 3. Comparison of the indirect determinations of m_W and m_t (LEP+SLD+ νN data)(solid contour) and the direct measurements (Tevatron and LEP II data) (dashed contour). In both cases the 68% CL contours are plotted. Also shown is the Standard Model relationship for the masses as a function of the Higgs mass.

The second method works at energies well above the threshold ($\geq$ 170 GeV). At those energies, the cross section has lost most of its sensitivity to the mass, but it is high ($\simeq$ 12 pb^{-1}). A significant statistical sample of W can be produced, and a direct reconstruction of the mass is accomplished through the invariant mass of its decay products. LEP provided 10 pb^{-1} to each experiment at $\sqrt{s}$=172 GeV in 1996. The 4 jets topology (both W's decaying hadronically), which accounts for 44% of the W^+W^- pairs, is selected by the experiments with efficiency around 70% and $\simeq$ 25% background. The major source of background comes from the process $e^+e^- \rightarrow \gamma/Z \rightarrow qq(\gamma)$, where the two quarks produce 4 jets in the detector (either by gluon radiation or misreconstruction). The topology of 2 jets and one lepton (one W decaying hadronically, the other semileptonically), which accounts for other 44% of the W^+W^- pairs, is selected by the experiments with efficiency around 85% and only few percent background. The procedure used to determine the W mass from a given event is the following. In a 4 jets event, the jets are coupled to form a candidate W, and the invariant mass is calculated from the reconstructed energy and direction of the jets. Two invariant masses per event, m_1 and m_2, are determined. In a 2 jet - lepton topology, the neutrino energy and direction is not reconstructed, but can be determined from the missing momentum of the event, and invariant masses can still be calculated. Events where both W's decay semileptonically, having thus two missing neutrinos, do not have enough constraints and are not used. A significant improvement in the mass resolution is obtained by using a constrained kinematic fit, namely imposing the conservation of energy and momentum of the event (4 constraints fit) and a possible additional constraint of equal masses $m_1 = m_2$ (5 constraints fit). Notice that while in the 2 jets-lepton topology the assignement of the decay products to the mother W is

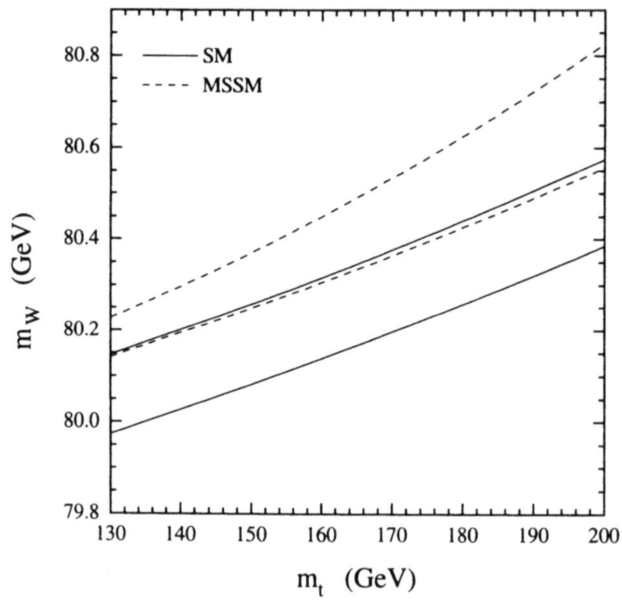

Figure 4. The W mass range in the Standard Model (solid line) and in the MSSM (dashed line). The bounds correspond to the possible situation that no Higgs bosons and SUSY particles are found at LEP II.

unique, in the 4-jets case three possible pairings are allowed between the 4 jets, with only one being correct. The experiments try to get as much information as possible from the event by using the two combinations with best χ^2 from the kinematical fit, or even using all the three possible pairings weighted by an ideogram tecnique. A collection of invariant mass distributions from the LEP experiments is presented in Figs. 6-7. An empirical parametrization of the mass distribution ($\simeq$ a Breit-Wigner) is used to fit the data, and the result is then calibrated from the study of Monte Carlo samples with different values of the W mass. Details of the analysis tecniques can be found in [13]. Preliminary measurements with the direct reconstruction method using 172 GeV data are presented in Table 4.

Table 4. Measurements of the mass of the W boson by the LEP Collaborations with the direct reconstruction method.

	m_W (GeV) 2 jets - lepton	m_W (GeV) 4 jets
ALEPH	$80.38 \pm 0.43 \pm 0.12$	$81.30 \pm 0.47 \pm 0.10$
DELPHI	$80.51 \pm 0.57 \pm 0.06$	$79.90 \pm 0.59 \pm 0.12$
L3	$80.42 \pm 0.54 \pm 0.08$	$80.91 \pm 0.42 \pm 0.13$
OPAL	$80.53 \pm 0.41 \pm 0.10$	$80.08 \pm 0.44 \pm 0.15$
LEP Average	80.46 ± 0.24	80.62 ± 0.26
χ^2/DoF	0.1/3	5.4/3

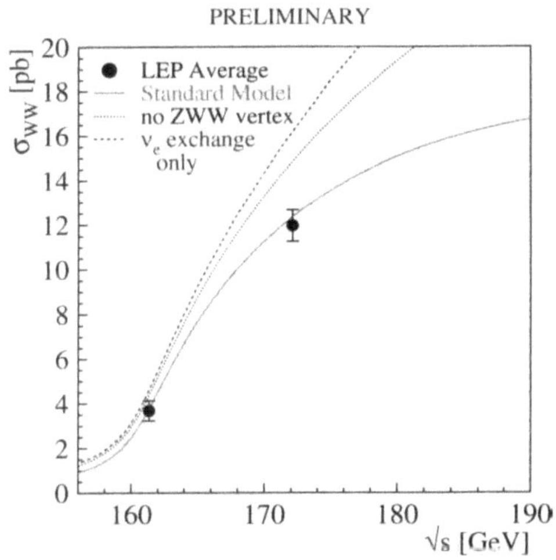

Figure 5. The W pair cross section as a function of the center of mass energy. The data points are the LEP averages. Also shown is the Standard Model prediction (solid line), and for comparison the cross section if the ZWW coupling did not exist (dotted line), or if only the t-channel ν_e exchange diagram existed (dashed line).

At the current level of statistics, the mass determined from 2 jets-lepton and that from 4-jets samples are consistent, and are combined to obtain:

$$m_W = 80.53 \pm 0.17 \pm 0.05 \text{ GeV} \quad \text{(Direct reconstruction).}$$

This result shows that the sensitivity of the direct reconstruction method is more than enough to reach the goal of 30 MeV error. In fact, a statistical error of 170 MeV obtained with 10 pb^{-1} scales below 30 MeV for 500 pb^{-1} expected at the end of LEP II. The experimental systematical uncertainties (detectors, fit, etc.) are already not dominant in the total systematic error, and most of them are of statistical origin and thus expected to become smaller with increasing statistics. The LEP energy calibration currently contributes with a systematic uncertainty of 30 MeV. The resonant depolarization method cannot work as well as at LEP I, since the beam polarization rapidly decreases with increasing beam energy. In fact, the calibration is performed up to $\simeq$50 GeV beam energy, and then an extrapolation is needed for higher energies. However, changes in the optics which allowed calibration at higher energies are expected to significantly improve the calibration for 1997 data (by a factor at least two), thus fulfilling the requirement for a final 30 MeV total error on the W mass. The measurement with the 4-jets sample is affected by a $\simeq$100 MeV theoretical uncertainty, due to Bose-Einstein and colour reconnection effects. In a simplified picture, the width of the W corresponds to a scale of 0.1 F, to be compared with a scale of 1 F typical of the hadronization process. Thus, the four quarks from the W decays can 'talk' to each other modifying the hadronization, and eventually the invariant mass distributions. However, 100 MeV error is quite conservative. Preliminary measurements performed by the LEP experiments show no evidence of large Bose-Einstein effects [14]. Also, large changes in the charge multiplicity of 4-jets events as predicted by some models of

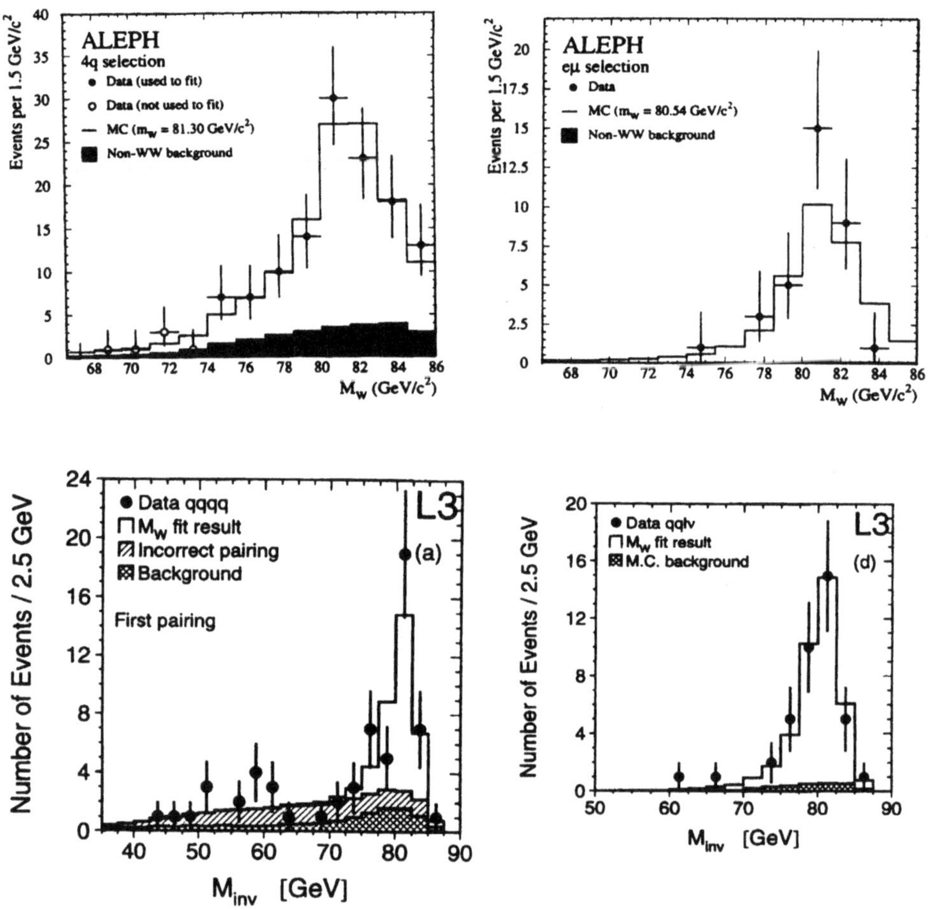

Figure 6. Reconstructed W mass distributions at $\sqrt{s}$=172 GeV: ALEPH and L3.

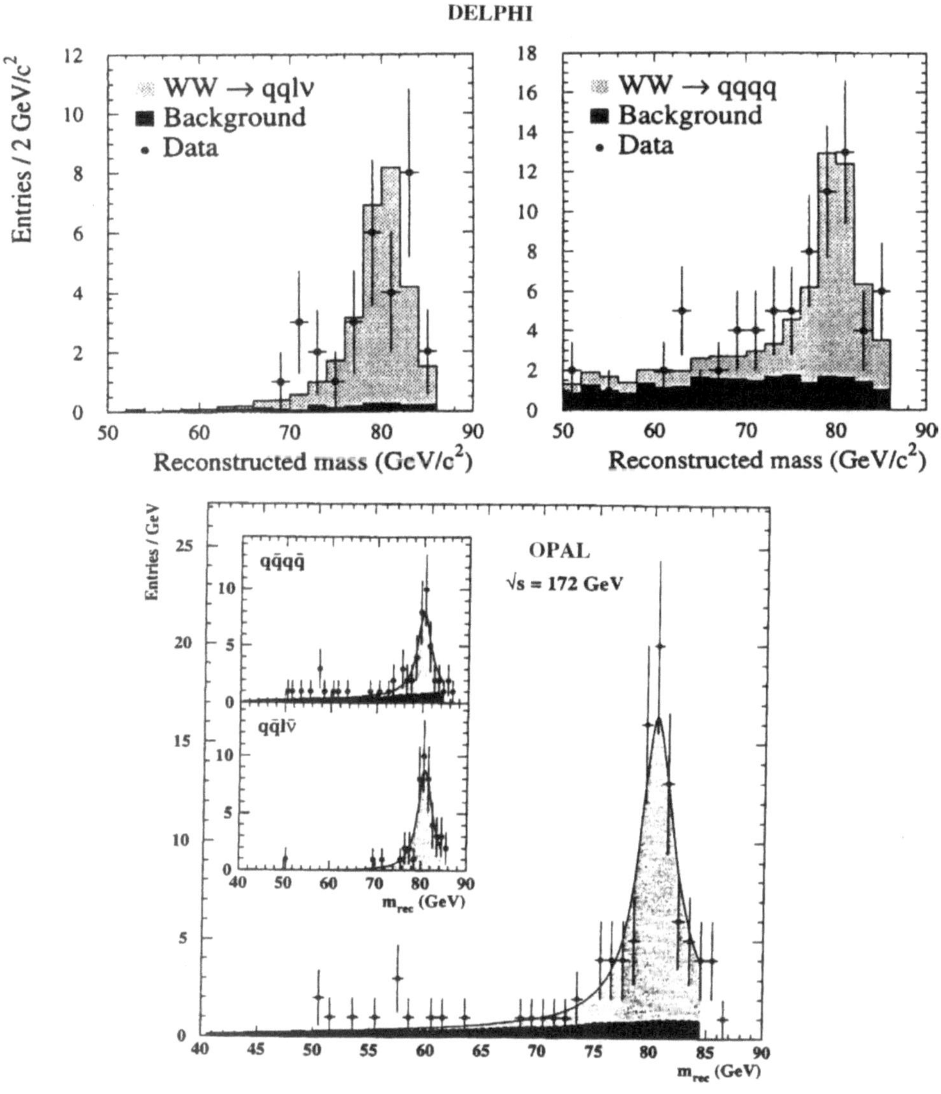

Figure 7. Reconstructed W mass distributions at $\sqrt{s}$=172 GeV: DELPHI and OPAL.

colour reconnection have not been observed [14]. The combined effort of experimental and theoretical physicists should bring the systematic uncertainty due to this source down to a level which fully exploits the potentiality of the 4-jets channel.

The current world average obtained from measurements performed at the Tevatron [15] and from the LEP measurements at threshold and at 172 GeV [12] is compared to the indirect determination obtained from the Standard Model fit in Table 5.

Table 5. Measurements of the mass of the W boson.

	W mass (GeV)
Tevatron	80.41 ± 0.09
LEP	80.48 ± 0.14
World Average	80.43 ± 0.08
χ^2/DoF	0.2/1
SM indirect determ.	80.329 ± 0.041

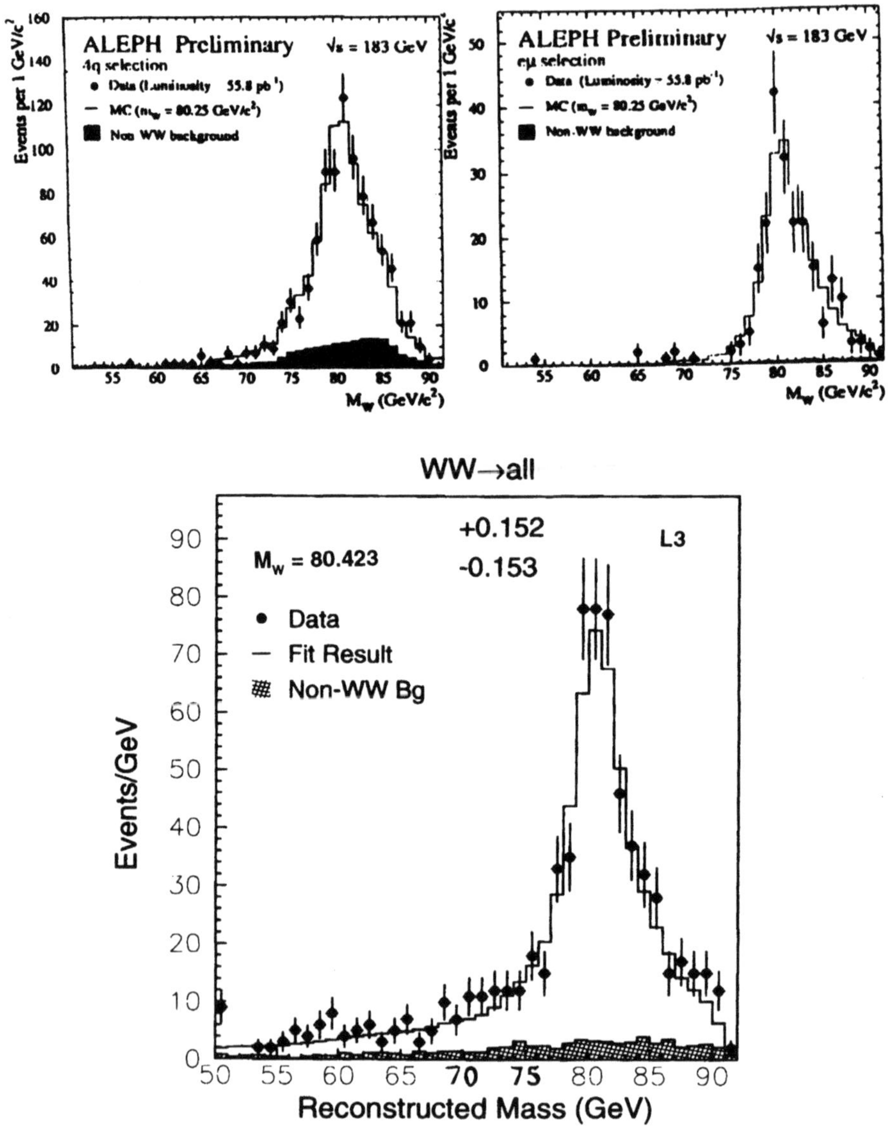

Figure 8. Reconstructed W mass distributions at $\sqrt{s}$=183 GeV: ALEPH and L3.

The 1997 data taking, just ended, has provided $\simeq$ 60 pb^{-1} per experiment at $\sqrt{s}$=183 GeV. In Figs. 8-9 very preliminary mass distributions are presented.

4 CONCLUSIONS

The LEP I program, concluded in 1995, has been a great success. The mass of the Z boson has been measured with a relative precision of $\simeq 2 \cdot 10^{-5}$, allowing stringent tests of the Standard Model. The full data sample has been analyzed, and results are almost final.

The first data collected at LEP II during 1996 have already provided a preliminary measurement of the W boson mass. The performance of the LEP machine and exper-

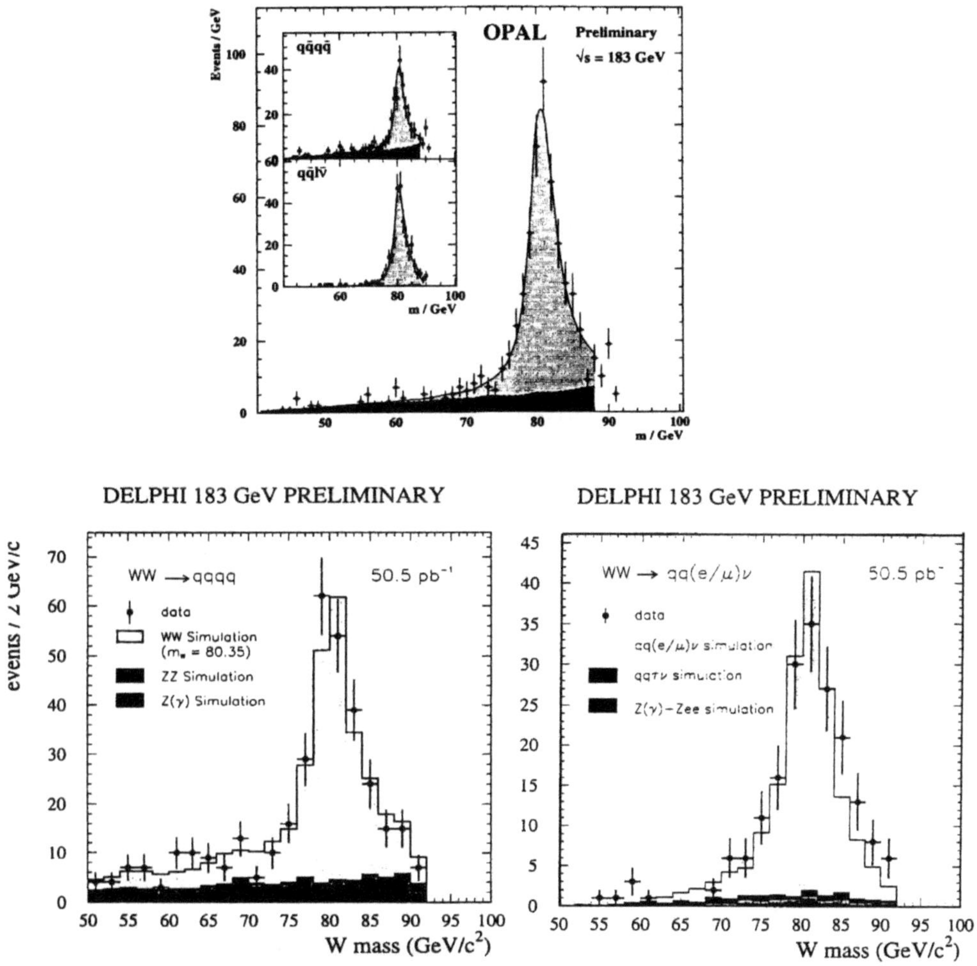

Figure 9. Reconstructed W mass distributions at $\sqrt{s}$=183 GeV: DELPHI and OPAL.

iments is consistent with the goal of $\simeq$30 MeV total error on m_W at the end of LEP II.

5 Acknowledgements

I would like to thank Dr. Kursunoglu for his invitation to this very nice and stimulating conference. This presentation has been possible thanks to the effort of the whole LEP scientific community who have been working so well during all these years.

REFERENCES

[1] S. L. Glashow, Nucl. Phys. **22** (1961) 579;
S. Weinberg, Phys. Rev. Lett. **19** (1967) 1264;
A. Salam, "Elementary Particle Theory", Ed. N. Svartholm, Stockholm, "Almquist and Wiksell" (1968), 367.

[2] C. Y. Prescott *et al.*, Phys. Lett. B77 (1978) 347

[3] N. Cabibbo and R. Gatto, Phys. Rev. **124** (1961) 1577.

[4] The LEP Electroweak Working Group and the SLD Heavy Flavour Group, CERN preprint CERN-PPE/97-154, December 1997.

[5] F. Ferroni and P. Privitera, Electroweak Physics at LEP, Proceedings of 'Les Rencontres de Physique de la Vallee d'Aoste' La Thuile March 1991, ed. M. Greco, Editions Frontieres (1991)

[6] A. Arbuzov, *et al.*, Phys. Lett. **B383** (1996) 238;
S. Jadach, *et al.*, Comp. Phys. Comm. **102** (1997) 229.

[7] L. Arnaudon *et al.*, Nucl. Instr. Meth. A357 (1995) 249.

[8] LEP Energy Working Group, private communication.

[9] See, for example, M. Consoli *et al.*, in "Z Physics at LEP 1", CERN Report CERN 89-08 (1989), eds G. Altarelli, R. Kleiss and C. Verzegnassi, Vol. 1.

[10] Reports of the working group on precision calculations for the Z resonance, eds. D. Bardin, W. Hollik and G. Passarino, CERN Yellow Report 95-03, Geneva, 31 March 1995.

[11] A. Borrelli, M. Consoli, L. Maiani, R. Sisto, Nucl. Phys. B333 (1990) 357;
R.G. Stuart, Phys. Lett. B272 (1991) 353;
A. Leike, T. Riemann, J. Rose, Phys. Lett. B273 (1991) 513;
T. Riemann, Phys. Lett. B293 (1992) 451;
S. Kirsch, T. Riemann, Comp. Phys. Comm. 88 (1995) 89.

[12] ALEPH Collaboration, R. Barate, *et al.*, Phys. Lett. B401 (1997) 347;
DELPHI Collaboration, P. Abreu, *et al.*, Phys. Lett. B397 (1997) 158;
L3 Collaboration, M. Acciarri, *et al.*, Phys. Lett. B398 (1997) 223;
OPAL Collaboration, K. Ackerstaff, *et al.*, Phys. Lett. B389 (1996) 416.

[13] ALEPH Collaboration, W mass measurement through direct reconstruction in ALEPH, contributed paper to EPS-HEP-97, Jerusalem, EPS-600;
DELPHI Collaboration, Measurement of the W-pair cross-section and of the W mass in e^+e^- interactions at 172 GeV, DELPHI 97-108 CONF 90, contributed paper to EPS-HEP-97, Jerusalem, EPS-347;
L3 Collaboration, M. Acciarri, *et al.*, Phys. Lett. B407 (1997) 419;
L3 Collaboration, M. Acciarri, *et al.*, Measurements of Mass, Width and Gauge Couplings of the W Boson at LEP, CERN-PPE/97-98, submitted to Phys. Lett. B;
OPAL Collaboration, Measurement of the Mass of the W Boson Mass and $W^+ W^-$ Production and Decay Properties in e^+e^- Collisions at $\sqrt{s}$ = 172 GeV, CERN-PPE/97-116, accepted by Zeit. f. Physik C.

[14] A. De Angelis, Interconnection effects in multiparticle production from WW events at LEP, talk presented at the 27th International Symposium on Multiparticle Dynamics, ISMD 97 Frascati Italy, 8-12 September 1997, to appear in the proceedings.

[15] Y.K. Kim, talk presented at the Lepton-Photon Symposium 1997, Hamburg, 28 July - 1 Aug, 1997, to appear in the proceedings.

POLARIZED PARITY VIOLATING ELECTRON SCATTERING ON ^3He FROM LOW TO HIGH ENERGY

S.L. Mintz, G.M. Gerstner, and M. A. Barnett

Physics Department
Florida International University
Miami, Florida 33199

M. Pourkaviani

M.P. Consulting Associates
714 Fox Valley Dr.
Longwood, Florida 32701

INTRODUCTION

From a practical point of view it has been difficult to study weak interactions in nuclei. At low energies ($q^2 \simeq 0$) a substantial number of beta decay process are available. At $q^2 \simeq -m_\mu^2$ both inclusive and exclusive muon capture reactions are observed. Above the q^2 range of muon capture the situation is much more difficult. Neutrino reactions in principle would be very valuable but neutrinos are generally available only over a spectrum thus washing out detail. In addition neutrino reactions have proven to be very difficult to observe experimentally and carry large error bars. Inverse beta decays which have been suggested and which could be run over a wide range of q^2 are usually[1] obscured by competing competing reactions.

However it has been suggested that polarized parity violating electron scattering might be useful for studying the weak interaction in nucleons and nuclei[1] and more recently for studying the role of strange quarks in the nuclear medium[2] and for tests of the standard model[3]. These reactions can be run at fixed incident electron energies at a number of machines. Furthermore, the asymmetry, the quantity usually determined can be found under suitable conditions to high accuracy. Thus reactions of this kind might be run over a wide range of q^2 values than are generally available currently for studying weak interactions. Also there is evidence that the strange quark weak current resulting from a term of the form, $\bar{s}\gamma_\mu(1-\gamma_5)s$ might make a measurable contribution[1] to the asymmetry and so the possibility of obtaining such a contribution should also be considered.

In this paper we consider the reaction,$e^- +^3 He \longrightarrow e^- +^3 He$, run with right handed and left handed polarized electrons. This process is different from many previously considered in that it takes place on a spin 1/2 nucleus. This, as we shall see leads to structure in the asymmetry which is not present in scattering from 0^+ nuclei[1] or in the $0^+ \leftrightarrow 1^+$ transitions usually studied[4]. This structure is q^2 dependent in a more complicated way than the simple parabolic structure in q^2 seen in the previously studied cases.

In the section on matrix elements of this paper we shall obtain the matrix elements necessary to calculate the asymmetry. In the section on results of this paper we shall obtain the asymmetry and figure-of-merit and plot these quantities for incident electron energies of 0.1 GeV,0.5 GeV,1.0 GeV,2.0 GeV, 4.0 GeV, and 6.0 GeV. Finally in the discussion section of this paper we shall discuss our results and determine the accuracy to which the asymmetry might be determined and present conclusions.

MATRIX ELEMENTS

The quantity which is central in studying polarized parity violating electron scattering is the asymmetry given by:

$$A = \frac{\frac{d\sigma(L)}{d\Omega} - \frac{d\sigma(R)}{d\Omega}}{\frac{d\sigma(L)}{d\Omega} + \frac{d\sigma(R)}{d\Omega}} \tag{1}$$

where L and R indicate left and right handed electron polarization respectively. The numerator of this quantity contains only parity violating terms which in the lowest order come from the interference between the one photon exchange and one Z-boson exchange diagrams. These terms are contained in the square of the matrix element:

$$M_{fi} = \frac{e^2}{q^2}\overline{u}'\gamma_\mu u {<^3} He|J^\mu_{\text{em}}(0)|^3 He > + \frac{G}{\sqrt{2}}\overline{u}'\gamma_\mu[g_V + g_A\gamma_5]u {<^3} He|J^\mu_{\text{weak}}(0)|^3 He > \tag{2}$$

with

$$g_V = -1 + 4\sin^2(\theta_W) \tag{3a}$$

$$g_A = 1. \tag{3b}$$

This quantity may be evaluated for left and right handed electrons and the result subtracted. Only parity violating terms remain and are of the general form:

$$\begin{aligned} |M|_{PV} = \frac{2e^2G}{\sqrt{2}q^2}&\{\overline{u}'\gamma_\mu u_L {<^3} He|J^\mu_{\text{em}}(0)|^3 He >\{\overline{u}'\gamma_\nu[g_V + g_A\gamma_5]u_L \\ &\times {<^3} He|J^\nu_{\text{weak}}(0)|^3 He >\}^\dagger \\ &+ \{\overline{u}'\gamma_\mu u_L {<^3} He|J^\mu_{\text{em}}(0)|^3 He >\}^\dagger \overline{u}'\gamma_\nu[g_V + g_A\gamma_5]u_L \\ &\times {<^3} He|J^\nu_{\text{weak}}(0)|^3 He >\} \end{aligned} \tag{4}$$

where we have exhibited the terms for left handed polarization. The terms for right handed polarization are similar. In Eqs. (2) and (4) we have put in the familiar lepton part of the matrix elements. however the hadronic part of these equations is not so well determined and we must obtain expressions for it.

In particular we must find explicit expressions for the two nuclear matrix elements, which occur in Eqs.(2) and (4), namely $<^3 He|J^{\mu}_{em}(0)|^3He>$ and additionally $<^3 He|J^{(3)\mu}_{weak}|^3He>$ where

$$<^3 He|J^{(3)\mu}_{weak}(0)|^3He>=<^3 He|V^{(3)\mu}(0) - A^{(3)\mu}(0) - 2\sin^2(\theta_W)J^{\mu}_{em}(0)|^3He> . \quad (5)$$

As given,Eq.(5) is not complete. This is due to the possibility that there may be strange quark contributions to the nuclear weak current. This would add a term to the weak nuclear current of the form $\bar{s}\gamma_\mu s$. We consider only the vector case because axial contributions to polarized parity violating electron scattering will be extraordinarily difficult to detect. In the nuclear medium such strange terms will manifest themselves as a vector current of the form:

$$<^3 He|V^s_\mu|^3He>= \bar{u}_f[\gamma_\mu F^s_V(q^2) + iF^s_M(q^2)\sigma_{\mu\nu}\frac{q^\nu}{2m_p}]u_i. \quad (6)$$

Thus in addition to the terms in Eq.(5) which consist of the usual I =1 and I= 0 (V^8_μ) components there are contributions from Eq.(6). These may be be absorbed into the V_μ of Eq.(5) and will appear as departures from the expected values for the form factors describing the vector current. We shall mention this point again.

We make use of an elementary particle model treatment. In this model, the ^3He nucleus is treated as an elementary particle[5] of spin 1/2 and the nuclear structure is contained in form factors. This model has given particularly accurate results[6,7] in describing the muon capture reaction $\mu^- +^3 He \longrightarrow \nu_\mu +^3 H$. We have previously[8,9] used this model in calculating the electroweak processes, $e^- + {}^3He \longrightarrow \nu_e + {}^3H$ and $\bar{\nu}_e + {}^3He \longrightarrow e^+ + {}^3H$ for which it produced reasonable results. The matrix elements in this model are well known[5,8] and are given by:

$$<^3 He|V_\mu(0)|^3He>= \bar{u}_f[\gamma_\mu F_V(q^2) + i\frac{F_M(q^2)\sigma_{\mu\nu}q^\nu}{2m_p}]u_i \quad (7a)$$

$$<^3 He|A_\mu(0)|^3He>= \bar{u}_f[\gamma_\mu\gamma_5 F_A(q^2) + \frac{q_\mu\gamma_5 F_P(q^2)}{m_\pi}]u_i \quad (7b)$$

$$< k|J^{em}_\mu|k >= \bar{u}_k[F^1_k(q^2)\gamma_\mu + \frac{i\sigma_{\mu\nu}q^\nu F^2_k(q^2)}{2m_p}]u_k \quad (7c)$$

where k stands for ^3He or ^3H. As noted above, from Eq.(6), the strange contributions can be immediately incorporated into Eq.(7a).

The nuclear structure for the weak current matrix element is contained in the four form factors,$F^{weak}_V(q^2)$,$F^{weak}_M(q^2)$, $F^{weak}_A(q^2)$, and $F^{weak}_P(q^2)$. Similarly that for the electromagnetic current matrix element is contained in the $F_1(q^2)$ and $F_2(q^2)$ form factors. It is convenient to combine the vector parts of Eq.(5), making use of Eqs.(7a) and (7c) to obtain

$$F^{weak}_V = F^{(3)}_V - 2\sin^2(\theta_W)F_1 \quad (8a)$$

and

$$F^{weak}_M = F^{(3)}_M - 2\sin^2(\theta_W)F_2. \quad (8b)$$

In which the strange quark current contributions are now also contained. Thus a knowledge of these form factors would enable us to evaluate all terms in Eq.(2) and hence the asymmetry itself.

The non-strange quark contributions to the form factors of the weak vector current matrix element needed above may be obtained from electron scattering data by making use of the conserved vector current hypothesis in the form $[I_i, J_\mu^{em}] = V_\mu^\dagger$, and $[I^+, J_\mu^\dagger] = 2J_\mu^{(3)}$ to obtain:

$$F_V(q^2) = F^{(1)}_{{}^3He}(q^2) - F^{(1)}_{{}^3H}(q^2) \tag{9a}$$

$$F_M(q^2) = F^{(2)}_{{}^3He}(q^2) - F^{(2)}_{{}^3H}(q^2) \tag{9b}$$

and

$$F_i^{(3)} = \frac{F_i(q^2)}{2} \tag{9c}$$

where i=V,M, and A respectively. The form factors, F_V,F_M, and F_1,F_2 have already been obtained from electron scattering for values of q^2 from zero to $-50m_\pi^2$ and are given by:

$$F_V(q^2) = F_V(0)\cos^2(\frac{-q^2}{17.1m_\pi^2})(1 - \frac{q^2}{6.25m_\pi^2})^{-2} \tag{10a}$$

for $|q^2| \leq 24.5\ m_\pi^2$;

$$F_V(q^2) = F_V(0)\cos^2(\frac{-q^2}{14.39M_\pi^2})(1 - \frac{q^2}{4.35m_\pi^2})^{-2} \tag{10b}$$

for $|q^2| > 24.5\ m_\pi^2$ and

$$F_M(q^2) = F_M(0)\cos^2(\frac{-q^2}{28.4M_\pi^2})(1 - \frac{q^2}{4.5m_pi^2})^{-2} \tag{11a}$$

for $|q^2| \leq 43.0\ m_\pi^2$;

$$F_M(q^2) = F_M(0)\cos^2(\frac{-q^2}{26.0m_\pi^2})(1 - \frac{q^2}{3.5m_\pi^2})^{-2} \tag{11b}$$

for $|q^2| > 43.0\ m_\pi^2$. Again the form factors obtained by electron scattering do not include strange contributions of the form given in Eq.(6). Therefore the presence of strange contributions will be seen in deviations from these values in the processes which we are describing here.

The axial current form factors are not as well known as the vector current form factors. The two axial current form factors are $F_A(q^2)$ and $F_P(q^2)$. However,because cross-sectional terms containing $F_P(q^2)$ are also proportional to the charged lepton mass squared, measurable contributions do not occur for the electron case. Thus we shall not consider F_P further.

The form factor, $F_A(q^2)$, can be determined at $q^2 = 0$ from the beta decay[6] ,$^3H \longrightarrow ^3 He + e^- + \bar{\nu}$, which takes place at $q^2 \simeq 0$. There is no direct way to determine the q^2 dependence of F_A. However a result by Kim and Primakoff[5,10] based upon the impulse approximation but not making use of the actual form of the nuclear wave functions yields:

$$\frac{F_A(q^2)}{F_A(0)} \simeq \frac{F_M(q^2)}{F_M(0)}. \tag{12}$$

The derivation of this result is in the literature[5,10]. It was originally derived by Kim and Primakoff from a nucleons only impulse approximation but they later extended the calculation to include some exchange current contributions. It has been found that Eq.(12) works extremely well for muon capture[6,7] in ^3He as was previously noted and for a large number of other weak processes[11,12,13] in which F_A is a leading term which increases our confidence in Eq.(12). As we shall see, the axial current contributions to the asymmetry are suppressed and so only under unusual circumstances will we be highly concerned about the form of F_A.

The character of these form factors are all the same and they may be written in the form:

$$F_i(q^2) = F_i(0)\frac{\cos^2(\frac{-q^2}{\alpha_i m_\pi^2})}{(1-\frac{q^2}{\beta_i m_\pi^2})^2} \tag{13}$$

where i = V,M,A,1,2. The structure of these form factors is extremely important. At low q^2 the cosine squared term in the numerator is approximately one and the familiar dipole form of these form factors is apparent. At higher q^2, these form factors all exhibit diffraction minima due to the numerator of Eq.(13). Crucially important however is the fact that α_i and β_i are different for $F_M^{(3)}$, $F_V^{(3)}$, F_1,and F_2. This means that all of these form factors have different q^2 dependences. Because the variation for large q^2 for these form factors is relatively rapid, substantial cancellation can occur in F_V^{weak} and F_M^{weak} which can experience large changes in their values, becoming zero, and then reversing sign as can be seen from Eqs.(8a) and (8b). As we shall see this leads to interesting structure in the asymmetry itself.

RESULTS

We now have all information necessary to evaluate the asymmetry as given by Eq.(1). The result may be written as:

$$\begin{aligned} A = \frac{\sqrt{2}e^2G}{q^2}\Big[& [8F_V^{weak}F_1EE'(2M_i + (1-\cos(\theta)))(E-E'-M_i) \\ & + 8(F_V^{weak}F_2 + F_M^{weak}F_1)\frac{M_i}{m_p}(EE'(1-\cos(\theta)))^2 \\ & + 4F_M^{weak}F_2EE'(1-\cos(\theta))M_i^2[E^2+E'^2 \\ & + \frac{E'-E}{M_i}(1-\cos(\theta)) + EE'(1-\cos(\theta))]]g_A \\ & + 16F_A^{weak}(F_1 + F_2\frac{M_i}{m_p})[EE'(1-\cos(\theta))(E+E')M_i]g_V\Big]/|M|^2_{em} \end{aligned} \tag{14}$$

where $|M|^2_{em}$ is given by:

$$
\begin{aligned}
|M|^2_{em} = \frac{e^4}{q^4}\Big[& 8F_1^2 M_i EE'(2M_i + (1-\cos(\theta))(E - E' - M_i) \\
& + 16F_1F_2\frac{M_i}{m_p}(EE'(1-\cos(\theta)))^2 \\
& + \frac{4F_2^2}{m_p^2}EE'(1-\cos(\theta))M_i^2 \\
& \times [E^2 + E'^2 + (\frac{E'-E}{M_i} + 1)EE'(1-\cos(\theta))]\Big]
\end{aligned} \quad (15)
$$

As previously noted the axial terms are suppressed in the numerator of the asymmetry by a factor of $g_V = -1 + 4\sin^2(\theta_W)$. This is a small number and normally the axial part does not make an important contribution to the asymmetry. Because $g_A = 1$, there is no suppresion of the vector part of the numerator.

The case considered here is very different from the 0^+ or $0^+ \longrightarrow 1^+$ transitions normally considered. In these cases there is only one independent form factor. As a result the q^2 dependence attributable to the form factors cancels between the numerator and denominator. Most of the remaining energy dependence also cancels[12] . This leaves only the net linear q^2 dependence (from the $\frac{1}{q^2}$ in the numerator divided by the $\frac{1}{q^4}$ in the denominator). Our case is quite different. Here there are two independent form factors of roughly similar size but with different q^2 dependence. Hence there can be and is substantial cancellation in F_V^{weak} and F_M^{weak} as can be seen from Eqs.(8a) and (8b) as we mentioned before. This means that these two form factors can be positive,zero, or negative and (because F_1 and F_2 are always positive) that A, the asymmetry itself can be positive, negative or zero. This is strikingly distinct from the 0^+ and $0^+ \longrightarrow 1^+$ cases mentioned above.

We plot the asymmetries,A, for incident electron energies of 0.1 GeV, 0.5 GeV, 1.0 GeV, 2.0 GeV, 4.0 GeV, and 6.0 GeV respectively in figures 1,2,3, 4,5, and 6 respectively. The asymmetry is given as a function of the outgoing electron laboratory angle. In order to be able to plot our results on a logarithmic scale with have plotted positive values of A with a solid line. We have treated negative values of A by plotting their absolute values but using a dashed line.

We also plot figures of merit for the above cases making use of the the standard relation:

$$F - O - M = A^2\, d\sigma/d\Omega_e. \quad (16).$$

This is done in figures 7,8,9,10,11,and 12 respectively.

We plot the variation $\delta A/A$ for two situations. In figure 13 we plot this number for an electron scattering angle of 4 degrees as a function of incident electron energy from 0.1 GeV to 6.0 GeV. We take this angle because it is well inside of the first minimum for this range of energy. In figure 14 we choose a point in first fluctuation in A, approximately midway between zero and the first minimum.

Finally in figure 15 we plot the axial contribution to A (i.e. that part containing F_A^{weak} as a factor) divided by the vector contribution to A (that part not containing F_A^{weak} as a factor). We do this only for the 1.0 GeV case but the others cases are similar. As expected this ratio is in general near zero except for those angles for which the asymmetry is small. For these angles, the vector part of A is near zero leaving only the axial term.

For the case we are considering in figure 13, at an angle of 4 degrees , we obtain the variation in A, i.e. $\delta A/A$, from the F-O-M, with the assumption of a running

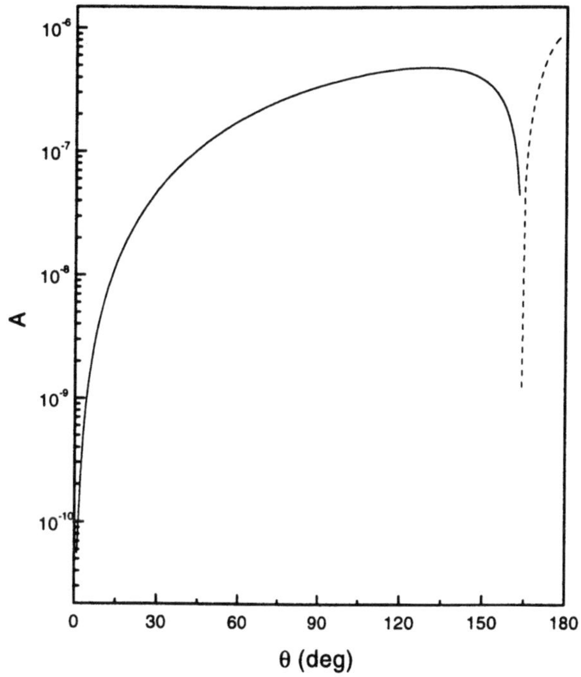

Fig. 1 Plot showing the asymmetry parameter,A,for the reaction $e^- +^3 He \longrightarrow e^- +^3 He$ as a function of the outgoing laboratory angle for an incident electron energy of 0.1 GeV. Positive values of A are indicated by a solid line. For negative values the absolute values are given and indicated by a dashed line.

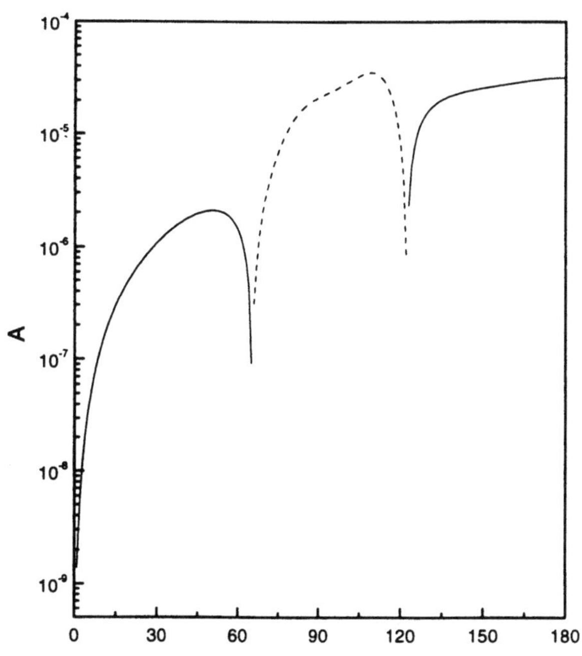

Fig. 2 Plot showing the asymmetry parameter,A,as in figure 1 for an incident electron energy of 0.5 GeV.

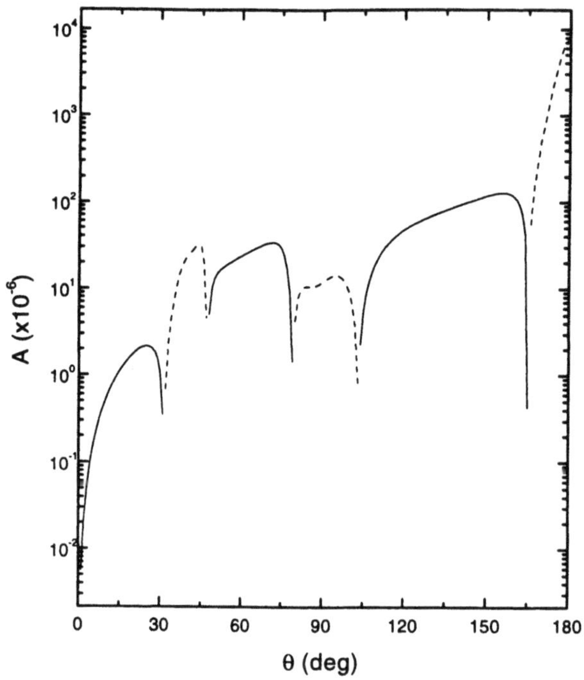

Fig. 3 Plot showing the asymmetry parameter, A, as in figure 1 for an incident electron energy of 1.0 GeV.

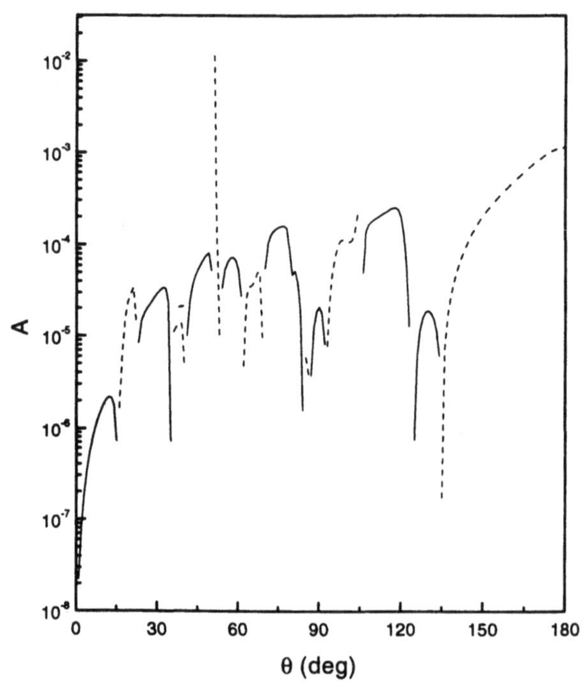

Fig. 4 Plot showing the asymmetry parameter, A, as in figure 1 for an incident electron energy of 2.0 GeV.

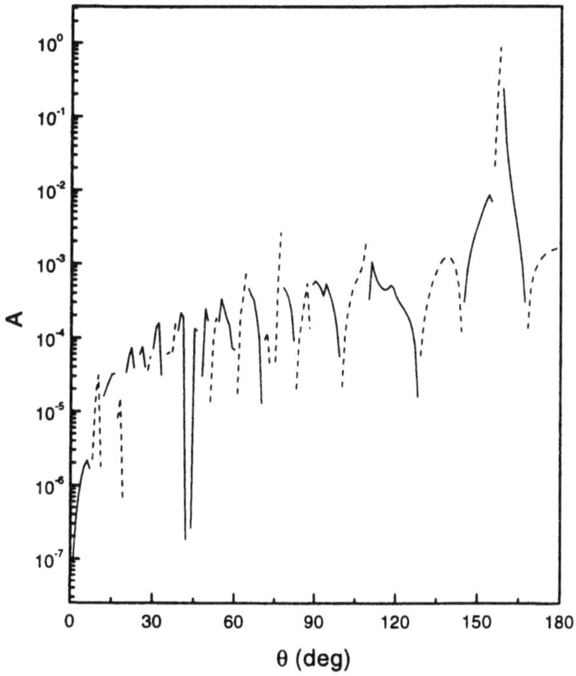

Fig. 5 Plot showing the asymmetry parameter, A, as in figure 1 for an incident electron energy of 4.0 GeV.

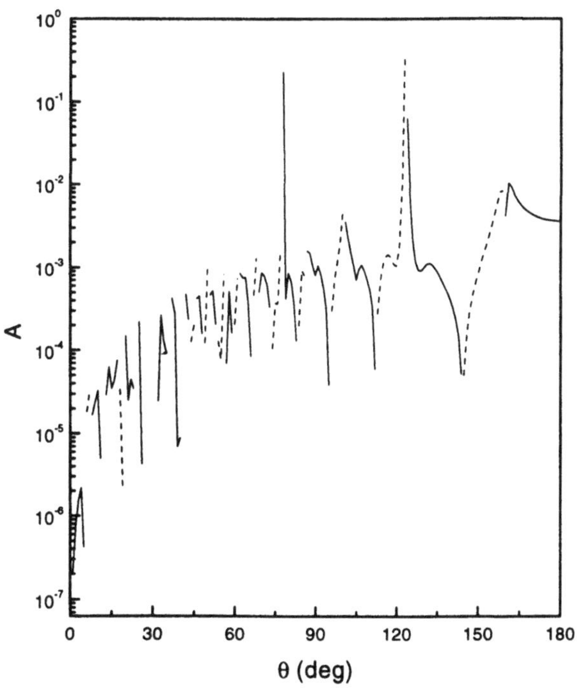

Fig. 6 Plot showing the asymmetry parameter, A, as in figure 1 for an incident electron energy og 6.0 GeV.

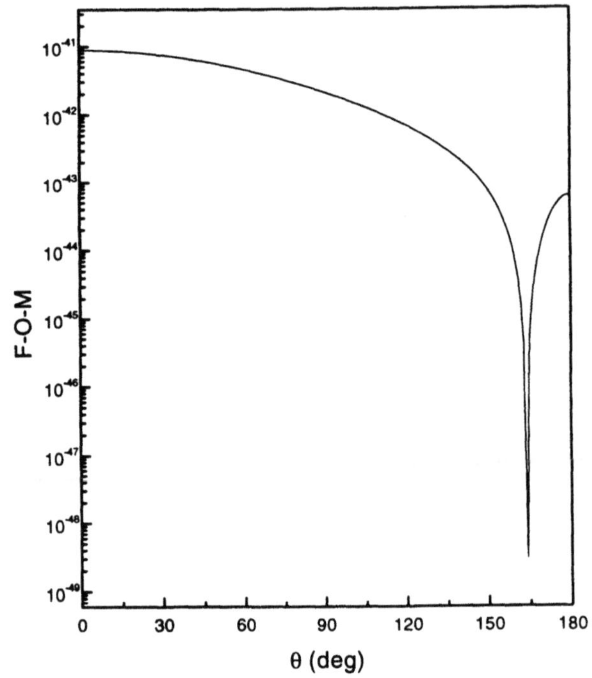

Fig. 7 Plot of the figure of merit (F-O-M) for the reaction, $e^- + ^3He \longrightarrow e^- + ^3He$, as a function of outgoing laboratory electron energy for an incident electron energy of 0.1 GeV.

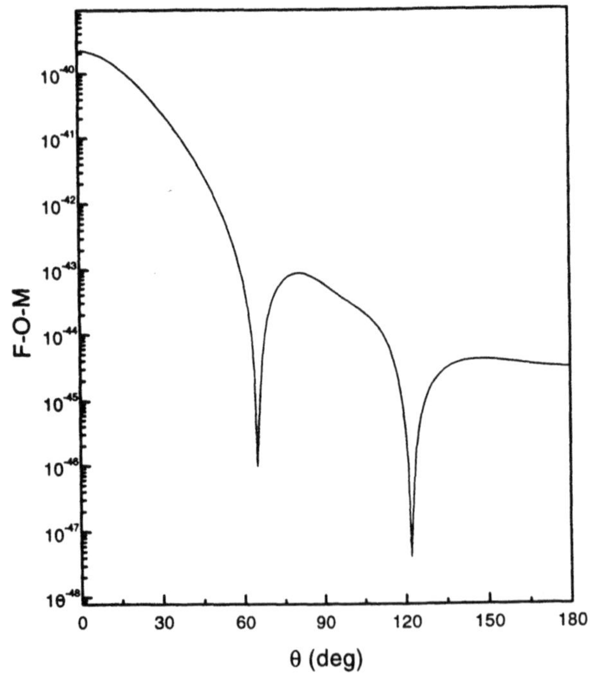

Fig. 8 Plot of the figure of merit as in figure 7 for an incident electron energy of 0.5 GeV.

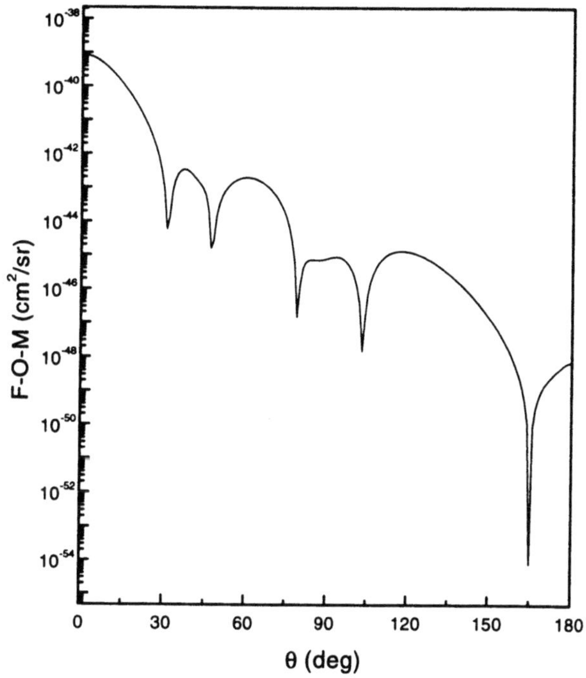

Fig. 9 Plot of the figure of merit as in figure 7 for an incident electron energy of 1.0 GeV.

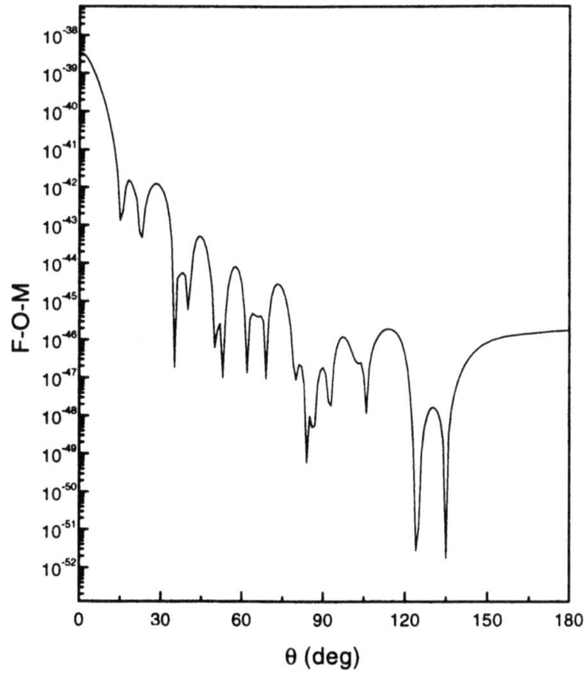

Fig. 10 Plot of the figure of merit as in figure 7 for an incident electron energy of 2.0 GeV.

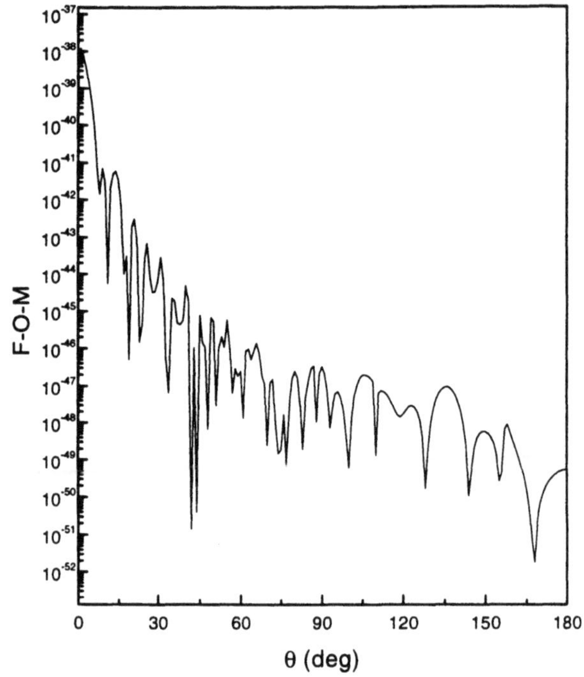

Fig. 11 Plot of the figure of merit as in figure 7 for an incident electron energy of 4.0 GeV.

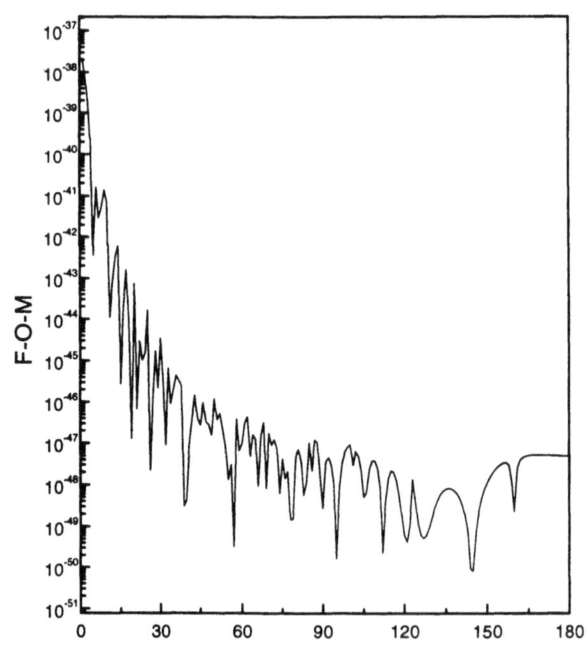

Fig. 12 Plot of the figure of merit as in figure 7 for an incident electron energy of 6.0 GeV.

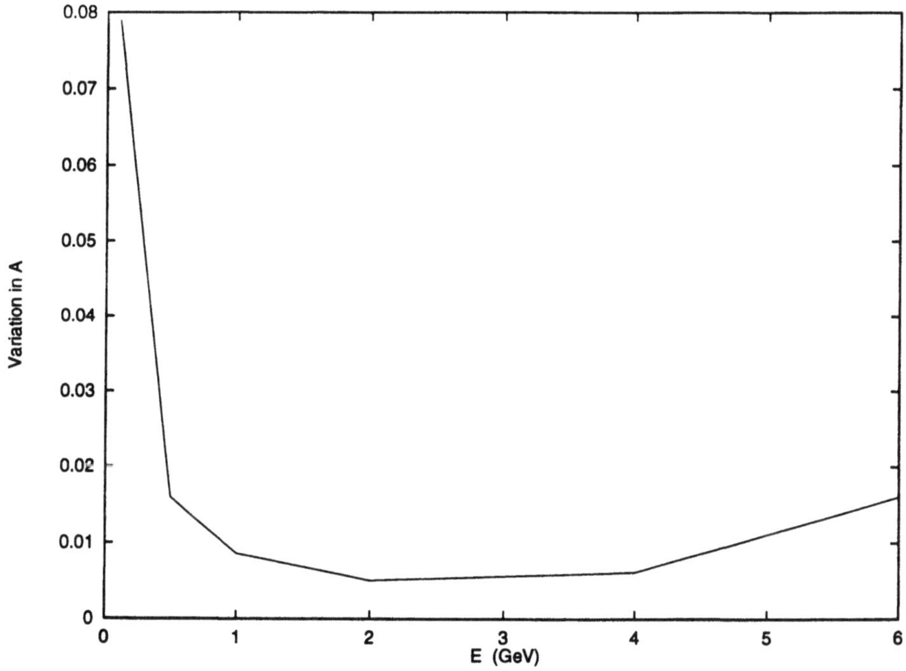

Fig. 13 Plot of the quantity $\delta A/A$ at an electron scattering angle of 4 degrees for incident electron energies from 0.1 GeV to 6.0 GeV

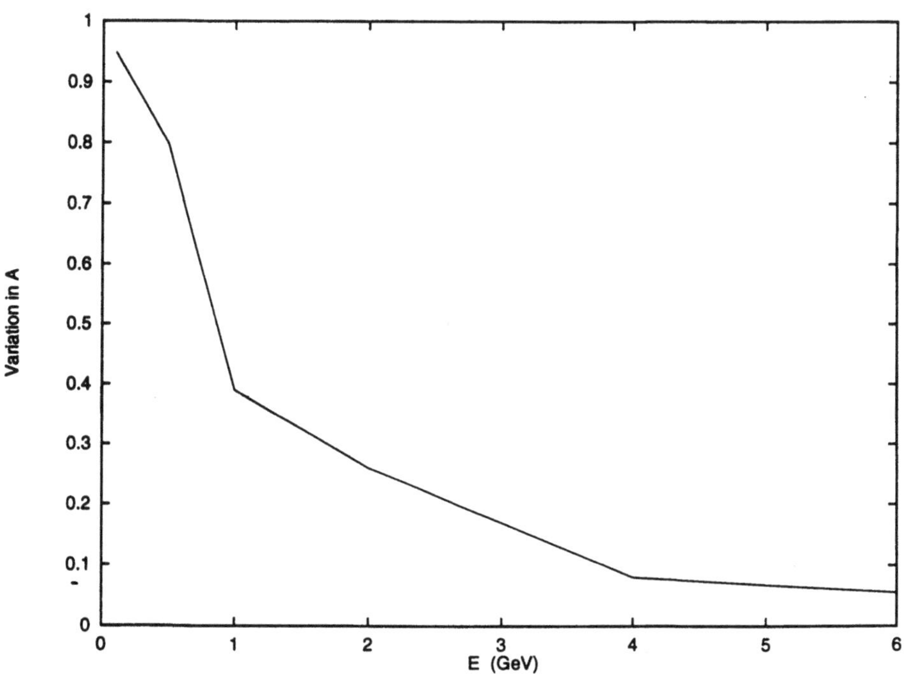

Fig. 14 Plot of the quantity $\delta A/A$ near the first zero in A for incident electron energies from 0.1 GeV to 6.0 GeV.

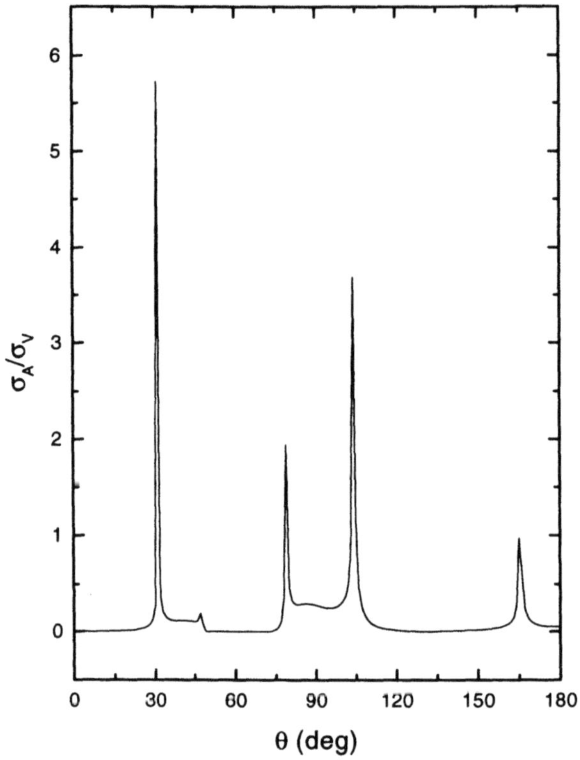

Fig. 15 Plot of the ratio of the axial contribution to the asymmetry (the part containg F_A^{weak} over the rest) and indicated by σ_A/σ_V. The results are given as a function of the outgoing electron laboratory angle for an incident electron energy of 1.0 GeV. Only the absolute values are given.

time of 1000 hours, a luminosity,L, of 5 $\times$ $10^{-38}cm^{-2}s^{-1}$, and a $\Delta\Omega$ of 10 msr,by making use of a standard[14] result,$\delta A/A = 1/\sqrt(FOM \times X)$,$X = L\Delta\Omega T$.

DISCUSSION

We find, as expected, from figures 13 and 14, that the variation in the asymmetry decreases as the energy of the incident electron increases. Even at 0.1 GeV, the uncertainty in A is of the order of 7.9 % and by 2.0 GeV it drops to around .51 %. Because calculations of weak nuclear process in this energy range are probably not more accurate[6] than 8 to 10 %, a discrepancy of the order of 12 to 15 % due to strange quark contributions should be determinable. This would be useful as some estimates for such contributions are at the 20 % level.

A region of interest is obviously in the neighborhood of the first zero of the asymmetry for the energies that we have studied. The angle at which the zero occurs decreases as the incident electron energy increases. We have chosen points representative of this neibhborhood at approximately the half-way point between the first zero and first minimum. These correspond to angles of 180, 81,37,20,9, and 6.5 degrees for incident electron energies of 0.1 GeV,0.5 GeV, 1.0 GeV, 2.0 GeV,4.0 GeV, and 6.0 GeV respectively. The values for $\delta A/A$ are plotted for this values in figure 14.

For this figure we have used a T of 2000 hours. We have discussed experiments of this length with a CEBAF experimentalist and such an experiment is not out of the question.

From figure 14 it can be seen that the uncertainty in A is too large to learn anything much for the 0.1 GeV, and 0.5 GeV cases. Useful information on strange quark contributions to A begins to become possible as early as 1.0 GeV where the uncertainty in A is of the order of 25%. However the situation is much better at 4 GeV or higher. At 4 GeV the uncertainty is only 5.65%. Even at 1000 hours of running time it is only 8 % which as we have noted above would be an acceptable uncertainty in A. Moreover because the curve for A is steep in the region on either side of the first zero, the uncertainties in A on either side of a zero are much larger than the uncertainties in the position of the zero. For example for the 1 GeV case uncertainties in A of approximately 25 % lead to an uncertainty of only $\pm.4$ degrees in the zero point angle. By direct calculation it is found that F_V^{weak} is largely responsible for the behavior of A near the zeros. This effect is even more striking at higher energies and could lead to a possible test for strange quark current contributions to F_V^{weak} at the 12 to 15 % level. If F_V^{weak} can be determined more accurately, contributions at the 8 to 10 % level from the strange quarks might be observable.

Finally we review what could be learned by observing this reaction at a variety of incident electron energies. First because there is a non-trivial q^2 dependence to the asymmetry,A, it might be possible to determine the weak vector form factors and thus to check for discrepancies that might represent strange quark contributions or possibly other phenomena. At small angles this test can probably be performed at or below the 10 % level over the entire range of energies considered. In addition near the zero values for A, F_V^{weak} dominates A and thus a displacement of the angles corresponding to these zero values from the expected values might be related to the strange quark current contributions to F_V^{weak} at the order of 12 % if a high enough incident electron energy (at least 4.0 GeV) is used. Although higher energies have a smaller variation in A, the rapid fluctuations in A might make the experiment more difficult.

Because q^2 values are relatively large in the region of high incident electron energy, the results might be interesting since we have no direct measurements of the weak form factors at large q^2. We again remark that the charge changing weak current form factors which are closely related to the neutral current $I = 1$ form factors work very well for muon capture but there are no higher q^2 tests. For the purpose of studying the form factors, the predicted structure is useful and unexpected and occurs at all energies considered. For these reasons experiments on ^3He could be very interesting.

REFERENCES

[1] G. Feinberg,Phys. Rev. **D 12**,3575(1975).
[2] M. J. Musolf and T.W. Donnelly,Nucl. Phys. **A 546**,509(1992).
[3] R. Carlini,talk, Workshop on Parity Violation, Seattle(1997).
[4] S.L. Mintz and M. Pourkaviani,Phys. Rev. **C 47**,873(1993).
[5] C.W.Kim and H. Primakoff,Phys. Rev. **140**,B566(1965).
[6] J.G. Congleton and H.W. Fearing, Nucl. Phys. **A552**, 534(1993).
[7] J. Frazier and C.W. Kim,Phys. Rev. **177**,2568(1968).

[8] S.L. Mintz,M.A. Barnett,G.M. Gerstner, and M. Pourkaviani, Nucl. Phys. **A598**,367(1995).
[9] "Anti-neutrino Reactions in ^3He",S.L. Mintz,G.M. Gerstner, M.A. Barnett, and M. Pourkaviani,preprint 1997.
[10] C.W. Kim and H. Primakoff,*Mesons in Nuclei*,edited by M. Rho and D.H. Wilkinson,p.68,North Holland, Amsterdam(1979).
[11] C.W. Kim and S.L. Mintz,Phys. Letters **31B**,503(1970).
[12] M. Pourkaviani and S.L. Mintz,J. Phys G: Nucl. Part. Phys.**17**,1139(1991).
[13] S.L. Mintz and M. Pourkaviani,Nucl. Phys. **A609**,441(1996).
[14] M.J. Musolf et al.,Physics Reports **239**,1(1994).
[15] Werner Boeglin,private communication.

GAUGE DYONIC STRINGS AND THEIR GLOBAL LIMIT

M.J. Duff*[1] James T. Liu†[2] H. Lü,‡ and C. N. Pope*[3]

*Center for Theoretical Physics, Department of Physics
Texas A&M University, College Station, Texas 77843–4242

†Department of Physics, The Rockefeller University
1230 York Avenue, New York, NY 10021-6399

‡Laboratoire de Physique Théorique de l'École Normale Supérieure[4]
24 Rue Lhomond - 75231 Paris CEDEX 05, France

ABSTRACT

We show that six-dimensional supergravity coupled to tensor and Yang-Mills multiplets admits not one but two different theories as global limits, one of which was previously thought not to arise as a global limit and the other of which is new. The new theory has the virtue that it admits a global anti-self-dual string solution obtained as the limit of the curved-space gauge dyonic string, and can, in particular, describe tensionless strings. We speculate that this global model can also represent the worldvolume theory of coincident branes. We also discuss the Bogomol'nyi bounds of the gauge dyonic string and show that, contrary to expectations, zero eigenvalues of the Bogomol'nyi matrix do not lead to enhanced supersymmetry and that negative tension does not necessarily imply a naked singularity.

1 Introduction

This paper is devoted to certain properties of the six-dimensional gauge dyonic string [1] and in particular to its global limit in which it becomes anti-self dual. An important special

[1] Research supported in part by NSF Grant PHY-9411543

[2] Research supported in part by the U.S. DOE under grant no. DOE-91ER40651-TASKB

[3] Research supported in part by the U.S. DOE under grant no. DE-FG03-95ER40917

[4] Unité Propre du Centre National de la Recherche Scientifique, associée à l'École Normale Supérieure et à l'Université de Paris-Sud

Physics of Mass
Edited by Kursunoglu *et al.*, Kluwer Academic / Plenum Publishers, New York, 1999

case corresponds to the tensionless string, which has been the subject of much interest lately [2–13,1], especially in the context of phase transitions [14,11,1,15].[1] This global limit is particularly interesting because one might then expect to be able to find an anti-self-dual string solution by directly solving the global supersymmetric theory in six-dimensions [11] describing an anti-self-dual tensor multiplet coupled to Yang-Mills. However, an apparently paradoxical claim was made in [18] that no such global limit exists. Here we resolve the paradox, and show that not only does the limit exist but that there are in fact two different limits, each giving different globally supersymmetric theories. One of these is the theory constructed in [18], which we shall refer to as the "BSS theory". The other flat-space theory, which for reasons described below we shall refer to as the "interacting theory", appears to be new, and admits an anti-self-dual string solution which can indeed be obtained as the flat-space limit of the dyonic string of the supergravity theory.

A surprising feature of the BSS theory constructed in [18] is that there is an asymmetry in the interactions between the Yang-Mills multiplet and the anti-self-dual tensor multiplet. In particular, the Yang-Mills multiplet satisfies free equations of motion, whereas the equations of motion for the tensor multiplet do involve couplings to the Yang-Mills fields. By contrast, the interactions in the "interacting" theory obtained in the present paper here are more symmetrical, in that they occur in all the equations of motion. Interestingly, however, the additional interaction terms of the new theory cancel in the special case of its anti-self-dual string solution, and so the same configuration is also a solution of the BSS theory. Curiously, however, it is not tensionless in that theory, and indeed the BSS theory is inappropriate for describing any tensionless string solution.

Another intriguing aspect of the gauge dyonic string concerns the counter-intuitive relations between its Bogomol'nyi bound, unbroken supersymmetry and its singularity structure [1]. We confirm:

(1) The dyonic string continues to preserve just half of the supersymmetry even in the tensionless limit, notwithstanding the standard Bogomol'nyi argument that a BPS state with vanishing central charge leads to completely unbroken supersymmetry.

(2) A solution with *negative* tension can be completely non-singular, contrary to the folk-wisdom that negative mass necessarily implies naked singularities.

Finally, six dimensional global models are also important as fivebrane worldvolume theories [19–21,18] and as the worlvolume theories of coincident higher-dimensional branes with six dimensions in common [25,24]. We speculate that the interacting anti-self-dual-tensor Yang-Mills system is indeed such a worldvolume theory. Hence the global gauge anti-self-dual string, and in particular the tensionless string, may also be regarded as a

[1]But note that, contrary to some claims in the literature, the tensionless string corresponds to the (quasi)-anti-self-dual limit of the dyonic string of [16], where the string couples dominantly to the 3-form field strength of the tensor matter multiplet, and not the self-dual string of [17] where the string couples only to the 3-form field strength of the gravity multiplet.

string on the worldvolume. In the case of the tensionless string, in the limit as the size ρ of the Yang-Mills instanton shrinks to zero, one recovers the global limit of the neutral tensionless string [16, 1] which is also a solution of the $(2,0)$ theory that resides on the worldvolume of the M-theory fivebrane. It is curious, therefore, that we find in this limit that the tension really is zero, as opposed to the infinite tension of the string solution of the free $(2,0)$ theory [22,23].

2 $N=1$ supergravity and the gauge dyonic string

The low-energy $D = 6$ $N = (1,0)$ supergravity is generated by a pair of symplectic Majorana-Weyl spinors ϵ transforming in the 2 of $Sp(2)$. This theory has the unusual feature in that the antisymmetric tensor breaks up into self-dual and anti-self-dual components. The basic supergravity theory consists of the graviton multiplet $(g_{\mu\nu}, \psi_\mu, B^+_{\mu\nu})$ coupled to n_T tensor multiplets $(B^-_{\mu\nu}, \chi, \phi)$. When $n_T = 1$, corresponding to the heterotic string compactified on $K3$, these multiplets may be combined, yielding a single ordinary antisymmetric tensor $B_{\mu\nu}$.

We are interested, however, in the general case with n_T tensor multiplets coupled to an arbitrary number of vector multiplets (A_μ, λ). Due to the presence of chiral antisymmetric tensor fields, there is no manifestly covariant Lagrangian formulation of this theory. Nevertheless, the equations of motion may be constructed, and were studied in [26,27]. With n_T tensor multiplets, there are n_T scalars parametrizing the coset $SO(1,n_T)/SO(n_T)$. This may be described in terms of a $(n_T+1)\times(n_T+1)$ vielbein transforming as vectors of both $SO(1,n_T)$ and $SO(n_T)$. Following the conventions of [27], the vielbein may be decomposed as

$$V = \begin{bmatrix} V_+ \\ V_- \end{bmatrix} = \begin{bmatrix} v_0 & v_M \\ x^m{}_0 & x^m{}_M \end{bmatrix} , \tag{2.1}$$

satisfying the condition $V^{-1} = \eta V^T \eta$ where η is the $SO(1,n_T)$ metric, $\eta = \mathrm{diag}(1, -I_{n_T})$. Below, we use indices $r, s, \ldots = \{0, M\}$ to denote $SO(1,n_T)$ vector indices. The composite $SO(n_T)$ connection is then given by

$$S^{[mn]}_\mu = (\partial_\mu V_- \eta V_-^T)^{[mn]} = -x^m{}_0 \partial_\mu x^n{}_0 + x^m{}_M \partial_\mu x^n{}_M , \tag{2.2}$$

so that the fully covariant derivative acting on $SO(n_T)$ vectors is given by $\mathcal{D}_\mu = \nabla_\mu + S_\mu$.

To describe the combined supergravity plus tensor system, we introduce (n_T+1) antisymmetric tensors $B_{\mu\nu}$ transforming as a vector of $SO(1,n_T)$. In the presence of Yang-Mills fields, the three-form field strengths pick up a Chern-Simons coupling

$$\mathcal{H} = dB + c\,\omega_3 , \tag{2.3}$$

where $\omega_3 = AdA + \frac{2}{3}A^3$, so that $d\mathcal{H} = c \operatorname{tr} F^2$. The constants c form a $(n_T+1)\times n_V$ matrix where n_V is the number of vector multiplets[2]. Note that this coupling of the vector

[2]For non-abelian gauge fields, instead of having n_V independent quantities, there is a single set of c's for each factor of the gauge group.

and tensor multiplets is dictated by supersymmetry and encompasses both tree-level and one-loop Yang-Mills corrections. Furthermore, the supersymmetry guarantees that there are no higher-loop corrections. The vielbein is then used to transform the field strengths $\mathcal{H}$ into their chiral components $H = v_r\mathcal{H}^r$ and $K^m = x^m{}_r\mathcal{H}^r$ so that the (anti-)self-duality conditions for the tensors become $H = *H$ and $K^m = - *K^m$.

With the above conventions, the bosonic equations of motion are

$$\begin{aligned}
G_{\mu\nu} \equiv R_{\mu\nu} - \tfrac{1}{2}g_{\mu\nu}R &= T_{\mu\nu} \\
\mathcal{D}_\mu P^{m\,\mu} &= -\tfrac{\sqrt{2}}{3}H_{\mu\nu\rho}K^{m\,\mu\nu\rho} - \tfrac{1}{\sqrt{2}}x^m{}_r c^r\,\mathrm{tr}\,(F_{\mu\nu}F^{\mu\nu}) \\
dH &= -\sqrt{2}P^m K^m + v_r c^r\,\mathrm{tr}\,F^2 \\
(d\delta^{mn} + S^{mn})K^n &= -\sqrt{2}P^m H + x^m{}_r c^r\,\mathrm{tr}\,F^2 \\
v_r c^r D^\mu F_{\mu\nu} &= \sqrt{2}P^{m\,\mu}x^m{}_r c^r F_{\mu\nu} + H_{\nu\rho\sigma}v_r c^r F^{\rho\sigma} + K^m_{\nu\rho\sigma}x^m{}_r c^r F^{\rho\sigma}\ ,
\end{aligned} \tag{2.4}$$

where

$$\begin{aligned}
P^m_\mu &= \tfrac{1}{\sqrt{2}}(\partial_\mu V_+ \eta V_-^T)^m \\
&= \tfrac{1}{\sqrt{2}}(x^m{}_0\partial_\mu v_0 - x^m{}_M\partial_\mu v_M)\ ,
\end{aligned} \tag{2.5}$$

and S and P are the 1-forms, $S = S_\mu dx^\mu$, $P = P_\mu dx^\mu$. The symmetric stress tensor is given by

$$T_{\mu\nu} = H_{\mu\rho\sigma}H_\nu{}^{\rho\sigma} + K^m_{\mu\rho\sigma}K^{m\,\rho\sigma}_\nu + 2[P^m_\mu P^m_\nu - \tfrac{1}{2}g_{\mu\nu}P^m_\rho P^{m\,\rho}] + 4v_r c^r\,\mathrm{tr}\,[F_{\mu\lambda}F_\nu{}^\lambda - \tfrac{1}{4}g_{\mu\nu}F_{\lambda\sigma}F^{\lambda\sigma}]\ . \tag{2.6}$$

For the antisymmetric tensors, Eqn. (2.4) along with the (anti-)self-duality constraint may be viewed as the equivalent of the combined Bianchi identities and equations of motion. Finally, the fermionic equations of motion are

$$\begin{aligned}
\gamma^{\mu\nu\rho}\nabla_\nu\psi_\rho &= -H^{\mu\nu\rho}\gamma_\nu\psi_\rho + \tfrac{i}{2}K^{m\,\mu\nu\rho}\gamma_{\nu\rho}\chi^m - \tfrac{i}{\sqrt{2}}P^m_\nu\gamma^\nu\gamma^\mu\chi^m - \tfrac{1}{\sqrt{2}}\gamma^{\sigma\tau}\gamma^\mu v_r c^r\,\mathrm{tr}\,F_{\sigma\tau}\lambda \\
\gamma^\mu\nabla_\mu\chi^m &= \tfrac{i}{2}K^{m\,\mu\nu\rho}\gamma_{\mu\nu}\psi_\rho + \tfrac{1}{12}H_{\mu\nu\rho}\gamma^{\mu\nu\rho}\chi^m + \tfrac{i}{\sqrt{2}}P^m_\nu\gamma^\mu\gamma^\nu\psi_\mu - \tfrac{i}{\sqrt{2}}\gamma^{\mu\nu}x^m{}_r c^r\,\mathrm{tr}\,F_{\mu\nu}\lambda \\
v_r c^r\gamma^\mu D_\mu\lambda &= \tfrac{1}{\sqrt{2}}P^m_\mu\gamma^\mu x^m{}_r c^r\lambda - \tfrac{1}{2\sqrt{2}}v_r c^r F_{\lambda\tau}\gamma^\mu\gamma^{\lambda\tau}\psi_\mu - \tfrac{i}{2\sqrt{2}}x^m{}_r c^r F_{\mu\nu}\gamma^{\mu\nu}\chi^m \\
&\quad -\tfrac{1}{12}K^m_{\mu\nu\rho}x^m{}_r c^r\gamma^{\mu\nu\rho}\lambda\ .
\end{aligned} \tag{2.7}$$

In order to examine the Bogomol'nyi bound, we need the supersymmetry variations for the fermionic fields:

$$\begin{aligned}
\delta\psi_\mu &= [\nabla_\mu + \tfrac{1}{4}H_{\mu\nu\rho}\gamma^{\nu\rho}]\epsilon \\
\delta\chi^m &= i[\tfrac{1}{\sqrt{2}}\gamma^\mu P^m_\mu + \tfrac{1}{12}K^m_{\mu\nu\rho}\gamma^{\mu\nu\rho}]\epsilon \\
\delta\lambda &= -\tfrac{1}{2\sqrt{2}}F_{\mu\nu}\gamma^{\mu\nu}\epsilon
\end{aligned} \tag{2.8}$$

(given to lowest order). For completeness, the bosonic fields transform according to

$$\begin{aligned}
\delta e_\mu{}^a &= -i\bar\epsilon\gamma^a\psi_\mu \\
\delta B^r_{\mu\nu} &= \eta^{rs}\bar\epsilon[iv_s\gamma_{[\mu}\psi_{\nu]} - \tfrac{1}{2}x^m{}_s\gamma_{\mu\nu}\chi^m] + 2c^r\,\mathrm{tr}\,A_{[\mu}\delta A_{\nu]} \\
\delta v_r &= x^m{}_r\bar\epsilon\chi^m \\
\delta A_\mu &= -\tfrac{i}{\sqrt{2}}\bar\epsilon\gamma_\mu\lambda\ .
\end{aligned} \tag{2.9}$$

Careful examination of Eqns. (2.8) and (2.9) reveals the intricate interplay between terms of various chiralities necessary to maintain $D = 6$ $N = (1,0)$ supersymmetry. In particular, ϵ is a chiral spinor satisfying $P_+\epsilon = 0$ where $P_\pm = \frac{1}{2}(1 \pm \gamma^7)$ is the chirality projection in six dimensions. As a consequence, H and K satisfy the identities

$$\begin{aligned}(H_{\mu\nu\lambda}\gamma^{\mu\nu\lambda})\epsilon &= 0 \\ (K^m_{\mu\nu\lambda}\gamma^{\mu\nu\lambda}\gamma_\alpha)\epsilon &= 0\,,\end{aligned} \tag{2.10}$$

which prove to be useful in manipulating Nester's form below.

2.1 The gauge dyonic string solution

It was shown in [1] that the equations of motion (2.4) admit a gauge dyonic string solution carrying both self-dual and anti-self-dual tensor charges. Under an appropriate $SO(n_T)$ rotation, the latter charge can be put in a single tensor component, so that we may focus on the theory with $n_T = 1$. In this case, corresponding to a compactified heterotic string, the self-dual and anti-self-dual three-forms in the graviton and tensor multiplets respectively may be combined together according to

$$H = \tfrac{1}{2}e^{-\phi}(*\mathcal{H} + \mathcal{H})\,, \qquad K = \tfrac{1}{2}e^{-\phi}(*\mathcal{H} - \mathcal{H})\,, \tag{2.11}$$

where we have chosen a vielbein

$$V = \begin{bmatrix} \cosh\phi & \sinh\phi \\ \sinh\phi & \cosh\phi \end{bmatrix}\,. \tag{2.12}$$

For a simple gauge group, we pick the coupling vector c to be

$$c = \frac{\alpha'}{16}\begin{bmatrix} v + \tilde{v} \\ -v + \tilde{v} \end{bmatrix}\,, \tag{2.13}$$

so that the $\mathcal{H}$ Bianchi identity and equation of motion, given in Eqn. (2.4), may be rewritten as

$$d\mathcal{H} = \tfrac{1}{8}\alpha'\, v\, \mathrm{tr}\, F \wedge F \qquad d(e^{-2\phi} * \mathcal{H}) = \tfrac{1}{8}\alpha'\, \tilde{v}\, \mathrm{tr}\, F \wedge F\,. \tag{2.14}$$

The gauge dyonic string is built around a single self-dual $SU(2)$ Yang-Mills instanton in transverse space, and is given in terms of three parameters, which are the electric and magnetic charges Q and P carried by the string, and ρ which is the scale parameter of the instanton. Splitting the six-dimensional space into longitudinal $\mu, \nu = 0, 1$ and transverse $m, n, \ldots = 2, 3, 4, 5$ components, the gauge dyonic string solution is given by [1]

$$\begin{aligned} ds^2 &= e^{2A}\eta_{\mu\nu}dx^\mu dx^\nu + e^{-2A}dy^m dy^m \\ \mathcal{H}_{mnp} &= \tfrac{1}{2}\epsilon_{mnpq}\partial_q H_1 \qquad \mathcal{H}_{\mu\nu m} = \tfrac{1}{2}\epsilon_{\mu\nu}\partial_m H_2^{-1} \\ e^{-\phi} &= \sqrt{H_2/H_1} \qquad e^{-2A} = \sqrt{H_1 H_2}\,, \end{aligned} \tag{2.15}$$

where $\epsilon_{01} = 1$, $\epsilon_{2345} = 1$. The functions H_1 and H_2 are

$$H_1 = e^{\phi_0} + \frac{P(2\rho^2 + r^2)}{(\rho^2 + r^2)^2} , \qquad H_2 = e^{-\phi_0} + \frac{Q(2\rho^2 + r^2)}{(\rho^2 + r^2)^2} , \tag{2.16}$$

and are determined by the effect of the instanton source

$$F^a = \frac{2\rho^2}{(\rho^2 + r^2)^2} \eta^a_{mn} dy^m \wedge dy^n , \tag{2.17}$$

on the three-form tensor according to (2.14). (Note that $\mathrm{tr}\,(F^2) = 2F^a_{mn} F^{amn}$.) In particular, the charges are thus given by $Q = 2\alpha' \tilde{v}$ and $P = 2\alpha' v$. The mass per unit length of the dyonic string is given by

$$2\pi\alpha'^2 \, m = Pe^{-\phi_0} + Qe^{\phi_0} . \tag{2.18}$$

This expression for the mass, and its relation to the Bogomol'nyi bound, will be examined in detail in the following section.

In the $\rho \to 0$ limit, we recover the neutral dyonic string obtained in [16].

3 The Bogomol'nyi bound in six dimensions

It is well known that the six-dimensional $N = (1, 0)$ supersymmetry algebra admits a single real string-like central charge, putting a lower bound on the tension of the six-dimensional string. Thus the tensionless string only arises in the limit of vanishing central charge. Before focusing on the tensionless string, we examine the Bogomol'nyi mass bound in general and determine the conditions for which it is satisfied.

For a string-like field configuration in six dimensions, we may construct the supercharge per unit length of the string from the behavior of the gravitino at infinity [28]

$$Q_\epsilon = \int_{\partial\mathcal{M}} \bar{\epsilon} \gamma^{\mu\nu\lambda} \psi_\lambda d\Sigma^{\mu\nu} , \tag{3.1}$$

where $\mathcal{M}$ is the four-dimensional space transverse to the string. We note that in writing the supercharge in terms of the gravitino, this expression holds only up to the equations of motion. It is for this reason that, unlike in the global case, saturation of the Bogomol'nyi bound alone is insufficient to guarantee that the bosonic background solves the supergravity equations of motion.

Using Nester's procedure [29, 28, 30], we may take the anticommutator of two supercharges to get

$$\{Q_\epsilon, Q_{\epsilon'}\} = \delta_\epsilon Q_{\epsilon'} = \int_{\partial\mathcal{M}} N^{\mu\nu} d\Sigma_{\mu\nu} , \tag{3.2}$$

where

$$N^{\mu\nu} = \overline{\epsilon'} \gamma^{\mu\nu\lambda} \delta_\epsilon \psi_\lambda = \overline{\epsilon'} \gamma^{\mu\nu\lambda} [\nabla_\lambda + \tfrac{1}{4} H_{\lambda\rho\sigma} \gamma^{\rho\sigma}] \epsilon \tag{3.3}$$

is a generalized Nester's form. Appealing to the supersymmetry algebra, we then see that the mass and central charge per unit length of the six-dimensional string is encoded in the surface integral of $N^{\mu\nu}$. For a string in the 0-1 direction, the ADM mass per unit length M of the string is given by the asymptotic behavior of the metric

$$ds^2 = (1 - \frac{GM}{2r^2} + \cdots)[-dt^2 + dz^2] + (1 + \frac{GM}{2r^2} + \cdots)dy^i dy^i \ , \tag{3.4}$$

where $r^2 = y^i y^i$ is the transverse radial distance from the string. Using this definition of the ADM mass, the surface integral of Nester's form becomes

$$\int_{\partial\mathcal{M}} N^{\mu\nu} d\Sigma_{\mu\nu} = 2\pi^2 \epsilon'^{\dagger}[\frac{GM}{2} - Z\gamma^0\gamma^1]\epsilon \ , \tag{3.5}$$

where the real string-like central charge Z is given by the self dual H charge

$$\int_{\partial\mathcal{M}} H = 2\pi^2 Z \ . \tag{3.6}$$

This reinforces the close relation between the central charges of a supergravity theory and the bosonic charges of the fields in the graviton multiplet.

From the point of view of the supersymmetry algebra, the left hand side of Eqn. (3.2) is non-negative for identical (commuting) spinors $\epsilon' = \epsilon$. Since $\gamma^0\gamma^1$ has eigenvalues ± 1, this gives rise to the Bogomol'nyi bound

$$GM \geq 2|Z| \ , \tag{3.7}$$

with saturation of the bound corresponding to (partially) unbroken supersymmetry. However an issue has arisen over the necessary conditions for this bound to apply. In particular, it has been noted that the gauge dyonic string [1] may have a tensionless limit without naked singularities when the instanton size in the gauge solution is sufficiently large. Corresponding to Eqn. (3.7), this tensionless string has vanishing central charge and is hence quasi-anti-self-dual. Nevertheless, examination of the Killing spinor equations indicates that it still breaks exactly half of the supersymmetries, in contrast to the expectation that $M = 0$ yields completely unbroken supersymmetry. In terms of singular four-dimensional solutions, this breakdown of the Bogomol'nyi argument has also been discussed in [31, 32].

In order to address the issue of where the Bogomol'nyi expression may break down, we take a closer look at the Witten-Nester proof of the positive energy theorem [33, 29]. Following [28], the charges at infinity may be related to the divergence of Nester's form:

$$\int_{\partial\mathcal{M}} N^{\mu\nu} d\Sigma_{\mu\nu} = \int_{\mathcal{M}} \nabla_\mu N^{\mu\nu} d\Sigma_\nu \ . \tag{3.8}$$

Proof of the Bogomol'nyi bound is then a matter of reexpressing this divergence in a manifestly non-negative form. Straightforward but tedious manipulations allow the divergence of Nester's form to be rewritten in terms of the supersymmetry variations of the fermionic fields given in Eqn. (2.8). Starting with

$$\begin{aligned}\nabla_\mu N^{\mu\nu} &= \overline{\delta_{\epsilon'}\psi_\mu}\gamma^{\mu\nu\rho}\delta_\epsilon\psi_\rho - \tfrac{1}{2}\overline{\epsilon'}G^\nu{}_\sigma\gamma^\sigma\epsilon \\ &\quad + \tfrac{1}{4}\overline{\epsilon'}\gamma^{\mu\nu\rho}\gamma^{\beta\gamma}(\nabla_\mu H_{\rho\beta\gamma})\epsilon + \tfrac{1}{16}\overline{\epsilon'}\gamma^{\beta\gamma}\gamma^{\mu\nu\rho}\gamma^{\lambda\sigma}H_{\mu\beta\gamma}H_{\rho\lambda\sigma}\epsilon \ ,\end{aligned} \tag{3.9}$$

it is apparent that the H equations of motion must enter the calculation. Working through these equations then gives the final result

$$\begin{aligned}\nabla_\mu N^{\mu\nu} &= \overline{\delta_{\epsilon'}\psi_\mu}\gamma^{\mu\nu\rho}\delta_\epsilon\psi_\rho + \overline{\delta_{\epsilon'}\chi}\gamma^\nu\delta_\epsilon\chi + v_r c^r \operatorname{tr}\overline{\delta_{\epsilon'}\lambda}\gamma^\nu\delta_\epsilon\lambda - \tfrac{1}{2}\overline{\epsilon'}[G^{\nu\sigma} - \mathcal{T}^{\nu\sigma}]\gamma_\sigma\epsilon \\ &\quad -\tfrac{1}{12}\overline{\epsilon'}\gamma^{\mu\rho}\gamma^\nu\gamma^{\beta\gamma}[\partial_{[\mu}H_{\rho\beta\gamma]} + \sqrt{2}P^m_{[\mu}K^m_{\rho\beta\gamma]} - \tfrac{3}{2}v_r c^r \operatorname{tr} F_{[\mu\rho}F_{\beta\gamma]}]\epsilon \\ &\quad +\tfrac{1}{2}\overline{\epsilon'}[\nabla_\alpha H^{\alpha\nu\sigma} - \sqrt{2}P^m_\alpha K^{m\,\alpha\nu\sigma} - \tfrac{1}{4}\epsilon^{\nu\sigma\alpha\beta\gamma\delta}v_r c^r F_{\alpha\beta}F_{\gamma\delta}]\gamma_\sigma\epsilon\,. \end{aligned} \tag{3.10}$$

We wish to point out that this is an exact expression, where only kinematics has been used in rewriting the divergence. The last two lines are related to the self-dual H equation of motion (in Bianchi identity and divergence form respectively), and hence vanish on-shell. In addition to the expected terms, this divergence has the unusual feature in that the full stress tensor $\mathcal{T}_{\mu\nu}$ arising from the supersymmetry manipulations is modified by the inclusion of an antisymmetric contribution

$$\begin{aligned}\mathcal{T}_{\mu\nu} &= T_{\mu\nu} + T'_{\mu\nu} \\ &= T_{\mu\nu} - 2v_r c^r \operatorname{tr}[F_{\mu\alpha}F_\nu{}^\alpha - \tfrac{1}{4}g_{\mu\nu}F_{\alpha\beta}F^{\alpha\beta} - \tfrac{1}{8}\epsilon_{\mu\nu\alpha\beta\gamma\delta}F^{\alpha\beta}F^{\gamma\delta}]\,, \end{aligned} \tag{3.11}$$

where $T_{\mu\nu}$, given in (2.6), is the symmetric stress tensor appearing in Einstein's equation. In particular, this antisymmetric component, which arises as a consequence of the $N = (1,0)$ supersymmetry algebra in six dimensions [34], is related to the fact that the classical equations of motion, Eqns. (2.4), are actually inconsistent in such a manner as to cancel the effects of the gauge anomalies when loop corrections are taken into account [35, 36]. As a result, the equations violate Bose symmetry in a way that would be impossible if they were derivable from a Lagrangian. There exists a Lagrangian, at least in the case $n_T = 1$, which automatically leads to Bose symmetric equations but which lacks gauge invariance [14]. As discussed in [35,36], these two formulations are related to the difference between consistent and covariant anomalies. It is interesting to note, however, that the gauge dyonic string solves both sets of equations, since the Bose non-symmetric terms vanish in this background.

We are now in a position to examine the conditions under which the Bogomol'nyi bound, Eqn. (3.7), may hold. Based on the rewriting of the Bogomol'nyi equation in terms of a volume integral, it is apparent that the mass bound will hold provided the divergence $\nabla_\mu N^{\mu\nu}$ is positive semi-definite over the entire transverse space $\mathcal{M}$. This gives rise to the following three conditions: *i*) the supergravity equations of motion must be satisfied[3], *ii*) Witten's condition must hold globally so the gravitino variation is non-negative, and *iii*) the Yang-Mills contributions from both the gaugino variation and the correction $T'_{\mu\nu}$ to the stress tensor must be non-negative. While the first condition is straightforward, the other two require further explanation. Witten's condition [33] is essentially a spatial Dirac equation, $\gamma^i\delta_\epsilon\psi_i = 0$, where $i = 1, \ldots, 5$. While this condition may be satisfied for a well behaved background, it is also important to ensure that such spinors are normalizable on all of $\mathcal{M}$ so that the divergence integral is well defined. In particular, this normalizability

[3]Only Einstein's equation and the H equation of motion are relevant for the Bogomol'nyi calculation. Note that when we refer to Einstein's equation, we do not include the correction $T'_{\mu\nu}$ which is accounted for separately.

condition apparently breaks down in the presence of naked singularities, as we subsequently verify for the gauge dyonic string solution. This leads us to believe that Witten's condition is essentially equivalent to demanding that the background contains no naked singularities.

We now turn to the conditions that need to be imposed on the Yang-Mills fields. Looking at the gaugino variation in Eqn. (3.10), it is natural to impose the condition that all components of the n_V dimensional vector $v_r c^r$ are to be non-negative. Since the n_T scalars encoded in the vielbein v_r act as gauge coupling constants, this condition simply states that the Yang-Mills fields must have the correct sign kinetic terms. Starting from a weakly coupled point in moduli space, it is apparent that the only way to generate a wrong sign term is to pass through infinite coupling. Since this corresponds to a phase transition [14], driven by tensionless strings [13,11,1], it indicates that the Bogomol'nyi results need to be applied with care when discussing the strong coupling dynamics of six dimensional strings.

Since the Yang-Mills fields lead to a modification of the stress tensor, it is also necessary to require that $T'_{\mu\nu}$ enters non-negatively into the divergence of Nester's form. For a string-like geometry in the 0-1 direction, this condition is equivalent to demanding that $-T'_{00} \geq |T'_{01}| \geq 0$, which is automatically satisfied for gauge fields living only in transverse space (again provided $v_r c^r$ is non-negative). To see this, note that for $\mu, \nu = 0, 1$ we may write

$$T'_{\mu\nu} = \tfrac{1}{2} v_r c^r \operatorname{tr}\left[g_{\mu\nu} F_{mn} F^{mn} + \epsilon_{\mu\nu} F_{mn} *_4 F^{mn}\right] , \tag{3.12}$$

and use the instanton argument, $\operatorname{tr}(F \pm *_4 F)^2 \geq 0$, to show that the T' conditions are satisfied. Therefore as long as the Yang-Mills fields vanish in the longitudinal directions of the string-like solution, no further condition is necessary. It is perhaps not coincidental that this vanishing of the gauge fields on the string also renders unimportant the inconsistency of the classical equations of motion.

3.1 Supersymmetry of the gauge dyonic string

It is instructive to see how the Bogomol'nyi equation breaks down in the various limits of the gauge dyonic string. For this string background, given by (2.15), the supersymmetry variations of the fermions, (2.8), become

$$\begin{aligned}
\delta\psi_\mu &= -\gamma^n \partial_n A \gamma_\mu \mathcal{P}_2^+ \epsilon \\
\delta\psi_m &= \gamma^n \partial_n A \gamma_m \mathcal{P}_2^+ \epsilon + e^{A/2} \partial_m (e^{-A/2} \epsilon) \\
\delta\chi &= -i \gamma^n \partial_n \phi \mathcal{P}_2^+ \epsilon \\
\delta\lambda &= -\frac{1}{2\sqrt{2}} F_{mn} \gamma^{mn} \mathcal{P}_2^+ \epsilon ,
\end{aligned} \tag{3.13}$$

where $\mathcal{P}_2^+ = \frac{1}{2}(1 + \gamma^{\overline{01}})$ is a projection onto the chiral two-dimensional world-sheet of the string-like solution (overlined symbols indicate tangent-space indices). This indicates, as noted in [1], that the Killing spinor equations are solved for spinors ϵ satisfying

$$\mathcal{P}_2^+ \epsilon = 0 , \qquad \epsilon = e^{A/2} \epsilon_0 . \tag{3.14}$$

On the other hand, the fermion zero modes are given by spinors ϵ surviving the projection, namely $\mathcal{P}_2^+\epsilon = \epsilon$. Note that for the zero modes there is no further condition on ϵ.

Based on the above supersymmetry variations, we may explicitly calculate the divergence of Nester's expression. Since this expression obviously vanishes for Killing spinors, we only concern ourselves with the fermion zero modes. For simplicity in working with the derivative term entering $\delta\psi_m$, we assume a simple scaling so that ϵ is given by

$$\epsilon = e^{\alpha A}\epsilon_0 \ , \qquad \mathcal{P}_2^+\epsilon = \epsilon \ , \tag{3.15}$$

where ϵ_0 is a constant spinor. Working out the divergence then gives

$$\begin{aligned} \int_{\mathcal{M}} \nabla_\mu N^{\mu\nu} d\Sigma_\nu &= 2\pi^2\, \epsilon_0^\dagger\epsilon_0 \int e^{2(\alpha-\frac{1}{2})A}[4(\alpha-\tfrac{1}{2})\partial_m A\partial_m A + 2\partial_m\partial_m A]r^3 dr \\ &= 2\pi^2\, \epsilon_0^\dagger\epsilon_0 \frac{1}{\alpha-\frac{1}{2}} \int r^3 dr\, \partial_m\partial_m e^{2(\alpha-\frac{1}{2})A} \ , \end{aligned} \tag{3.16}$$

where the last line holds for $\alpha \neq \frac{1}{2}$ and is in fact a total derivative, which is not surprising considering the origin of this expression. Substituting in the explicit function $A(r)$, we then find

$$2\pi^2\, \epsilon_0^\dagger\epsilon_0[\frac{GM}{2} - Z] = \int_{\mathcal{M}} \nabla_\mu N^{\mu\nu} d\Sigma_\nu = 2\pi^2\, \epsilon_0^\dagger\epsilon_0[Pe^{-\phi_0} + Qe^{\phi_0}] \ , \tag{3.17}$$

which is independent of α as expected. Combining this with $[\frac{GM}{2} + Z] = 0$ appropriate to Killing spinors then gives an explicit derivation of the Bogomol'nyi bound,

$$GM = -2Z = Pe^{-\phi_0} + Qe^{\phi_0} \ , \tag{3.18}$$

for the gauge dyonic string.

So far we have not addressed the issue of what conditions are necessary to ensure the validity of the Bogomol'nyi bound. While the equations of motion are satisfied by construction, both Witten's condition and the positivity of the gauge function are not guaranteed. Examining first Witten's condition, we find

$$\gamma^i\delta_\epsilon\psi_i = \gamma^n\partial_n A[\alpha - \tfrac{1}{2} - \mathcal{P}_2^+]\epsilon \ . \tag{3.19}$$

Therefore, for Killing spinors, we choose $\alpha = \frac{1}{2}$ as noted previously in order to satisfy Witten's condition. On the other hand, we must choose $\alpha = \frac{3}{2}$ for the case of the fermion zero modes. Provided there are no naked singularities, this value of α gives rise to a well-behaved integral, so that there is no problem satisfying Witten's condition. However this is no longer the case whenever there are naked singularities. To see this, we note that such naked singularities develop whenever $2Pe^{-\phi_0} \leq -\rho^2$ or $2Qe^{\phi_0} \leq -\rho^2$ so that e^{-2A} vanishes for some $r^2 \geq 0$ [1]. Convergence of the volume integral near the singularity then requires $\alpha < -\frac{3}{2}$ (or $\alpha < -\frac{1}{2}$ for the case $Pe^{-\phi_0} = Qe^{\phi_0}$) which clearly indicates the incompatibility of Witten's condition with normalizable fermion zero modes whenever naked singularities are present.

Note that for any value of the mass given by Eqn. (3.18), it is always possible to avoid naked singularities in the gauge dyonic string by choosing a sufficiently large instanton size ρ. Therefore evasion of the Bogomol'nyi bound, Eqn. (3.7), is possible even without singularities. Whenever $M < 0$ we may see that the breakdown in Bogomol'nyi occurs because the Yang-Mills couplings have the wrong sign (this is already obvious because M itself is related to the gauge coupling at infinity). A quick check shows that this breakdown is also present for the tensionless ($M = 0$) quasi-anti-self-dual string where there is an exact cancellation between the contributions from the graviton and tensor multiplet fields and the wrong sign Yang-Mills fields. As shown below, this cancellation continues to hold when examining the energy integral for the tensionless string in the flat-space limit.

4 The flat-space limit

If the charges P and Q are such that $P = Qe^{2\phi_0}$, the anti-self-dual 3-form field strength and the dilaton decouple, *i.e.* $K^m_{\mu\nu\rho} = 0$, $\phi = \phi_0$. In other words, the matter multiplet decouples in this case, and we recover the self-dual string of [17]. On the other hand if $P = -Qe^{2\phi_0}$, the dyonic string becomes massless, as can be seen from (2.18). At first sight, one might think that in this case the self-dual 3-form $H_{\mu\nu\rho}$ and the metric of the gravity multiplet would be decoupled. However, this is not in fact what happens. This can easily be seen from the fact that the metric (2.15) does not become flat: indeed the $1/r^2$ terms cancel asymptotically, as they must since the solution is now massless, but the metric still has non-vanishing asymptotic deviations from Minkowski spacetime of order $1/r^4$. Similarly, the self-dual 3-form $H_{\mu\nu\rho}$ falls off as $1/r^4$. On the other hand, the fields $K^m_{\mu\nu\rho}$ and $\phi - \phi_0$ fall off as $1/r^2$ at large r. For this reason, the dyonic string in this limit should more appropriately be called quasi-anti-self-dual [1], rather than anti-self-dual. However, the solution becomes anti-self-dual asymptotically, since the self-dual part of the 3-form falls off faster by a factor of $1/r^2$.

The above discussion suggests that it should be possible to take the flat-space limit of the $N = (1,0)$ supergravity theory, and the quasi-anti-self-dual solution, where Newton's constant is set to zero. In fact, as we shall show below, there are actually two distinct limits that can be taken, yielding two inequivalent flat-space theories. To show this, we shall first construct the flat-space limit of the more general $N = 1$ supergravity coupled to an arbitrary number of anti-self-dual fields $K^m_{\mu\nu\rho}$. To do this, it is convenient to re-introduce Newton's constant κ in the supergravity theory, by rescaling the fields of the tensor multiplet in the following manner:

$$\begin{aligned} V = \begin{bmatrix} v_0 & v_M \\ x^m{}_0 & x^m{}_M \end{bmatrix} &\longrightarrow \begin{bmatrix} v_0 & \kappa v_M \\ \kappa x^m{}_0 & x^m{}_M \end{bmatrix}, \\ B^M_{\mu\nu} &\longrightarrow \kappa B^M_{\mu\nu}, \\ \chi^m &\longrightarrow \kappa \chi^m, \end{aligned} \tag{4.1}$$

while the fields of the Yang-Mills multiplet have not been rescaled, the coupling constants c^r are naturally dimensionless in the global limit, and hence must be rescaled according to

$$c^r \longrightarrow \kappa c^r \; . \tag{4.2}$$

Note that $P^m_\mu \to \kappa P^m_\mu$ under the rescalings. As a result of this rescaling, the equations of motion for the supergravity fields become

$$\begin{aligned}
G_{\mu\nu} - H_{\mu\rho\sigma} H_\nu{}^{\rho\sigma} &= \kappa^2 [K^m_{\mu\rho\sigma} K^{m\,\rho\sigma}_\nu + 2(P^m_\mu P^m_\nu - \tfrac{1}{2} g_{\mu\nu} P^m_\rho P^{m\,\rho}) \\
&\quad + \kappa[4(v_0 c^0 + \kappa v_M c^M)\,\mathrm{tr}\,(F_{\mu\lambda} F_\nu{}^\lambda - \tfrac{1}{4} g_{\mu\nu} F_{\lambda\sigma} F^{\lambda\sigma})] \\
dH &= -\kappa^2 \sqrt{2} P^m K^m + \kappa(v_0 c^0 + \kappa v_M c^M)\,\mathrm{tr}\,F^2 \\
\gamma^{\mu\nu\rho} \nabla_\nu \psi_\rho + H^{\mu\nu\rho} \gamma_\nu \psi_\rho &= \kappa^2 [\tfrac{i}{2} K^{m\,\mu\nu\rho} \gamma_{\nu\rho} \chi^m - \tfrac{i}{\sqrt{2}} P^m_\nu \gamma^\mu \gamma^\nu \chi^m] \\
&\quad - \kappa[\tfrac{1}{\sqrt{2}} \gamma^{\sigma\tau} \gamma^\mu (v_0 c^0 + \kappa v_M c^M)\,\mathrm{tr}\,F_{\sigma\tau}\lambda] \; ,
\end{aligned} \tag{4.3}$$

where now $H = v_0 \mathcal{H}^0 + \kappa^2 v_M \mathcal{H}^M$, indicating that in the limit $\kappa \to 0$ we may consistently set the gravity fields to their flat-space backgrounds,

$$g_{\mu\nu} \to \eta_{\mu\nu}, \qquad B^0_{\mu\nu} \to 0, \qquad \psi_\mu \to 0 \; . \tag{4.4}$$

Note that the terms proportional to c^0 (the coupling of Yang-Mills to the self-dual H) enter at $O(\kappa)$. This suggests the possibility that two different limits can arise; one where c^0/κ is held fixed, and the other where c^0 is non-vanishing and held fixed, as κ goes to zero. This may be made more transparent by examining the Yang-Mills equation of motion

$$D^\mu[(v_0 c^0 + \kappa v_M c^M) F_{\mu\nu}] = H_{\nu\rho\sigma}(v_0 c^0 + \kappa v_M c^M) F^{\rho\sigma} + \kappa K^m_{\nu\rho\sigma}(\kappa x^m{}_0 c^0 + x^m{}_M c^M) F^{\rho\sigma} \; , \tag{4.5}$$

from which we see that the $O(\kappa^0)$ terms survive only in the second limit, whilst the equation is of order κ in the first limit. Before proceeding with the flat-space limits, we note that the constrained vielbein matrix V simplifies greatly in the $\kappa \to 0$ limit, and the n_T degrees of freedom can be parametrized by scalar fields ϕ^m defined by $\delta^m_M v_M = x^m{}_0 = \phi^m$. The other components of V simply become $v_0 = 1$ and $x^m{}_M = \delta^m_M$. The two flat-space limits arise as follows:

Flat-space limit with c^0/κ fixed:

In this limit, it is natural to define $\tilde{c}^0 = c^0/\kappa$ before taking the flat-space limit. We see that there are now no κ-independent terms in (4.5), and we obtain a Yang-Mills equation that includes interactions with the anti-self-dual matter multiplets. We find that the complete set of flat-space equations is

$$\begin{aligned}
\Box \phi^m &= c^m \; \mathrm{tr}\,(F_{\mu\nu} F^{\mu\nu}) \; , \\
\partial^\mu K^m_{\mu\nu\rho} &= -\tfrac{1}{4} c^m \, \epsilon_{\mu\nu\rho\alpha\beta\gamma} \; \mathrm{tr}\,(F^{\mu\alpha} F^{\beta\gamma}) \; , \\
\gamma^\mu \partial_\mu \chi^m &= -\tfrac{i}{\sqrt{2}} c^m \; \mathrm{tr}\,(F_{\mu\nu} \gamma^{\mu\nu} \lambda) \; , \\
D^\mu[(\tilde{c}^0 + c^m \phi^m) F_{\mu\nu}] &= c^m F^{\rho\sigma} K^m_{\nu\rho\sigma} \; ,
\end{aligned} \tag{4.6}$$

$$(\tilde{c}^0 + c^m\phi^m)\gamma^\mu D_\mu\lambda \;=\; -\tfrac{1}{2}c^m(\partial_\mu\phi^m)\gamma^\mu\lambda - \tfrac{i}{2\sqrt{2}}c^m\, F_{\mu\nu}\gamma^{\mu\nu}\chi^m - \tfrac{1}{12}c^m K^m_{\mu\nu\rho}\gamma^{\mu\nu\rho}\lambda\,.$$

Note that the anti-self-dual field strengths are given by

$$K^m = dB^m + c^m\,\omega_3\;, \tag{4.7}$$

where $\omega_3 = AdA + \frac{2}{3}A^3$. The supersymmetry transformation rules in this flat-space limit become

$$\begin{aligned}
&\delta\phi^m = \bar{\epsilon}\chi^m\;, \qquad \delta A_\mu = -\tfrac{i}{\sqrt{2}}\bar{\epsilon}\gamma_\mu\lambda\;, \qquad \delta B^m_{\mu\nu} = \tfrac{1}{2}\bar{\epsilon}\gamma_{\mu\nu}\chi^m + 2c^m\,\mathrm{tr}\,(A_{[\mu}\delta A_{\nu]})\;,\\
&\delta\lambda = -\tfrac{1}{2\sqrt{2}}F_{\mu\nu}\gamma^{\mu\nu}\epsilon\;, \qquad \delta\chi^m = -\tfrac{i}{2}\partial_\mu\phi^m\,\gamma^\mu\epsilon + \tfrac{i}{12}K^m_{\mu\nu\rho}\,\gamma^{\mu\nu\rho}\epsilon\,.
\end{aligned} \tag{4.8}$$

The energy-momentum tensor for this flat-space theory may be obtained simply by applying the same limiting procedure to the right-hand side of the Einstein equation of the original supergravity theory, given in (4.3). By this means we obtain the flat-space expression

$$T_{\mu\nu} = K^m_{\mu\rho\sigma}\,K^{m\rho\sigma}_\nu + \partial_\mu\phi^m\,\partial_\nu\phi^m - \tfrac{1}{2}\eta_{\mu\nu}\,(\partial\phi^m)^2 + 4(\tilde{c}^0 + c^m\,\phi^m)\,\mathrm{tr}\,(F_{\mu\lambda}\,F_\nu{}^\lambda - \tfrac{1}{4}\eta_{\mu\nu}\,F_{\lambda\sigma}\,F^{\lambda\sigma})\,. \tag{4.9}$$

This is the theory that we refer to as the "interacting theory". It is interesting to note that the bosonic equations of motion of (4.6) can be derived from the Lagrangian

$$\mathcal{L} = -(\partial\phi^m)^2 - \tfrac{1}{3}K^2 - 2(\tilde{c}^0 + c^m\,\phi^m)\;\mathrm{tr}\,(F^2) - 2c^m *(B^m \wedge \mathrm{tr}\,(F\wedge F))\;, \tag{4.10}$$

where K is taken to be unconstrained, with its anti-self-duality being imposed only after having obtained the equations of motion.

Flat-space limit with c^0 held fixed:

The situation is different when c^0 is non-vanishing and is held fixed when κ goes to zero. As can be seen from (4.5), the leading-order terms in the Yang-Mills equation are now independent of κ, and in fact there are now no interactions with the anti-self-dual multiplets in the $\kappa \to 0$ limit. All equations of motion, and supersymmetry transformation rules, remain the same as in the previous $c^0 \sim \kappa$ limit with the exception of the Yang-Mills equations and the gaugino equation, which are now source-free and given by

$$\begin{aligned}
&D^\mu\,F_{\mu\nu} = 0\;,\\
&\gamma^\mu\,D_\mu\,\lambda = 0\,.
\end{aligned} \tag{4.11}$$

Note however that the energy-momentum tensor, again obtained from the right-hand side of the Einstein equation in (4.3) by applying the limiting procedure, is now simply given by

$$T_{\mu\nu} = 4c^0\;\mathrm{tr}\,(F_{\mu\lambda}\,F_\nu{}^\lambda - \tfrac{1}{4}\eta_{\mu\nu}\,F_{\lambda\sigma}\,F^{\lambda\sigma})\,. \tag{4.12}$$

In the case of a single self-dual tensor multiplet, this is the theory that we refer to as the "BSS theory".

A word of explanation is in order here. Firstly, it should be noted that this energy-momentum tensor arose as a term of order κ in the Einstein equation, rather than the usual order κ^2 for matter fields. Consequently the Yang-Mills contribution dominates the $O(\kappa^2)$ contributions from the tensor multiplets, and so they are absent in this flat-space limit. Indeed, it is evident that if one were to add "standard" contributions for the fields of the tensor multiplets, one would find that the resulting energy-momentum tensor was not conserved upon using the equations of motion. Effectively the tensor multiplets describe "test fields" in a Yang-Mills background, whose energy-momentum tensor is negligible in comparison to that of the Yang-Mills field. For the same reason, they do not affect the Yang-Mills equation. The energy-momentum tensor (4.12) would cease to be appropriate in a configuration where the Yang-Mills field was zero, since now the previously-neglected matter contributions would become important. This rather pathological feature of the BSS theory is reflected also in the fact that it cannot be described by an analogue of the Lagrangian (4.10), owing to the inherent asymmetry between the occurrence of interaction terms in the matter and Yang-Mills equations.

A number of further comments are also in order. Firstly, it should be emphasised that the higher-order fermi terms are not included in the equations of motion and supersymmetry transformation rules (4.6) and (4.8); they were not included in [26,27], and indeed they have only recently been computed [37]. (See also [36,38].) Nevertheless, one can see on general grounds that the inclusion of the higher-order terms in the supergravity theory will not present any obstacle in the taking of the two inequivalent flat-space limits. Alternatively, the higher-order completion of the supersymmetry transformations may be determined in either of the flat-space theories by demanding the closure of the supersymmetry algebra on the fermi fields. For the interacting theory the supersymmetry transformation rules for the bosons remain unchanged, while the complete transformation rules for the fermions are

$$\begin{aligned}\delta\chi^m &= -\tfrac{i}{2}[\partial_\mu\phi^m\gamma^\mu - \tfrac{1}{6}K^m_{\mu\nu\rho}\gamma^{\mu\nu\rho}]\epsilon - \tfrac{1}{2}c^m\,\mathrm{tr}\,[\gamma_\mu\lambda(\bar\epsilon\gamma^\mu\lambda)]\ , \\ \delta\lambda &= -\frac{1}{2\sqrt{2}}F_{\mu\nu}\gamma^{\mu\nu}\epsilon + \frac{c^m}{(\tilde c^0 + c^n\phi^n)}[-\tfrac{1}{2}(\bar\chi^m\lambda)\epsilon - \tfrac{1}{4}(\bar\chi^m\epsilon)\lambda + \tfrac{1}{8}(\bar\chi^m\gamma_{\mu\nu}\epsilon)\gamma^{\mu\nu}\lambda]\ ,\end{aligned} \tag{4.13}$$

and agree with the flat-space limit of the transformations in the supergravity theory [37]. On the other hand, in the BSS theory the gaugino variation remains unmodified, and only $\delta\chi^m$ picks up a higher-order correction (identical to that of the interacting theory). Note that it is straightforward to see that this must be the case, since the lowest-order transformation for the Yang-Mills multiplet, (4.8), already closes on the source-free gaugino equation of motion, (4.11). So in fact we see that the only difference in the supersymmetry transformation rules in the two flat-space limits is in the higher-order terms in the gaugino variation, consistent with the difference in the equations of motion for the Yang-Mills multiplet between the two limits.

The complete gaugino transformation rule in the interacting theory is somewhat unusual, in that it contains a possibly singular denominator, $(\tilde c^0 + c^n\phi^n)$ (which was also noted in

[36,37]). As in the supergravity situation, this singular denominator is just a manifestation of the strong coupling singularity already present in the lowest-order Yang-Mills equations. This form of the denominator also shows up in the complete equations of motion, given for the fermi fields in the interacting theory by[4]

$$
\begin{aligned}
\gamma^\mu \partial_\mu \chi^m &= -\tfrac{i}{\sqrt{2}} c^m \operatorname{tr}(F_{\mu\nu}\gamma^{\mu\nu}\lambda) + \frac{i c^m c^n}{(\tilde{c}^0 + c^p \phi^p)} \operatorname{tr}[\tfrac{3}{2}(\bar{\chi}^n\lambda)\lambda - \tfrac{1}{4}(\bar{\chi}^n\gamma_{\mu\nu}\lambda)\gamma^{\mu\nu}\lambda] , \\
(\tilde{c}^0 + c^m\phi^m)\gamma^\mu D_\mu \lambda &= -\tfrac{1}{2} c^m (\partial_\mu \phi^m)\gamma^\mu \lambda - \tfrac{i}{2\sqrt{2}} c^m F_{\mu\nu}\gamma^{\mu\nu}\chi^m - \tfrac{1}{12} c^m K^m_{\mu\nu\rho}\gamma^{\mu\nu\rho}\lambda \\
&\quad -i \frac{c^m c^n}{(\tilde{c}^0 + c^p\phi^p)} [\tfrac{3}{4}(\bar{\lambda}\chi^m)\chi^n - \tfrac{1}{8}(\bar{\lambda}\gamma_{\mu\nu}\chi^m)\gamma^{\mu\nu}\chi^n] \\
&\quad + i\alpha c^m c^{m\prime} \operatorname{tr}'[(\bar{\lambda}\gamma_\mu\lambda')\gamma^\mu\lambda'] , \qquad (4.14)
\end{aligned}
$$

where the primes in the last line indicate the quantities involved in the trace. (Recall that there can be different c^m constants for each factor in a semi-simple group.) Note that α is an arbitrary parameter that is not fixed by the supersymmetry algebra [37], and appears to be related to the gauge anomaly (see [37] for a more complete discussion).

We also note that the flat-space limit when c^0 is non-vanishing and held fixed, if we specialise to the case where there is only a single anti-self-dual multiplet, coincides with the BSS theory, constructed in [18]. It was argued in [18] that this theory could not be obtained as a $\kappa \to 0$ limit of the supergravity theory, on the grounds that the Chern-Simons form ω enters the 3-form field strengths in (2.3) with a factor of κ (after restoring Newton's constant, as in (4.2)), and thus it would disappear in the flat-space limit. However, while this is indeed the case for the self-dual field of the gravity multiplet, the potentials $B^m_{\mu\nu}$ for the anti-self-dual matter fields also acquire factors of κ, with the net result that the Chern-Simons terms are of the same order, and hence they survive in the $\kappa \to 0$ limit, as we saw in (4.7) above. The non-standard dimensions of the energy-momentum tensor (4.12) is a reflection of the need for a dimensionful free parameter, which was also seen in [18]. Finally, we remark that the more general flat-space theory we obtained in the limit where $c^0 \sim \kappa$, does not conflict with the results in [18] which found only the free Yang-Mills equations (4.11), since in [18] it was assumed that the kinetic term for the Yang-Mills multiplet was described by the standard superspace free action. Note also that the BSS theory can be obtained from the interacting flat-space theory by taking k to zero after making the following rescalings of the fields of the interacting theory:

$$
\phi^m \to k\,\phi^m , \qquad B^m_{\mu\nu} \to k\,B^m_{\mu\nu} , \qquad \chi^m \to k\,\chi^m , \qquad (4.15)
$$

together with the rescaling $c^m \to k\,c^m$. Thus the interacting flat-space theory encompasses the BSS theory as a singular limiting case.

Let us now consider the flat-space limit of the quasi-anti-self-dual dyonic string solution (2.15) of the supergravity theory. This solution is massless, and hence from (2.18) it follows

[4]While these equations of motion were obtained by taking the flat-space limit of [37], they equally well follow from closure of the supersymmetry algebra, (4.13).

that the magnetic charge is related to the electric charge by $P = -Q\, e^{2\phi_0}$. Consequently, the parameters $c^0 = (Q+P)/32$ and $c^1 = (Q-P)/32$ are given by

$$c^0 = -\tfrac{1}{16} Q\, e^{\phi_0} \sinh\phi_0 \,, \qquad c^1 = \tfrac{1}{16} Q\, e^{\phi_0} \cosh\phi_0 \,. \tag{4.16}$$

In the flat-space limit, where in particular ϕ was rescaled by κ, we see that $c^0 = -\frac{1}{16} Q\, \kappa\, \phi_0$ prior to sending κ to zero, and hence we are in the regime of the "interacting theory", corresponding to the first of the two limits discussed above. We find that the flat-space solution is given by

$$\begin{aligned} \phi &= \phi_0 - \frac{Q(2\rho^2 + r^2)}{(\rho^2 + r^2)^2} \\ K_{mnp} &= -\tfrac{1}{2}\epsilon_{mnpq}\partial_q \phi \,, \qquad K_{\mu\nu m} = -\tfrac{1}{2}\epsilon_{\mu\nu}\partial_m \phi \,, \\ F^a &= \frac{2\rho^2}{(\rho^2 + r^2)^2}\, \eta^a_{mn}\, dy^m \wedge dy^n \,. \end{aligned} \tag{4.17}$$

Note that ϕ_0 no longer has physical significance as a coupling constant, and it can be eliminated by making a constant shift of ϕ.

Since this solution has been obtained as the flat-space limit of a tensionless string, we expect that it should have vanishing energy. This might at first sight seem surprising, since it is described by a non-trivial field configuration. However, a straightforward calculation of T_{00} given by (4.9) yields

$$\begin{aligned} T_{00} &= K_{00}^2 + \tfrac{1}{2}(\partial\phi)^2 + \tfrac{1}{16} Q\, (\phi - \phi_0)\, \mathrm{tr}\,(F^2) \,, \\ &= \frac{4Q^2\, r^2\, (3\rho^2 + r^2)^2}{(\rho^2 + r^2)^6} - \frac{24 Q^2\, \rho^4\, (2\rho^2 + r^2)}{(\rho^2 + r^2)^6} \,, \end{aligned} \tag{4.18}$$

where the first term in the second line comes from the (equal) contributions from K and ϕ, and the second term comes from F. It is easily verified that while T_{00} itself is non-vanishing, the integral $\int_0^\infty T_{00}\, r^3\, dr$ is equal to zero. Clearly the Yang-Mills field is giving a negative contribution to the energy, in precisely such a way that the total energy is zero. This is the flat-space analogue of the cancellation that occurs in the supergravity theory, with its associated subtleties in the Bogomol'nyi analysis, which we discussed at the end of section 3.

It should be emphasised that the vanishing energy of the flat-space tensionless string occurs for arbitrary scale size ρ of the Yang-Mills instanton. However, if we consider instead the neutral tensionless string, which can be achieved by setting $\rho = 0$ so that the instanton is not present, then the expression (4.18) becomes $T_{00} = 4Q^2/r^6$, whose integral over the transverse space diverges at the core of the string. Thus the Yang-Mills instanton in the gauge-dyonic string can be viewed as a regulator for the total energy.

There are also massive string solutions to the interacting flat-space theory, which can also be obtained as flat-space limits of the curved-space gauge dyonic string. They arise by taking the ADM mass, as given by (2.18), to be non-zero and of the form $m_0\, \kappa$. Upon taking the $\kappa \to 0$ flat-space limit, this gives a solution of the same form as (4.17), but with

ϕ shifted by the constant m_0. From (4.18), this gives an extra term in T_{00} which gives rise to an energy m_0 per unit length for the flat-space string.

It is interesting to note that while the flat-space limit of the tensionless string always results in the $c^0 \sim \kappa$ limit of the interacting theory, the final solution itself, as given in (4.17), also satisfies the equations of motion of the BSS theory, where c^0 is held fixed in the flat-space limit[5]. To see this, we note that for a bosonic background, only the Yang-Mills equation differs between the two flat-space theories. In particular, both Yang-Mills equations may be expressed as $D^\mu F_{\mu\nu} = J_\nu$, where the current is

$$J_\mu = [c^m F_{\mu\nu} \partial^\nu \phi^m + c^m F^{\rho\sigma} K^m_{\mu\rho\sigma}]/[\tilde{c}^0 + c^n \phi^n] \ , \qquad (4.19)$$

for the first theory, and vanishes for the latter. Because of the form of the solution, (4.17), we see that J_μ identically vanishes, and hence the background is indeed a solution to both flat-space limits. Furthermore, examination of the BPS conditions arising from (4.8) indicates that $J_\mu = 0$ for any string-like background preserving half of the supersymmetries. It should be remarked, however, that when interpreted as a solution to the c^0 fixed flat-space limit, the string no longer has vanishing energy per unit length, since in this case the stress tensor (4.12) has only a positive contribution from the Yang-Mills instanton. Thus only the interacting theory from the first flat-space limit, (4.6), provides a suitable description of the tensionless string in flat space.

Finally, we note that by taking the divergence of (4.19), we obtain

$$D^\mu J_\mu = \tfrac{1}{8} c^m c^{m\prime} \epsilon_{\mu\nu\rho\sigma\eta\lambda} F^{\mu\nu} \,\mathrm{tr}' F^{\rho\sigma\prime} F^{\eta\lambda\prime}/[\tilde{c}^0 + c^n \phi^n] \ , \qquad (4.20)$$

indicating that the current is not conserved classically. Thus the inconsistency of the supergravity theory, which we discussed in section 3, survives in the "interacting" flat-space limit. Nevertheless since, as for the gauge dyonic string in curved space, J_μ vanishes identically for the global gauge string, this classical inconsistency does not spoil the solution. On the other hand, since the BSS theory is free of this inconsistency it is possible that such a classical inconsistency, necessary for anomaly cancellation in the quantum theory, is an integral part of a fully interacting theory.

We have not paid much attention in this paper to the question of gravitational anomalies which is always an important issue when dealing with *chiral* theories. In particular, we have for simplicity ignored the presence of hypermultiplets. A coupled supergravity-matter theory which is initailly free of gravitational anomalies theory will not remain so when the gravity multiplet is switched off because the contributions from the gravitino and self-dual 2-form, necessary for the anomaly cancellation, are no longer present. Naively, of course, one could argue that gravitational anomalies are no longer of any concern in the flat space limit. However, it may be that subtleties arise when one tries to take the global limit of a

[5]The solution (4.17) has also been obtained in the BSS theory by directly solving its first-order BPS equations [39].

worldvolume theory which relies for its anomaly freedom on anomaly inflow from the bulk. This is deserving of further study.

In conclusion, we note that six-dimensional global models are also important as fivebrane worldvolume theories [19–21, 18]. In the absence of Yang-Mills fields, the (1, 0) multiplet is the only one available to describe the worldvolume theory of the $D = 7$, $N = 1$ fivebrane solution found in [40]. Six-dimensional global models also arise from configurations of higher-dimensional branes with six worldvolume dimensions in common. Indeed, the brane configurations yielding (1, 0) theories with tensor multiplets, vector multiplets and hypermultiplets have been identified in [24, 25], although no field equations were written down. Here we speculate further that the interacting anti-self-dual-tensor Yang-Mills system given in this paper (together with hypermultiplets where necessary) is the appropriate one to describe these global models. The global gauge anti-self-dual string, and in particular the tensionless string, could then be regarded as strings on the worldvolume.

Note added

Global $D = 6$, (1, 0) models of the type discussed in this paper have recently been shown to arise from configurations of NS fivebranes, Dirichlet sixbranes and eightbranes [41]

Acknowledgment.

MJD thanks E. Sezgin, E. Verlinde and E. Witten, and JTL thanks V.P. Nair, for useful discussions.

References

[1] M.J. Duff, H. Lü and C.N. Pope, *Heterotic phase transitions and singularities of the gauge dyonic string*, Phys. Lett. **B378** (1996) 101, hep-th/9603037.

[2] E. Witten, *Some comments on string dynamics, Contributed to STRINGS 95: Future Perspectives in String Theory, Los Angeles, CA, March 1995*, hep-th/9507121.

[3] M.J. Duff, Electric/magnetic duality and its stringy origins, Int. J. Mod. Phys. **A11** (1996) 4031, hep-th/9509106

[4] H. Lü and C.N. Pope, *p-brane solitons in maximal supergravities*, Nucl. Phys. **B465** (1996) 127, hep-th/9512012.

[5] A. Strominger, *Open p-branes*, Phys. Lett. **B383** (1996) 44, hep-th/9512059.

[6] P.K. Townsend, *D-branes from M-branes*, Phys. Lett. **B373** (1996) 68, hep-th/9512062.

[7] M. Douglas, *Branes within branes*, hep-th/9512077.

[8] M.B. Green, *Worldvolumes and string target spaces*, Fortsch. Phys. **44** (1996) 551, hep-th/9602061.

[9] K. Becker and M. Becker, *Boundaries in M-theory*, Nucl. Phys. **B472** (1996) 221, hep-th/9602071.

[10] E. Witten, *Fivebranes and M-theory on a orbifold*, Nucl. Phys. **B474** (1996) 122, hep-th/9602120.

[11] N. Seiberg and E. Witten, *Comments on string dynamics in six dimensions*, Nucl. Phys. **B471** (1996) 121, hep-th/9603003.

[12] O. Aharony, J. Sonnenschein and S. Yankielowicz, *Interactions of strings and D-branes from M-theory*, Nucl. Phys. **B474** (1996) 309, hep-th/9603009.

[13] O. Ganor and A. Hanany, *Small E_8 instantons and tensionless non-critical strings*, Nucl. Phys. **B476** (1996) 437, hep-th/9603161.

[14] M.J. Duff, R. Minasian and E. Witten, *Evidence for heterotic/heterotic duality*, Nucl. Phys. **B465** (1996) 413, hep-th/9601036

[15] L. Ibanez and A. M. Uranga , *D=6 N=1 string vacua and duality*, hep-th/9707075.

[16] M.J. Duff, S. Ferrara, R.R. Khuri and J. Rahmfeld, *Supersymmetry and dual string solitons*, Phys. Lett. **B356** (1995) 479, hep-th/9506057.

[17] M.J. Duff and J.X. Lu, *Black and super p-branes in diverse dimensions*, Nucl. Phys. **B416** (1994) 301, hep-th/9306052.

[18] E. Bergshoeff, E. Sezgin and E. Sokatchev, *Couplings of selfdual tensor multiplet in six-dimensions*, Class. Quant. Grav. **13** (1996) 2875, hep-th/9605087.

[19] R. Dijkgraaf, E. Verlinde and H. Verlinde, *BPS quantization of the fivebrane*, Nucl. Phys. **B486** (1996) 89, hep-th/9604055.

[20] N. Seiberg, *New theories in six dimensions and matrix description of M-theory on T^5 and T^5/Z_2*, Phys. Lett. **B408** (1997) 98, hep-th/9705221.

[21] J.S. Schwarz, *Self-dual string in six dimensions*, hep-th/9604171.

[22] M. Perry and J.S. Schwarz, *Interacting chiral gauge fields in six dimensions and Born-Infeld theory*, Nucl. Phys. **B489** (1997) 47, hep-th/9611065.

[23] P.S. Howe, N.D. Lambert and P.C. West, *The self-dual string soliton*, hep-th/9709014.

[24] A. Hanany and A. Zaffaroni, *Chiral symmetry from Type IIA branes*, hep-th/9706047.

[25] K. Intriligator, *New string theories in six dimensions via branes at orbifold singularities*, hep-th/9708117.

[26] L.J. Romans, *Self-duality for interacting fields*, Nucl. Phys. **B276** (1986) 71.

[27] A. Sagnotti, *A note on the Green-Schwarz mechanism in open string theories*, Phys. Lett. **B294** (1992) 196, hep-th/9210127.

[28] A. Dabholkar, G.W. Gibbons, J.A. Harvey and F. Ruiz-Ruiz, *Superstrings and solitons*, Nucl. Phys. **B340** (1990) 33.

[29] J.M. Nester, *A new gravitational energy expression with a simple positivity proof*, Phys. Lett. **A83** (1981) 241.

[30] J.A. Harvey and J. Liu, *Magnetic monopoles in $N = 4$ supersymmetric low-energy superstring theory*, Phys. Lett. **B268** (1991) 40.

[31] M. Cvetič and D. Youm, *Singular BPS saturated states and enhanced symmetries of four-dimensional $N = 4$ supersymmetric string vacua*, Phys. Lett. **B359** (1995) 87, hep-th/9507160.

[32] K.-L. Chan and M. Cvetič, *Massless BPS-saturated states on the two-torus moduli sub-space of heterotic string*, Phys. Lett. **B375** (1996) 98, hep-th/9512188.

[33] E. Witten, *A new proof of the positive energy theorem*, Commum. Math. Phys. **80** (1981) 381.

[34] D. Olive, *The electric and magnetic charges as extra components of four-momentum*, Nucl. Phys. **B153** (1979) 1.

[35] S. Ferrara, R. Minasian and A. Sagnotti, *Low-energy analysis of M and F theories on Calabi-Yau threefolds*, Nucl. Phys. **B474** (1996) 323, hep-th/9604097.

[36] H. Nishino and E. Sezgin, *New couplings of six-dimensional supergravity*, Nucl. Phys. **B505** (1997) 497, hep-th/9703075.

[37] S. Ferrara, F. Riccioni and A. Sagnotti, *Tensor and vector multiplets in six-dimensional supergravity*, hep-th/9711059.

[38] G. Dall'Agata, K. Lechner and M. Tonin, *Covariant actions for $N = 1, D = 6$ supergravity theories with chiral bosons*, hep-th/9710127.

[39] V. P. Nair and S. Randjbar-Daemi, *Solitons in a six-dimensional super Yang-Mills-tensor system and non-critical strings*, hep-th/9711125.

[40] H. Lü, C.N. Pope, E. Sezgin and K.S.Stelle, *Dilatonic p-brane solitons*, Phys. Lett. **B371** (1996) 46, hep-th/9511203.

[41] A. Hanany and A. Zaffaroni, *Branes and six dimensional supersymmetric theories*, hep-th/9712145.

SECTION IV
Particle Masses

QUARK MASS HIERARCHY AND FLAVOR MIXING

Harold Fritzsch

Sektion Physik, Universität München,
D–80333 München, Germany

ABSTRACT

In view of the observed strong hierarchy of the quark and lepton masses and of the flavor mixing angles it is argued that the description of flavor mixing must take this into account. One particular interesting way to describe the flavor mixing, which, however, is not the one used today, emerges, which is best suited for models of quark mass matrices based on flavor symmetries. We conclude that the unitarity triangle important for B physics should be close to or identical to a rectangular triangle. CP violation is maximal in this sense.

Invited talk given at the
International Conference on Orbis Scientiae 1997:
Physics of Mass
Miami Beach, Florida (December 12–15, 1997)

The phenomenon of flavor mixing, which is intrinsically linked to CP–violation, is an important ingredient of the Standard Model of Basic Interactions. Yet unlike other features of the Standard Model, e. g. the mixing of the neutral electroweak gauge bosons, it is a phenomenon which can merely be described. A deeper understanding is still lacking, but most theoreticians would agree that it is directly linked to the mass spectrum of the quarks – the possible mixing of lepton flavors will not be discussed here. Furthermore there is a general consensus that a deeper dynamical understanding would require to go beyond the physics of the Standard Model. In this talk I shall not go thus far. Instead I shall demonstrate that the observed properties of the flavor mixing, combined with our knowledge about the quark mass spectrum, suggest specific symmetry properties which allow to fix the flavor mixing parameters with high precision, thus predicting the outcome of the experiments which will soon be performed at the B–meson factories.

In the standard electroweak theory, the phenomenon of flavor mixing of the quarks is described by a 3×3 unitary matrix, the Cabibbo-Kobayashi-Maskawa (CKM) matrix[1,2]. This matrix can be expressed in terms of four parameters, which are usually taken as three rotation angles and one phase. A number of different parametrizations have been proposed in the literature[2,3,4,5]. Of course, adopting a particular parametrization of flavor mixing is arbitrary and not directly a physics issue. Nevertheless it is quite likely that the actual values of flavor mixing parameters (including the strength of CP violation), once they are known with high precision, will give interesting information about the physics beyond the standard model. Probably at this point it will turn out that a particular description of the CKM matrix is more useful and transparent than the others. For this reason, let me first analyze all possible parametrizations and point out their respective advantages and disadvantages.

In the standard model the quark flavor mixing arises once the up- and down-type mass matrices are diagonalized. The generation of quark masses is intimately related to the phenomenon of flavor mixing. In particular, the flavor mixing parameters do depend on the elements of quark mass matrices. A particular structure of the underlying mass matrices calls for a particular choice of the parametrization of the flavor mixing matrix. For example, in it was noticed[6] that a rather special form of the flavor mixing matrix results, if one starts from Hermitian mass matrices in which the (1,3) and (3,1) elements vanish. This has been subsequently observed again in a number of papers[7]. Recently we have studied the exact form of such a description from a general point of view and pointed out many advantages of this type of representation in the discussion of flavor mixing and CP-violating phenomena[5], which will be discussed later.

In the standard model the weak charged currents are given by

$$\overline{(u,\ c,\ t)_L} \begin{pmatrix} V_{ud} & V_{us} & V_{ub} \\ V_{cd} & V_{cs} & V_{cb} \\ V_{td} & V_{ts} & V_{tb} \end{pmatrix} \begin{pmatrix} d \\ s \\ b \end{pmatrix}_L , \tag{1}$$

where u, c, ..., b are the quark mass eigenstates, L denotes the left-handed fields, and V_{ij} are elements of the CKM matrix V. In general V_{ij} are complex numbers, but their absolute values are measurable quantities. For example, $|V_{cb}|$ primarily determines the lifetime of B mesons. The phases of V_{ij}, however, are not physical, like the phases of quark fields. A phase transformation of the u quark ($u \rightarrow u\ e^{i\alpha}$), for example, leaves the quark mass term invariant but changes the elements in the first row of V (i.e., $V_{uj} \rightarrow V_{uj}\ e^{-i\alpha}$). Only a common phase transformation of all quark fields leaves all elements of V invariant, thus there is a five-fold freedom to adjust the phases of V_{ij}.

In general the unitary matrix V depends on nine parameters. Note that in the absence of complex phases V would consist of only three independent parameters, corresponding to three (Euler) rotation angles. Hence one can describe the complex matrix V by three angles and six phases. Due to the freedom in redefining the quark field phases, five of the six phases in V can be absorbed, and we arrive at the well-known result that the CKM matrix V can be parametrized in terms of three rotation angles and one CP-violating phase. The question about how many different ways to describe V may exist was raised some time ago[8]. Recently the problem was reconsidered and brought in connection with the mass hierarchy[5].

In our view the best possibility to describe the flavor mixing in the standard model is to adopt the parametrization discussed in ref. (5). This parametrization has a number of significant advantages. In the following part I shall show that this parametrization follows automatically if we impose the constraints from the chiral symmetries and the hierarchical structure of the mass eigenvalues[9,10,11]. We take the point of view that the quark mass eigenvalues are dynamical entities, and one could change their values in order to study certain symmetry limits, as it is done in QCD. In the standard electroweak model, in which the quark mass matrices are given by the coupling of a scalar field to various quark fields, this can certainly be done by adjusting the related coupling constants. Whether it is possible in reality is an open question. It is well-known that the quark mass matrices can always be made hermitian by a suitable transformation of the right-handed fields. Without loss of generality, we shall suppose in this paper that the quark mass matrices are hermitian. In the limit where the masses of the u and d quarks are set to zero, the quark mass matrix $\tilde{M}$ (for both charge $+2/3$ and charge $-1/3$ sectors) can be arranged such that its elements $\tilde{M}_{i1}$ and $\tilde{M}_{1i}$ ($i = 1, 2, 3$) are all zero[9,10]. Thus the quark mass matrices have the form

$$\tilde{M} = \begin{pmatrix} 0 & 0 & 0 \\ 0 & \tilde{C} & \tilde{B} \\ 0 & \tilde{B}^* & \tilde{A} \end{pmatrix} . \tag{2}$$

The observed mass hierarchy is incorporated into this structure by denoting the entry which is of the order of the t-quark or b-quark mass by $\tilde{A}$, with $\tilde{A} \gg \tilde{C}, |\tilde{B}|$. It can easily be seen (see, e.g., ref. (12) that the complex phases in the mass matrices (1) can be rotated away by subjecting both $\tilde{M}_{\rm u}$ and $\tilde{M}_{\rm d}$ to the same unitary transformation. Thus we shall take $\tilde{B}$ to be real for both up- and down-quark sectors. As expected, CP violation cannot arise at this stage. The diagonalization of the mass matrices leads to a mixing between the second and third families, described by an angle $\tilde{\theta}$. The flavor mixing matrix is then given by

$$\tilde{V} = \begin{pmatrix} 1 & 0 & 0 \\ 0 & \tilde{c} & \tilde{s} \\ 0 & -\tilde{s} & \tilde{c} \end{pmatrix} , \tag{3}$$

where $\tilde{s} \equiv \sin\tilde{\theta}$ and $\tilde{c} \equiv \cos\tilde{\theta}$. In view of the fact that the limit $m_u = m_d = 0$ is not far from reality, the angle $\tilde{\theta}$ is essentially given by the observed value of $|V_{cb}|$ ($= 0.039 \pm 0.002$)[13,14]; i.e., $\tilde{\theta} = 2.24° \pm 0.12°$.

At the next and final stage of the chiral evolution of the mass matrices, the masses of the u and d quarks are introduced. The Hermitian mass matrices have in general the form:

$$M = \begin{pmatrix} E & D & F \\ D^* & C & B \\ F^* & B^* & A \end{pmatrix} \tag{4}$$

with $A \gg C, |B| \gg E, |D|, |F|$. By a common unitary transformation of the up- and down-type quark fields, one can always arrange the mass matrices $M_{\rm u}$ and $M_{\rm d}$ in such a way that $F_{\rm u} = F_{\rm d} = 0$; i.e.,

$$M = \begin{pmatrix} E & D & 0 \\ D^* & C & B \\ 0 & B^* & A \end{pmatrix} . \qquad (5)$$

This can easily be seen as follows. If phases are neglected, the two symmetric mass matrices M_u and M_d can be transformed by an orthogonal transformation matrix O, which can be described by three angles such that they assume the form (5). The condition $F_u = F_d = 0$ gives two constraints for the three angles of O. If complex phases are allowed in M_u and M_d, the condition $F_u = F_u^* = F_d = F_d^* = 0$ imposes four constraints, which can also be fulfilled, if M_u and M_d are subjected to a common unitary transformation matrix U. The latter depends on nine parameters. Three of them are not suitable for our purpose, since they are just diagonal phases; but the remaining six can be chosen such that the vanishing of F_u and F_d results.

The basis in which the mass matrices take the form (5) is a basis in the space of quark flavors, which in our view is of special interest. It is a basis in which the mass matrices exhibit two texture zeros, for both up- and down-type quark sectors. These, however, do not imply special relations among mass eigenvalues and flavor mixing parameters (as pointed out above). In this basis the mixing is of the "nearest neighbour" form, since the (1,3) and (3,1) elements of M_u and M_d vanish; no direct mixing between the heavy t (or b) quark and the light u (or d) quark is present (see also ref. (15). In certain models (see, e.g., ref. (16), this basis is indeed of particular interest, but we shall proceed without relying on a special texture models for the mass matrices.

A mass matrix of the type (5) can in the absence of complex phases be diagonalized by a rotation matrix, described by two angles only. At first the off-diagonal element B is rotated away by a rotation between the second and third families (angle θ_{23}); at the second step the element D is rotated away by a transformation of the first and second families (angle θ_{12}). No rotation between the first and third families is required. The rotation matrix for this sequence takes the form

$$\begin{aligned} R = R_{12} R_{23} &= \begin{pmatrix} c_{12} & s_{12} & 0 \\ -s_{12} & c_{12} & 0 \\ 0 & 0 & 1 \end{pmatrix} \begin{pmatrix} 1 & 0 & 0 \\ 0 & c_{23} & s_{23} \\ 0 & -s_{23} & c_{23} \end{pmatrix} \\ &= \begin{pmatrix} c_{12} & s_{12}c_{23} & s_{12}s_{23} \\ -s_{12} & c_{12}c_{23} & c_{12}s_{23} \\ 0 & -s_{23} & c_{23} \end{pmatrix} , \end{aligned} \qquad (6)$$

where $c_{12} \equiv \cos\theta_{12}$, $s_{12} \equiv \sin\theta_{12}$, etc. The flavor mixing matrix V is the product of two such matrices, one describing the rotation among the up-type quarks, and the other describing the rotation among the down-type quarks:

$$V = R_{12}^u R_{23}^u (R_{23}^d)^{-1} (R_{12}^d)^{-1} . \qquad (7)$$

The product $R_{23}^u (R_{23}^d)^{-1}$ can be written as a rotation matrix described by a single angle θ. In the limit $m_u = m_d = 0$, this is just the angle $\tilde{\theta}$ encountered in eq. (6). The angle which describes the R_{12}^u rotation shall be denoted by θ_u; the corresponding angle for the R_{12}^d rotation by θ_d. Thus in the absence of CP-violating phases the flavor mixing matrix takes the following specific form:

$$
\begin{aligned}
V &= \begin{pmatrix} c_u & s_u & 0 \\ -s_u & c_u & 0 \\ 0 & 0 & 1 \end{pmatrix} \begin{pmatrix} 1 & 0 & 0 \\ 0 & c & s \\ 0 & -s & c \end{pmatrix} \begin{pmatrix} c_d & -s_d & 0 \\ s_d & c_d & 0 \\ 0 & 0 & 1 \end{pmatrix} \\
&= \begin{pmatrix} s_u s_d c + c_u c_d & s_u c_d c - c_u s_d & s_u s \\ c_u s_d c - s_u c_d & c_u c_d c + s_u s_d & c_u s \\ -s_d s & -c_d s & c \end{pmatrix} , \qquad (8)
\end{aligned}
$$

where $c_u \equiv \cos\theta_u$, $s_u \equiv \sin\theta_u$, etc.

We proceed by including the phase parameters of the quark mass matrices in eq. (5). It can easily be seen that, by suitable rephasing of the quark fields, the flavor mixing matrix can finally be written in terms of only a single phase φ as follows:

$$
\begin{aligned}
V &= \begin{pmatrix} c_u & s_u & 0 \\ -s_u & c_u & 0 \\ 0 & 0 & 1 \end{pmatrix} \begin{pmatrix} e^{-i\varphi} & 0 & 0 \\ 0 & c & s \\ 0 & -s & c \end{pmatrix} \begin{pmatrix} c_d & -s_d & 0 \\ s_d & c_d & 0 \\ 0 & 0 & 1 \end{pmatrix} \\
&= \begin{pmatrix} s_u s_d c + c_u c_d e^{-i\varphi} & s_u c_d c - c_u s_d e^{-i\varphi} & s_u s \\ c_u s_d c - s_u c_d e^{-i\varphi} & c_u c_d c + s_u s_d e^{-i\varphi} & c_u s \\ -s_d s & -c_d s & c \end{pmatrix} . \qquad (9)
\end{aligned}
$$

Note that the three angles θ_u, θ_d and θ in eq. (12) can all be arranged to lie in the first quadrant through a suitable redefinition of quark field phases. Consequently all s_u, s_d, s and c_u, c_d, c are positive. The phase φ can in general take values from 0 to 2π; and CP violation is present in the weak interactions if $\varphi \neq 0, \pi$ and 2π.

This particular representation of the flavor mixing matrix is the main result of this paper. In comparison with all other parametrizations discussed previously[2,3], it has a number of interesting features which in our view make it very attractive and provide strong arguments for its use in future discussions of flavor mixing phenomena, in particular, those in B-meson physics (see also refs. (17, 18)). We shall discuss them below.

a) The flavor mixing matrix V in eq. (12) follows directly from the chiral expansion of the mass matrices. Thus it naturally takes into account the hierarchical structure of the quark mass spectrum.

b) The complex phase φ describing CP violation appears only in the (1,1), (1,2), (2,1) and (2,2) elements of V, i.e., in the elements involving only the quarks of the first and second families. This is a natural description of CP violation since in our hierarchical approach CP violation is not directly linked to the third family, but rather to the first and second ones, and in particular to the mass terms of the u and d quarks.

It is instructive to consider the special case $s_u = s_d = s = 0$. Then the flavor mixing matrix V takes the form

$$
V = \begin{pmatrix} e^{-i\varphi} & 0 & 0 \\ 0 & 1 & 0 \\ 0 & 0 & 1 \end{pmatrix} . \qquad (10)
$$

This matrix describes a phase change in the weak transition between u and d, while no phase change is present in the transitions between c and s as well as t and b. Of course, this effect can be absorbed in a phase change of the u- and d-quark fields, and no CP violation is present. Once the angles $\theta_{\rm u}$, $\theta_{\rm d}$ and θ are introduced, however, CP violation arises. It is due to a phase change in the weak transition between u' and d', where u' and d' are the rotated quark fields, obtained by applying the corresponding rotation matrices given in eq. (9) to the quark mass eigenstates (u': mainly u, small admixture of c; d': mainly d, small admixture of s).

Since the mixing matrix elements involving the t or b quarks are real in the representation (9), one can find that the phase parameter of B_q^0-$\bar{B}_q^0$ mixing ($q = d$ or s), dominated by the box-diagram contributions in the standard model[19], is essentially unity:

$$\left(\frac{q}{p}\right)_{B_q} = \frac{V_{tb}^* V_{tq}}{V_{tb} V_{tq}^*} = 1 \, . \tag{11}$$

In most other parametrizations of the flavor mixing matrix, however, the two rephasing-variant quantities $(q/p)_{B_d}$ and $(q/p)_{B_s}$ take different (maybe complex) values.

c) The dynamics of flavor mixing can easily be interpreted by considering certain limiting cases in eq. (9). In the limit $\theta \rightarrow 0$ (i.e., $s \rightarrow 0$ and $c \rightarrow 1$), the flavor mixing is, of course, just a mixing between the first and second families, described by only one mixing angle (the Cabibbo angle $\theta_{\rm C}$[1]).
It is a special and essential feature of the representation (9) that the Cabibbo angle is *not* a basic angle, used in the parametrization. The matrix element V_{us} (or V_{cd}) is indeed a superposition of two terms including a phase. This feature arises naturally in our hierarchical approach, but it is not new. In many models of specific textures of mass matrices, it is indeed the case that the Cabibbo-type transition V_{us} (or V_{cd}) is a superposition of several terms. At first, it was obtained by one of the authors in the discussion of the two-family mixing[20].

In the limit $\theta = 0$ considered here, one has $|V_{us}| = |V_{cd}| = \sin\theta_{\rm C} \equiv s_{\rm C}$ and

$$s_{\rm C} = \left| s_{\rm u} c_{\rm d} - c_{\rm u} s_{\rm d} e^{-i\varphi} \right| . \tag{12}$$

This relation describes a triangle in the complex plane, as illustrated in Fig. 1, which we shall denote as the "LQ– triangle" ("light quark triangle"). This triangle is a feature of the mixing of the first two families (see also ref. (20)). Explicitly one has (for $s = 0$):

$$\tan\theta_{\rm C} = \sqrt{\frac{\tan^2\theta_{\rm u} + \tan^2\theta_{\rm d} - 2\tan\theta_{\rm u}\tan\theta_{\rm d}\cos\varphi}{1 + \tan^2\theta_{\rm u}\tan^2\theta_{\rm d} + 2\tan\theta_{\rm u}\tan\theta_{\rm d}\cos\varphi}} \, . \tag{13}$$

Certainly the flavor mixing matrix V cannot accommodate CP violation in this limit. However, the existence of φ seems necessary in order to make eq. (16) compatible with current data, as one can see below.

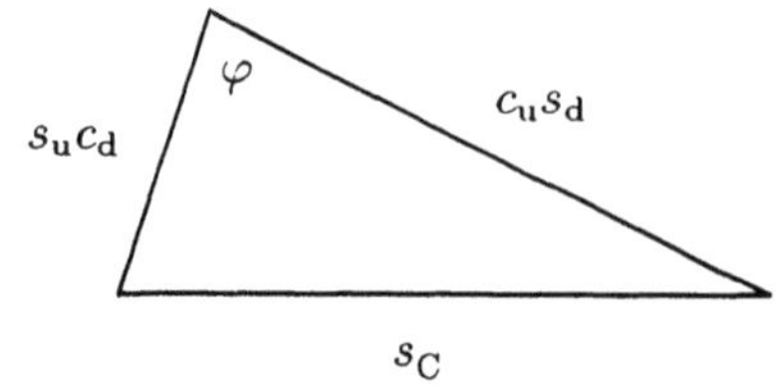

Abbildung 1: The LQ–triangle in the complex plane.

d) The three mixing angles θ, θ_u and θ_d have a precise physical meaning. The angle θ describes the mixing between the second and third families, which is generated by the off-diagonal terms B_u and B_d in the up and down mass matrices of eq. (9). We shall refer to this mixing involving t and b as the "heavy quark mixing". The angle θ_u, however, primarily describes the u-c mixing, corresponding to the D_u term in M_u. We shall denote this as the "u-channel mixing". The angle θ_d primarily describes the d-s mixing, corresponding to the D_d term in M_d; this will be denoted as the "d-channel mixing". Thus there exists an asymmetry between the mixing of the first and second families and that of the second and third families, which in our view reflects interesting details of the underlying dynamics of flavor mixing. The heavy quark mixing is a combined effect, involving both charge $+2/3$ and charge $-1/3$ quarks, while the u- or d-channel mixing (described by the angle θ_u or θ_d) proceeds solely in the charge $+2/3$ or charge $-1/3$ sector. Therefore an experimental determination of these two angles would allow to draw interesting conclusions about the amount and perhaps the underlying pattern of the u- or d-channel mixing.

e) The three angles θ, θ_u and θ_d are related in a very simple way to observable quantities of B-meson physics.
For example, θ is related to the rate of the semileptonic decay $B \to D^* l \nu_l$; θ_u is associated with the ratio of the decay rate of $B \to (\pi, \rho) l \nu_l$ to that of $B \to D^* l \nu_l$; and θ_d can be determined from the ratio of the mass difference between two B_d mass eigenstates to that between two B_s mass eigenstates. From eq. (9) we find the following exact relations:

$$\sin\theta = |V_{cb}| \sqrt{1 + \left|\frac{V_{ub}}{V_{cb}}\right|^2} , \tag{14}$$

and

$$\begin{aligned} \tan\theta_u &= \left|\frac{V_{ub}}{V_{cb}}\right| , \\ \tan\theta_d &= \left|\frac{V_{td}}{V_{ts}}\right| . \end{aligned} \tag{15}$$

These simple results make the parametrization (9) uniquely favorable for the study of B-meson physics.

By use of current data on $|V_{ub}|$ and $|V_{cb}|$, i.e., $|V_{cb}| = 0.039 \pm 0.002$[13,14] and $|V_{ub}/V_{cb}| = 0.08 \pm 0.02$[14], we obtain $\theta_u = 4.57° \pm 1.14°$ and $\theta = 2.25° \pm 0.12°$. Taking $|V_{td}| = (8.6 \pm 2.1) \times 10^{-3}$[14], which was obtained from the analysis of current data on B_d^0-$\bar{B}_d^0$ mixing, we get $|V_{td}/V_{ts}| = 0.22 \pm 0.07$, i.e., $\theta_d = 12.7° \pm 3.8°$. Both the heavy quark mixing angle θ and the u-channel mixing angle θ_u are relatively small. The smallness of θ implies that Eqs. (11) and (12) are valid to a high degree of precision (of order $1 - c \approx 0.001$).

f) According to eq. (12), as well as eq. (11), the phase φ is a phase difference between the contributions to V_{us} (or V_{cd}) from the u-channel mixing and the d-channel mixing. Therefore φ is given by the relative phase of D_d and D_u in the quark mass matrices (4), if the phases of B_u and B_d are absent or negligible.

The phase φ is not likely to be 0° or 180°, according to the experimental values given above, even though the measurement of CP violation in K^0-$\bar{K}^0$ mixing[19] is not taken into account. For $\varphi = 0°$, one finds $\tan\theta_C = 0.14 \pm 0.08$; and for $\varphi = 180°$, one gets $\tan\theta_C = 0.30 \pm 0.08$. Both cases are barely consistent with the value of $\tan\theta_C$ obtained from experiments ($\tan\theta_C \approx |V_{us}/V_{ud}| \approx 0.226$).

g) The CP-violating phase φ in the flavor mixing matrix V can be determined from $|V_{us}|$ ($= 0.2205 \pm 0.0018$)[19] through the following formula, obtained easily from eq. (8):

$$\varphi = \arccos\left(\frac{s_u^2 c_d^2 c^2 + c_u^2 s_d^2 - |V_{us}|^2}{2 s_u c_u s_d c_d c}\right) . \tag{16}$$

The two–fold ambiguity associated with the value of φ, coming from $\cos\varphi = \cos(2\pi - \varphi)$, is removed if one takes $\sin\varphi > 0$ into account (this is required by current data on CP violation in K^0-$\bar{K}^0$ mixing (i.e., ϵ_K). More precise measurements of the angles θ_u and θ_d in the forthcoming experiments of B physics will reduce the uncertainty of φ to be determined from eq. (19). This approach is of course complementary to the direct determination of φ from CP asymmetries in some weak B-meson decays into hadronic CP eigenstates[21].

Considering the presently known phenomenological constraints (see e.g. Ref.[22]) the value of φ is most likely in the range 40° to 120°; the central value is $\varphi \approx 81°$. Note that φ is essentially independent of the angle θ, due to the tiny observed value of the latter. Once $\tan\theta_d$ is precisely measured, one shall be able to fix the magnitude of φ to a satisfactory degree of accuracy.

h) It is well–known that CP violation in the flavor mixing matrix V can be rephasing–invariantly described by a universal quantity $\mathcal{J}$[23]:

$$\mathrm{Im}\left(V_{il}V_{jm}V_{im}^{*}V_{jl}^{*}\right) = \mathcal{J}\sum_{k,n=1}^{3}\left(\epsilon_{ijk}\epsilon_{lmn}\right] . \tag{17}$$

In the parametrisation (9), $\mathcal{J}$ reads

$$\mathcal{J} = s_u c_u s_d c_d s^2 c sin\varphi \tag{18}$$

Obviously $\varphi = 90°$ leads to the maximal value of $\mathcal{J}$. Indeed $\varphi = 90°$, a particularly interesting case for CP violation, is quite consistent with current data. In this case the mixing term D_d in eq. (5) can be taken to be real, and the term D_u to be imaginary, if $\mathrm{Im}(B_u) = \mathrm{Im}(B_d) = 0$ is assumed. Since in our description of the flavor mixing the complex phase φ is related in a simple way to the phases of the quark mass terms, the case $\varphi = 90°$ is especially interesting. It can hardly be an accident, and this case should be studied further. The possibility that the phase φ describing CP violation in the standard model is given by the algebraic number $\pi/2$ should be taken seriously. It may provide a useful clue towards a deeper understanding of the origin of CP violation and of the dynamical origin of the fermion masses.

In ref. (5) the case $\varphi = 90°$ has been denoted as "maximal" CP violation. It implies in our framework that in the complex plane the u-channel and d-channel mixings are perpendicular to each other. In this special case (as well as $\theta \to 0$), we have

$$\tan^2\theta_C = \frac{\tan^2\theta_u + \tan^2\theta_d}{1 + \tan^2\theta_u \tan^2\theta_d} . \tag{19}$$

To a good approximation (with the relative error $\sim 2\%$), one finds $s_C^2 \approx s_u^2 + s_d^2$.

i) At future B-meson factories, the study of CP violation will concentrate on measurements of the unitarity triangle.
The unitzarity triangle (a) and its rescaled counterpart (b) in the complex plane.

$$S_u + S_c + S_t = 0 , \tag{20}$$

where $S_i \equiv V_{id}V_{ib}^*$ in the complex plane (see Fig. 2(a)). The inner angles of this triangle. are denoted as[19].

$$\begin{aligned} \alpha &\equiv \arg(-S_t S_u^*) , \\ \beta &\equiv \arg(-S_c S_t^*) , \\ \gamma &\equiv \arg(-S_u S_c^*) . \end{aligned} \tag{21}$$

In terms of the parameters θ, $\theta_{\rm u}$, $\theta_{\rm d}$ and φ, we obtain

$$\begin{aligned} \sin(2\alpha) &= \frac{2c_{\rm u}c_{\rm d}\sin\varphi\,(s_{\rm u}s_{\rm d}c + c_{\rm u}c_{\rm d}\cos\varphi)}{s_{\rm u}^2 s_{\rm d}^2 c^2 + c_{\rm u}^2 c_{\rm d}^2 + 2s_{\rm u}c_{\rm u}s_{\rm d}c_{\rm d}c\cos\varphi} , \\ \sin(2\beta) &= \frac{2s_{\rm u}c_{\rm d}\sin\varphi\,(c_{\rm u}s_{\rm d}c - s_{\rm u}c_{\rm d}\cos\varphi)}{c_{\rm u}^2 s_{\rm d}^2 c^2 + s_{\rm u}^2 c_{\rm d}^2 - 2s_{\rm u}c_{\rm u}s_{\rm d}c_{\rm d}c\cos\varphi} . \end{aligned} \tag{22}$$

To an excellent degree of accuracy, one finds $\alpha \approx \varphi$. In order to illustrate how accurate this relation is, let us input the central values of θ, $\theta_{\rm u}$ and $\theta_{\rm d}$ (i.e., $\theta = 2.25°$, $\theta_{\rm u} = 4.57°$ and $\theta_{\rm d} = 12.7°$) to eq. (22). Then one arrives at $\varphi - \alpha \approx 1°$ as well as $\sin(2\alpha) \approx 0.34$ and $\sin(2\beta) \approx 0.65$. It is expected that $\sin(2\alpha)$ and $\sin(2\beta)$ will be directly measured from the CP asymmetries in $B_d \rightarrow \pi^+\pi^-$ and $B_d \rightarrow J/\psi K_S$ modes at a B-meson factory.

Note that the three sides of the unitarity triangle (21) can be rescaled by $|V_{cb}|$. In a very good approximation (with the relative error $\sim 2\%$), one arrives at

$$|S_u| : |S_c| : |S_t| \approx s_{\rm u}c_{\rm d} : s_{\rm C} : s_{\rm d} . \tag{23}$$

Equivalently, one can obtain

$$s_\alpha : s_\beta : s_\gamma \approx s_{\rm C} : s_{\rm u}c_{\rm d} : s_{\rm d} , \tag{24}$$

where $s_\alpha \equiv \sin\alpha$, etc. Comparing the unitarity triangle with the LQ–triangle in Fig. 1, we find that they are indeed congruent with each other to a high degree of accuracy. The congruent relation between these two triangles is particularly interesting, since the LQ–triangle is essentially a feature of the physics of the first two quark families, while the unitarity triangle is linked to all three families. In this connection it is of special interest to note that in models which specify the textures of the mass matrices the Cabibbo triangle and hence three inner angles of the unitarity triangle can be fixed by the spectrum of the light quark masses and the CP-violating phase φ.

j) It is worth pointing out that the u-channel and d-channel mixing angles are related to the so-called Wolfenstein parameters[23] in a simple way:

$$\begin{aligned} \tan\theta_{\rm u} &= \left|\frac{V_{ub}}{V_{cb}}\right| \approx \lambda\sqrt{\rho^2 + \eta^2} , \\ \tan\theta_{\rm d} &= \left|\frac{V_{td}}{V_{ts}}\right| \approx \lambda\sqrt{(1-\rho)^2 + \eta^2} , \end{aligned} \tag{25}$$

where $\lambda \approx s_C$ measures the magnitude of V_{us}. Note that the CP-violating parameter η is linked to φ through

$$\sin\varphi \approx \frac{\eta}{\sqrt{\rho^2+\eta^2}\sqrt{(1-\rho)^2+\eta^2}} \tag{26}$$

in the lowest-order approximation. Then $\varphi = 90°$ implies $\eta^2 \approx \rho(1-\rho)$, on the condition $0 < \rho < 1$. In this interesting case, of course, the flavor mixing matrix can fully be described in terms of only three independent parameters.

k) Compared with the standard parametrization of the flavor mixing matrix V the parametrization (9) has an additional advantage: the renormalization-group evolution of V, from the weak scale to an arbitrary high energy scale, is to a very good approximation associated only with the angle θ. This can easily be seen if one keeps the t and b Yukawa couplings only and neglects possible threshold effect in the one loop renormalization group equations of the Yukawa matrices[24]. Thus the parameters $\theta_{\rm u}$, $\theta_{\rm d}$ and φ are essentially independent of the energy scale, while θ does depend on it and will change if the underlying scale is shifted, say from the weak scale ($\sim 10^2$ GeV) to the grand unified theory scale (of order 10^{16} GeV). In short, the heavy quark mixing is subject to renormalization-group effects; but the u- and d-channel mixings are not, likewise the phase φ describing CP violation and the LQ–triangle as a whole.

We have presented a new description of the flavor mixing phenomenon, which is based on the phenomenological fact that the quark mass spectrum exhibits a clear hierarchy pattern. This leads uniquely to the interpretation of the flavor mixing in terms of a heavy quark mixing, followed by the u-channel and d-channel mixings. The complex phase φ, describing the relative orientation of the u-channel mixing and the d-channel mixing in the complex plane, signifies CP violation, which is a phenomenon primarily linked to the physics of the first two families. The Cabibbo angle is not a basic mixing parameter, but given by a superposition of two terms involving the complex phase φ. The experimental data suggest that the phase φ, which is directly linked to the phases of the quark mass terms, is close to 90°. This opens the possibility to interpret CP violation as a maximal effect, in a similar way as parity violation.

Our description of flavor mixing has many clear advantages compared with other descriptions. We propose that it should be used in the future description of flavor mixing and CP violation, in particular, for the studies of quark mass matrices and B-meson physics.

The description of the flavor mixing phenomenon given above is of special interest if for the U and D channel mixing the quark mass textures discussed first in[20] are applied (see also[5]). In that case one finds[25] (apart from small corrections)

$$\tan\Theta_d = \sqrt{\frac{m_d}{m_s}} \tag{27}$$

$$\tan\Theta_u = \sqrt{\frac{m_u}{m_c}} \, .$$

The experimental value for $\tan \Theta_u$ given by the ratio V_{ub}/V_{cb} is in agreement with the observed value for $(m_u/m_c)^{1/2} \approx 0.07$, but the errors for both $(m_u/m_c)^{1/2}$ and V_{ub}/V_{cb} are the same (about 25%). Thus from the underlying texture no new information is obtained.

This is not true for the angle Θ_d, whose experimental value is due to a large uncertainty.: $\Theta_d = 12.7° \pm 3.8°$. If Θ_d is given indeed by the square root of the quark mass ratio m_d/m_s, which is known to a high accuracy, we would know Θ_d and therefore all four parameters of the CKM matrix with high precision.

As emphasized in ref. (5), the phase angle φ is very close to 90°, implying that the LQ–triangle and the unitarity triangle are essentially rectangular triangles. In particular the angle β which is likely to be measured soon in the study of the reaction $B^\circ \rightarrow J/\psi K_s^\circ$ is expected to be close to 20°.

It will be very interesting to see whether the angles Θ_d and Θ_u are indeed given by the square roots of the light quark mass ration m_d/m_s and m_u/m_c, which imply that the phase φ is close to or exactly 90°. This would mean that the light quarks play the most important rôle in the dynamics of flavor mixing and CP violation and that a small window has been opened allowing the first view accross the physics landscape beyond the mountain chain of the Standard Model.

REFERENCES

1. N. Cabibbo, *Phys. Rev. Lett.* 10 (1963) 531.
2. M. Kobayashi and T. Maskawa, *Prog. Theor. Phys.* 49 (1973) 652.
3. See, e.g., L. Maiani, *in: Proc. 1977 Int. Symp. on Lepton and Photon Interactions at High Energies (DESY, Hamburg,* (1977), 867;
   L.L. Chau and W.Y. Keung, *Phys. Rev. Lett.* 53 (1984) 1802;
   H. Fritzsch, *Phys. Rev.* D32 (1985) 3058;
   H. Harari and M. Leurer, *Phys. Lett.* B181 (1986) 123;
   H. Fritzsch and J. Plankl, *Phys. Rev.* D35 (1987) 1732.
4. F.J. Gilman, Kleinknecht, and B. Renk, *Phys. Rev.* D 54 (1996) 94.
5. H. Fritzsch and Z.Z. Xing, *Phys. Lett.* B353 (1995) 114.
6. H. Fritzsch, *Nucl. Phys.* B155 (1979) 189.
7. S. Dimopoulos, L.J. Hall, and S. Raby, *Phys. Rev.* D45 (1992) 4192;
   L.J. Hall and A. Rasin, Phys. Lett. B315 (1993) 164;
   R. Barbieri, L.J. Hall, and A. Romanino, *Phys. Lett.* B401 (1997) 47.
8. C. Jarlskog, *in CP Violation, edited by C. Jarlskog (World Scientific, 1989)*, p. 3.
9. H. Fritzsch, *Phys. Lett.* B184 *(1987)* 391.
10. H. Fritzsch, *Phys. Lett.* B189 *(1987)* 191.
11. L.J. Hall and S. Weinberg, *Phys. Rev.* D48 (1993) 979.
12. H. Lehmann, C. Newton, and T.T. Wu, *Phys. Lett.* B384 (1996) 249.

13. M. Neubert, *Int. J. Mod. Phys.* A11 (1996) 4173.

14. R. Forty, talk given at the Second International Conference on B Physics and CP Violation, Honolulu, Hawaii, March 24 - 27, 1997.

15. G.C. Branco, L. Lavoura, and F. Mota, *Phys. Rev.* D39 (1989) 3443.

16. S. Dimopoulos, L.J. Hall, and S. Raby, *Phys. Rev. Lett.* 68 (1992) 1984; Phys. R. Barbieri, L.J. Hall, and A. Romanino, Phys. Lett. B401 (1997) 47.

17. H. Fritzsch and X. Xing, *Phys. Rev.* D57 (1998) 594–597

18. A. Rasin, Report No. hep-ph/9708216.

19. Particle Data Group, R.M. Barnett *et al.*, *Phys. Rev.* D54 (1996) 1.

20. H. Fritzsch, *Phys. Lett.* B70 (1977) 436; B73 (1978) 317.

21. A.B. Carter and A.I. Sanda, *Phys. Rev. Lett.* 45 (1980) 952; I.I. Bigi and A.I. Sanda, *Nucl. Phys.* B193 (1981) 85.

22. A.J. Buras, *Report No. MPI-PhT/96-111*; and references therein.

23. L. Wolfenstein, Phys. *Rev. Lett.* 51 (1983) 1945.

24. See, e.g., K.S. Babu and Q. Shafi, *Phys. Rev.* D47 (1993) 5004; and references therein.

25. H. Fritzsch and Z. Xing, in preparation.

QUARK MASSES, B-PARAMETERS, AND CP VIOLATION PARAMETERS ϵ AND ϵ'/ϵ

Rajan Gupta

Group T-8, Mail Stop B-285,
Los Alamos National Laboratory
Los Alamos, NM 87545, U. S. A
Email: rajan@qcd.lanl.gov

1 Introduction

The least well quantified parameters of the Standard Model (SM) are the masses of light quarks and the ρ and η parameters in the Wolfenstein representation of the CKM mixing matrix. A non-zero value of η signals CP violation. The important question is whether the CKM ansatz explains all observed CP violation. This can be addressed by comparing the SM estimates of the two CP violating parameters ϵ and ϵ'/ϵ against experimental measurements. The focus of this talk is to evaluate the dependence of these parameters on the light quark masses and on the bag parameters B_K, $B_6^{1/2}$, and $B_8^{3/2}$. I will therefore provide a status report on the estimates of these quantities from lattice QCD (LQCD).

Since this is the only talk presenting results obtained using LQCD at this conference, I have been asked to give some introduction to the subject. The only way I can cover my charter, introduce LQCD, summarize the results, and make contact with phenomenology is to skip details. I shall try to overcome this shortcoming by giving adequate pointers to relevant literature.

2 Lattice QCD

LQCD calculations are a non-perturbative implementation of field theory using the Feynman path integral approach. The calculations proceed exactly as if the field theory was being solved analytically had we the ability to do the calculations. The starting point is the partition function in Euclidean space-time

$$Z = \int \mathcal{D}A_\mu \, \mathcal{D}\psi \, \mathcal{D}\overline{\psi} \, e^{-S} \tag{1}$$

Physics of Mass
Edited by Kursunoglu *et al.*, Kluwer Academic / Plenum Publishers, New York, 1999

where S is the QCD action

$$S = \int d^4x \, (\frac{1}{4} F_{\mu\nu} F^{\mu\nu} - \overline{\psi} M \psi) \, . \tag{2}$$

and M is the Dirac operator. The fermions are represented by Grassmann variables ψ and $\overline{\psi}$. These can be integrated out exactly with the result

$$Z = \int \mathcal{D} A_\mu \, \det M \, e^{\int d^4x \, (-\frac{1}{4} F_{\mu\nu} F^{\mu\nu})} . \tag{3}$$

The fermionic contribution is now contained in the highly non-local term $\det M$, and the partition function is an integral over only background gauge configurations. One can write the action, after integration over the fermions, as $S = S_{gauge} + S_{quarks} = \int d^4x \, (\frac{1}{4} F_{\mu\nu} F^{\mu\nu}) - \sum_i \mathrm{Ln}(\mathrm{Det} M_i)$ where the sum is over the quark flavors distinguished by the value of the bare quark mass. Results for physical observables are obtained by calculating expectation values

$$\langle \mathcal{O} \rangle = \frac{1}{Z} \int \mathcal{D} A_\mu \, \mathcal{O} \, e^{-S} \, . \tag{4}$$

where $\mathcal{O}$ is any given combination of operators expressed in terms of time-ordered products of gauge and quark fields. The quarks fields in $\mathcal{O}$ are, in practice, re-expressed in terms of quark propagators using Wick's theorem for contracting fields. In this way all dependence on quarks as dynamical fields is removed. The basic building block for the fermionic quantities, the Feynman propagator, is given by

$$S_F(y, j, b; x, i, a) = (M^{-1})^{y,j,b}_{x,i,a} \, , \tag{5}$$

where M^{-1} is the inverse of the Dirac operator calculated on a given background field. A given element of this matrix $(M^{-1})^{y,j,b}_{x,i,a}$ is the amplitude for the propagation of a quark from site x with spin-color i, a to site-spin-color y, j, b.

So far all of the above is standard field theory. The problem we face in QCD is how to actually calculate these expectation values and how to extract physical observables from these. I will illustrate the second part first by using as an example the mass and decay constant of the pion.

Consider the 2-point correlation function, $\langle 0 | \sum_x \mathcal{O}_f(\vec{x}, t) \mathcal{O}_i(\vec{0}, 0) | 0 \rangle$, where the operators $\mathcal{O}$ are chosen to be the fourth component of the axial current $\mathcal{O}_f = \mathcal{O}_i = A_4 = \overline{\psi} \gamma_4 \gamma_5 \psi$ as these have a large coupling to the pion. The 2-point correlation function then gives the amplitude for creating a state with the quantum numbers of the pion out of the vacuum at space-time point 0 by the "source" operator $\mathcal{O}_i$; the evolution of this state to the point $(\vec{x}, t)$ via the QCD Hamiltonian; and finally the annihilation by the "sink" operator $\mathcal{O}_f$ at $(\vec{x}, t)$. The rules of quantum mechanics tell us that $\mathcal{O}_i$ will create a state that is a linear combination of all possible eigenstates of the Hamiltonian that have the same quantum numbers as the pion, *i.e.* the pion, radial excitations of the pion, three pions in $J = 0$ state, ... The second rule is that on propagating for Euclidean time t, a given eigenstate with energy E picks up a weight e^{-Et}. Thus, the 2-point function can be written in terms of a sum over all possible intermediate states

$$\langle 0 | \sum_x \mathcal{O}_f(\vec{x}, t) \mathcal{O}_i(0) | 0 \rangle = \sum_n \frac{\langle 0 | \mathcal{O}_f | n \rangle \langle n | \mathcal{O}_i | 0 \rangle}{2 E_n} \, e^{-E_n t} \, . \tag{6}$$

To study the properties of the pion at rest we need to isolate this state from the sum over n. To do this, the first simplification is to use the Fourier projection $\sum_{\vec{x}}$ as it restricts the sum over states to just zero-momentum states, so $E_n \to M_n$. (Note that it is sufficient to make the Fourier projection over either $\mathcal{O}_i$ or $\mathcal{O}_f$.) The second step to isolate the pion, *i.e.* project in the energy, consists of a combination of two strategies. One, make a clever choice of the operators $\mathcal{O}_i$ to limit the sum over states to a single state (the ideal choice is to set $\mathcal{O}_i$ equal to the quantum mechanical wave-function of the pion), and two, examine the large t behavior of the 2-point function where only the contribution of the lowest energy state that couples to $\mathcal{O}_i$ is significant due to the exponential damping. Then

$$\langle 0|\sum_x \mathcal{O}_f(x,t)\mathcal{O}_i(0)|0\rangle \underset{t\to\infty}{=} \frac{\langle 0|\mathcal{O}_f|\pi\rangle\langle\pi|\mathcal{O}_i|0\rangle}{2M_\pi} e^{-M_\pi t}. \tag{7}$$

The right hand side is now a function of the two quantities we want since $\langle 0|A_4|\pi\rangle = M_\pi f_\pi$. In this way, the mass and the decay constant are extracted from the rate of exponential fall-off in time and from the amplitude.

Let me now illustrate how the left hand side is expressed in terms of the two basic quantities we control in the path integral – the gauge fields and the quark propagator. Using Wick contractions, the correlation function can be written in terms of a product of two quark propagators S_F,

$$\langle 0|\sum_x \overline{\psi}\gamma_4\gamma_5\psi(x,t)\overline{\psi}\gamma_4\gamma_5\psi(0,0)|0\rangle \equiv \langle 0|\sum_x S_F(0;\vec{x},t)\gamma_4\gamma_5 S_F(\vec{x},t;0)\gamma_4\gamma_5|0\rangle. \tag{8}$$

This correlation function is illustrated in Fig. 1. It is important to note that one recovers the 2-point function corresponding to the propagation of the physical pion only after the functional integral over the gauge fields, as defined in Eq. 4, is done. To illustrate this Wick contraction procedure further, consider using gauge invariant non-local operators, for example using $\mathcal{O} = \overline{\psi}(x,t)\gamma_4\gamma_5(\mathcal{P}e^{\int_x^y dz igA_\mu(z)})\psi(y,t)$ where $\mathcal{P}$ stands for path-ordered. After Wick contraction the correlation function reads

$$\langle 0|\sum_x S_F(0;\vec{x},t)\gamma_4\gamma_5(\mathcal{P}e^{\int_z^x igA_\mu})S_F(\vec{z},t;\vec{y},0)\gamma_4\gamma_5(\mathcal{P}e^{\int_0^y igA_\mu})|0\rangle. \tag{9}$$

and involves both the gauge fields and quark propagators. This correlation function would have the same long t behavior as shown in Eq. 6, however, the amplitude will be different and consequently its relation to f_π will no longer be simple. The idea of improving the projection of $\mathcal{O}$ on to the pion is to construct a suitable combination of such operators that approximates the pion wave-function.

To implement such calculations of correlation functions requires the following steps. A way of generating the background gauge configurations and calculating the action S associated with each; calculating the Feynman propagator on such background fields; constructing the desired correlation functions; doing the functional integral over the gauge fields to get expectation values; making fits to these expectation values, say as a function of t as in Eq. 6 to extract the mass and decay constant; and finally including any renormalization factors needed to properly define the physical quantity. It turns out that at present the only first principles approach that allows us to perform these steps is LQCD. Pedagogical expose to LQCD can be found in [1, 2, 3, 4], and I shall only give a very brief description here.

Lattice QCD – QCD defined on a finite space-time grid – serves two purposes. One, the discrete space-time lattice serves as a non-perturbative regularization scheme. At finite values of the lattice spacing a, which provides the ultraviolet cutoff, there are no infinities. Furthermore, renormalized physical quantities have a finite well behaved limit as $a \to 0$. Thus, in principle, one could do all the standard perturbative calculations using lattice regularization, however, these calculations are far more complicated and have no advantage over those done in a continuum scheme. The pre-eminent utility of transcribing QCD on the lattice is that LQCD can be simulated on the computer using methods analogous to those used in Statistical Mechanics. These simulations allow us to calculate correlation functions of hadronic operators and matrix elements of any operator between hadronic states in terms of the fundamental quark and gluon degrees of freedom following the steps discussed above.

The only tunable input parameters in these simulations are the strong coupling constant and the bare masses of the quarks. Our belief is that these parameters are prescribed by some yet more fundamental underlying theory, however, within the context of the standard model they have to be fixed in terms of an equal number of experimental quantities. This is what is done in LQCD. Thereafter all predictions of LQCD have to match experimental data if QCD is the correct theory of strong interactions.

A summary of the main points in the calculations of expectation values via simulations of LQCD are as follows.

- The Yang-Mills action for gauge fields and the Dirac operator for fermions has to be transcribed on to the discrete space-time lattice in such a way as to preserve all the key properties of QCD – confinement, asymptotic freedom, chiral symmetry, topology, and a one-to-one relation between continuum and lattice fields. This step is the most difficult, and even today we do not have a really satisfactory lattice formulation that is chirally symmetric in the $m_q = 0$ limit and preserves the one-to-one relation between continuum and lattice fields, *i.e.* no doublers. In fact, the Nielson-Ninomiya theorem states that for a translationally invariant, local, hermitian formulation of the lattice theory one cannot simultaneously have chiral symmetry and no doublers [5]. One important consequence of this theorem is that, in spite of tremendous effort, there is no viable formulation of chiral fermions on the lattice. For a review of the problems and attempts to solve them see [6, 7, 8]. A second problem is encountered when approximating derivatives in the action by finite differences. As is well known this introduces discretization errors proportional to the lattice spacing a. These errors can be reduced by either using higher order difference schemes with coefficients adjusted to take into account effects of renormalization, or equivalently, by adding appropriate combinations of irrelevant

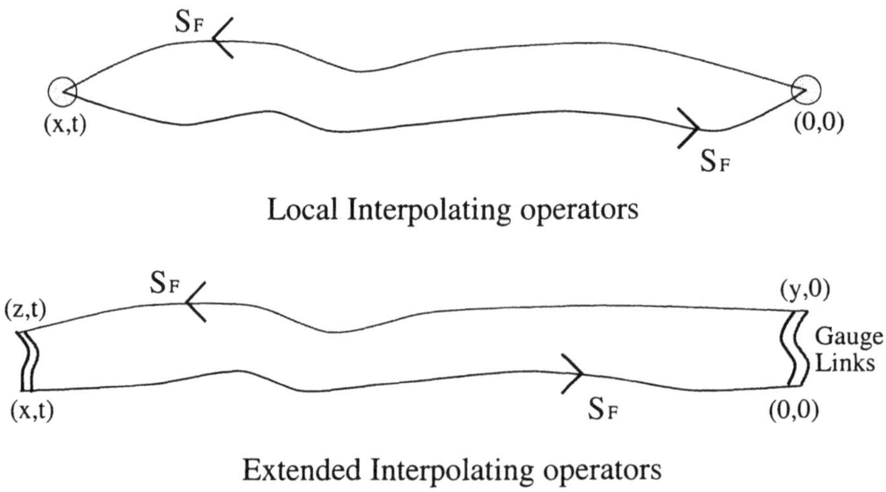

Figure 1. A schematic of the pion 2-point correlation function for local and non-local interpolating operators.

operators to the action that cancel the errors order by order in a. The various approaches to improving the fermion and gauge actions are discussed in [9, 10, 11]. Here I simply list the three most frequently used discretizations of the Dirac action – Wilson [12], Sheikholeslami-Wohlert (clover) [13], and staggered [14], which have errors of $O(a)$, $O(\alpha_s a) - O(a^2)$ depending on the value of the coefficient of the clover term, and $O(a^2)$ respectively. The important point to note is that while there may not yet exist a perfect action (no discretization errors) for finite a, improvement of the action is very useful but not necessary. Even the simplest formulation, Wilson's original gauge and fermion action [12], gives the correct results in the $a = 0$ limit. It is sufficient to have the ability to reliably extrapolate to $a = 0$ to quantify and remove the discretization errors.

- The Euclidean action $S = \int d^4x(\,\frac{1}{4}F_{\mu\nu}F^{\mu\nu} - \mathrm{TrLn}M)$ for QCD at zero chemical potential is real and bounded from below. Thus e^{-S} in the path integral is analogous to the Boltzmann factor in the partition function for statistical mechanics systems, *i.e.* it can be regarded as a probability weight for generating configurations. Since S is an extensive quantity the configurations that dominate the functional integral are those that minimize the action. The "importance sampled" configurations (configurations with probability of occurrence given by the weight e^{-S}) can be generated by setting up a Markov chain in exact analogy to say simulations of the Ising model. For a discussion of the methods used to update the configurations see [1] or the lectures by Creutz and Sokal in [2].
- The correlation functions are expressed as a product of quark propagators and path ordered product of gauge fields using Wick contractions. This part of the calculation is standard field theory. The only twist is that the calculation is done in Euclidean space-time.
- For a given background gauge configuration, the Feynman quark propagator is a matrix labeled by three indices – site, spin and color. A given element of this matrix gives the amplitude for the propagation of a quark with some spin, color, and space-time point to another space-time point, spin, and color. Operationally, it is simply the inverse of the Dirac operator. Once space-time is made discrete and finite, the Dirac matrix is also finite and its inverse can be calculated numerically. The gauge fields live on links between the sites with the identification $U_\mu(x, x+\hat{\mu}) = e^{iagA_\mu(x+\hat{\mu}/2)}$, *i.e.* the link at site x in the μ direction is an SU(3) matrix $U_\mu(x, x+\hat{\mu})$ denoting the average gauge field between x and $x+\hat{\mu}$ and labeled by the point $x + \hat{\mu}/2$. Also $U_\mu(x, x-\hat{\mu}) \equiv U_\mu^\dagger(x-\hat{\mu}, x)$. The links and propagators can be contracted to form gauge invariant correlation functions as discussed above in the case of the pion.
- On the "importance sampled" configurations, the expectation values reduce to simple averages of the correlation functions. The problem is that the set of background gauge configurations is infinite. Thus, while it is possible to calculate the correlation functions for specified background gauge configurations, doing the functional integral exactly is not feasible. It is, therefore, done numerically using monte carlo methods.

The simplest way to understand the numerical aspects of LQCD calculations is to gain familiarity with the numerical treatment of any statistical mechanics system, for example the Ising model. The differences are: (i) the degrees of freedom in LQCD are much more complicated – SU(3) link matrices rather than Ising spins, and quark propagators given by the inverse of the Dirac operator; (ii) The action involves the highly nonlocal term Ln Det M which makes the update of the gauge configurations very expensive; and (iii) the correlation functions are not simple products of spin variables like the specific heat or magnetic susceptibility, but complicated functions of the link variables and quark propagators.

The subtleties arising due to the fact that LQCD is a renormalizable field theory and not a classical statistical mechanics system come into play in the behavior of the correlation functions as the lattice spacing a is varied, and in the quantum corrections that renormalize the input parameters (quark and gluon masses and fields) and the composite operators used in the study of correlation functions. At first glance it might seem that one has introduced an additional parameter in LQCD, the lattice spacing a, however, recall that the coupling α_s and the cutoff a are not independent quantities but are related by the renormalization group

$$\Lambda_{QCD} = \frac{1}{a} e^{-1/2\beta_0 g^2(a)} \left(\beta_0 g^2(a)\right)^{-\beta_1/2\beta_0^2} + \dots \, , \tag{10}$$

where Λ_{QCD} is the non-perturbative scale of QCD, and $\beta_0 = (11 - 2n_f/3)/16\pi^2$ and $\beta_1 = (102 - 38n_f/3)/(16\pi^2)^2$ are the first two, scheme independent, coefficients of the β-function. In statistical mechanics systems, the lattice spacing a is a physical quantity – the intermolecular separation. In QFT it is simply the ultraviolet regulator that must eventually be taken to zero keeping physical quantities, like the renormalized coupling, spectrum, etc, fixed.

The reason that lattice results are not exact is because in numerical simulations we have to make a number of approximations. The size of these is dictated by the computer power at hand. They are being improved steadily with computer technology, better numerical algorithms, and better theoretical understanding. To evaluate the reliability of current lattice results, it is important to understand the size of the various systematic errors and what is being done to control them. I, therefore, consider it important to discuss these next before moving on to results.

3 Systematic Errors in Lattice Results

The various sources of errors in lattice calculations are as follows.

Statistical errors: The monte carlo method for doing the functional integral employs statistical sampling. The results, therefore, have statistical errors. The current understanding, based on agreement of results from ensembles generated using different algorithms and different initial starting configuration in the Markov process, is that the functional integral is dominated by a single global minimum. Also, configurations with non-trivial topology are properly represented in an ensemble generated using a Markov chain based on small changes to link variables. Another way of saying this is that the data indicate that the energy landscape is simple. As a result, the statistical accuracy can be improved by simply generating more statistically independent configurations with current update methods.

Finite Size errors: Using a finite space-time volume with (anti-)periodic boundary conditions introduces finite size effects. On sufficiently large lattices these effects can be analyzed in terms of interactions of the particle with its mirror images. Lüscher has shown that in this regime these effects vanish exponentially [15]. Current estimates indicate that for $L \gtrsim 3$ fermi and $M_\pi L \geq 6$ the errors are $\lesssim 1\%$, and decrease exponentially with increasing L.

Discretization errors: The discretization of the Euclidean action on a finite discrete lattice with spacing a leads, in general, to errors proportional to a, $\alpha_s^n a$, a^2, ... The precise form of the leading term depends on the choice of the lattice action and operators [16]. For example, lattice artefacts in the fermion action modify the quark propagator M^{-1} at large p from its continuum form. Numerical data show that the coefficients of the leading term are large, consequently the corrections for $1/a \approx 2$GeV are significant in many quantities, 10-30% [17]. The reliability of lattice results, with respect to $O(a)$ errors, is being improved by a two pronged strategy. First, for a given action extrapolations to the continuum limit $a = 0$ are performed by fitting data at a number of values of a using leading order corrections. Second, these extrapolations are being done for different types of actions (Wilson, Clover, staggered) that have significantly different discretization errors. We consider the consistency of the results in the $a = 0$ limit as a necessary check of the reliability of the results.

Extrapolations in Light Quark Masses: The physical u and d quark masses are too light to simulate on current lattices. For $1/a = 2$ GeV, realistic simulations require $L/a \gtrsim 90$ to avoid finite volume effects, *i.e.* keeping $\tilde{M}_\pi L \geq 6$ where $\tilde{M}_\pi$ is the lightest pseudoscalar meson mass on the lattice. Current best lattice sizes are $L/a = 32$ for quenched and $L/a = 24$ for unquenched. Thus, to get results for quantities involving light quarks, one typically extrapolates in $m_u = m_d$ from the range $m_s/3 - 2m_s$ using simple polynomial fits based on chiral perturbation theory. For quenched simulations there are additional problems for $m_q \lesssim m_s/3$ as discussed below in the item on quenching errors.

Discretization of heavy quarks: Simulations of heavy quarks (c and b) have discretization errors of $O(ma)$ and $O(pa)$. This is because quark masses measured in lattice units, $m_c a$ and $m_b a$, are of order unity for $2\text{GeV} \leq 1/a \leq 5\text{GeV}$. It turns out that these discretization errors are large even for m_c. Extrapolations of lattice data from lighter masses to m_b using HQET have also not been very reliable as the corrections are again large. The three most promising approaches to control these errors are non-relativistic QCD, $O(a)$ improved heavy Dirac, and HQET. These are discussed in [19, 20, 21]. There will not be any discussion of heavy quark physics in this talk.

Matching between the lattice and the continuum schemes (renormalization constants): Experimental data are analyzed using some continuum renormalization scheme like $\overline{MS}$, so results in the lattice scheme have to be converted to this scheme. The perturbative relation between renormalized quantities in say $\overline{MS}$ and the lattice scheme, are in almost all cases, known only to 1-loop. Data show that the $O(\alpha_s^2)$ corrections can be large, $\sim 10-50\%$ depending on the quantity at hand, even after implementation of the improved perturbation theory technique of Lepage-Mackenzie [18]. Recently, the technology to calculate these factors non-perturbatively has been developed and is now being exploited [22]. As a result, the reliance on perturbation theory for these matching factors will be removed.

Operator mixing: The lattice operators that arise in the effective weak Hamiltonian can, in general, mix with operators of the same, higher, and *lower* dimensions because at finite a the symmetries of the lattice and continuum theories are not the same. Perturbative estimates of this mixing can have an even more serious problem than the uncertainties discussed above in the matching coefficients. In cases where there is mixing with lower dimensional operators, the mixing coefficients have to be known very accurately otherwise the power divergences overwhelm the signal as $a \to 0$. In cases where there is mixing, due to the explicit chiral symmetry breaking in Wilson like actions, with operators of the same dimension but with different tensor structures, the chiral behavior may again be completely overwhelmed by the artefacts. In both of these cases a non-perturbative calculation of the mixing coefficients is essential.

Quenched approximation: The fermionic contribution, Ln det(M), in the Boltzmann factor e^{-S} for the generation of background gauge configurations increases the computational cost by a factor of $10^3 - 10^5$. The strategy, therefore, has been to initially neglect this factor, and to bring all other above mentioned sources of errors under quantitative control. The justification is that the quenched theory retains a number of the key features of QCD – confinement, asymptotic freedom, and the spontaneous breaking of chiral symmetry – and is expected to be good to within $10 - 20\%$ for a number of quantities. One serious drawback is that the quenched theory is not unitary and χPT analysis of it shows the existence of unphysical singularities in the chiral limit. For example, the chiral expansions of pseudoscalar masses and decay constants in the quenched theory are modified in two ways. One, the normal chiral coefficients are different in the quenched theory, and second there are additional terms that are artefacts and are singular in the limit $m_q = 0$ [23, 24]. These artefacts are expected to start becoming significant for $m_q \lesssim m_s/3$ [25, 26]. Thus, in quenched simulations one of the strategies for extrapolations in the light quark masses is to use fits based on χPT, keep only the normal coefficients, and restrict the data to the range $m_s/3 - 2m_s$ where the artifacts are expected to be small. In this talk I shall use this procedure to "define" the quenched results.

The above mentioned systematic errors are under varying degrees of control depending on the quantity at hand. Of the systematics effects listed above, quenching errors are by far the least well quantified, and are, to first approximation, unknown. Of the remaining sources the two most serious are the discretization errors and the matching of renormalized operators between the lattice and continuum theories. An example of the latter is the connection between the quark mass in lattice scheme and in a perturbative scheme like $\overline{MS}$. We shall discuss the status of control over these errors in more details when discussing data.

4 Light Quark Masses from χPT

The masses of light quarks cannot directly be measured in experiments as quarks are not asymptotic states. One has to extract the masses from the pattern of the observed hadron spectrum. Three approaches have been used to estimate these – chiral perturbation theory (χPT), QCD sum-rules, and lattice QCD.

χPT relates pseudoscalar meson masses to m_u, m_d, and m_s. However, due to the presence of an overall unknown scale in the chiral Lagrangian, χPT can predict only two ratios amongst the three light quark masses. The current estimates are [27, 28, 29]

	Lowest order	Next order
$2m_s/(m_u + m_d)$	$24.2 - 25.9$	24.4(1.5)
m_u/m_d	0.55	0.553(43) .

These ratios have been calculated neglecting the Kaplan-Manohar symmetry [30]. The subtle point here is that the masses μ_i extracted from low energy phenomenology are related to the fundamental parameters m_i, defined at some high scale by the underlying theory, as $\mu_i = m_i + \beta * \det M/m_i\Lambda_{\chi SB}$, where the second, correction, term is an instanton induced additive renormalization. For the u quark, the magnitude of this term is roughly equal to the χPT estimates of μ_u for $\beta \approx 2$ [31]. Consequently, using χPT, one cannot estimate the size of isospin breaking from low energy phenomenology alone.

At next to leading order, only one combination of ratios ($Q^2 = (m_s^2 - \overline{m}^2)/(m_d^2 - m_u^2)$ as defined in [29]) can be determined unambiguously from χPT. Even if one ignores the Kaplan-Manohar subtlety (such an approach has been discussed by Leutwyler under the assumption that the higher order terms are small [29]) one still needs input from sum-rules or LQCD to get absolute values of quark masses.

5 Light Quark Masses from LQCD

The most extensive and reliable results from LQCD have been obtained in the quenched approximation. In the last year the statistical quality of the quenched data has been improved dramatically especially by the work of the two Japanese Collaborations CP-PACS and JLQCD (see [32] for a recent review). Simultaneously, the lattice sizes have been pushed to $\gtrsim 3$ fermi for the lattice spacing in the range $0.5 - 0.25$ GeV^{-1}. In Fig. 2 we show the CP-PACS data obtained using Wilson fermions. To highlight the statistical improvement we show data at $\beta = 6.0$ from the next best calculation (with respect to both statistics and lattice size) [33].

To reliably extrapolate the lattice data to the continuum limit one needs control over discretization errors and over the matching relations between the lattice scheme and the continuum scheme, say $\overline{\text{MS}}$. The first issue has been addressed by the community by simulating three different discretizations of the Dirac action – Wilson, SW clover, and staggered – which have discretization errors of $O(a)$, $O(\alpha_s(a)a)$, and $O(a^2)$ respectively. The second issue, reliability of the 1-loop perturbative matching relations, is being checked by using non-perturbative estimates.

For Wilson and SW clover formulations, the internal consistency of the lattice calculations can be checked by calculating the quark masses two different ways. The first is based on methods of χPT, *i.e.* the calculated hadron masses are expressed as functions of the quark masses as in χPT. This method, based on hadron spectroscopy, is labeled HS for brevity. In the second method, labeled WI, quark masses are defined using the ward identity $\partial_\mu \overline{\psi}\gamma_5\gamma_\mu\psi = (m_1 + m_2)\overline{\psi}\gamma_5\psi$. An example of such checks is shown in Fig. 2. The solid lines are fits to the HS and WI estimates using the Wilson action and on the same statistical sample of configurations. The very close agreement of the extrapolated values is probably fortuitous since the linear extrapolation in a, shown in Fig. 2, neglects both the higher order discretization errors and the $O(\alpha_s(a)^2)$ errors in the 1-loop perturbative matching relations. The figure also shows preliminary results for the same WI data but now with non-perturbative Z's. The correction is large, note the large change in the slope in a, yet the extrapolated value is $\lesssim 4$ MeV. The final analysis using non-perturbative Z's will be available soon, and it is unlikely that the central value presented below will shift significantly.

Lastly, in the quenched approximation, estimates of quark masses can and do depend on the hadronic states used to fix them. The estimates of $\overline{m}$ given below were extracted using the pseudoscalars mesons, *i.e.* pions, with the scale a set by M_ρ. Using either the nucleon or the Δ to fix $\overline{m}$ would give $\sim 10\%$ smaller estimates. Similarly, extracting m_s using M_K gives estimates that are $\sim 20\%$ smaller that those using M_{K^*} or M_ϕ as shown below. While these estimates of quenching errors are what one would expect naively, we really have to wait for sufficient unquenched data to quantify these more precisely.

A summary of the quenched results in MeVat scale $\mu = 2$ GeV, based on an analysis of the 1997 world data [32], is

The difference in estimates between Wilson, tadpole improved (TI) clover, and staggered results could be due to the neglected higher order discretization errors and/or due to the difference between non-perturbative and 1-loop estimates of Z's. This uncertainty is currently $\approx 15\%$. Similarly, the $\approx 20\%$ variation in m_s with the state

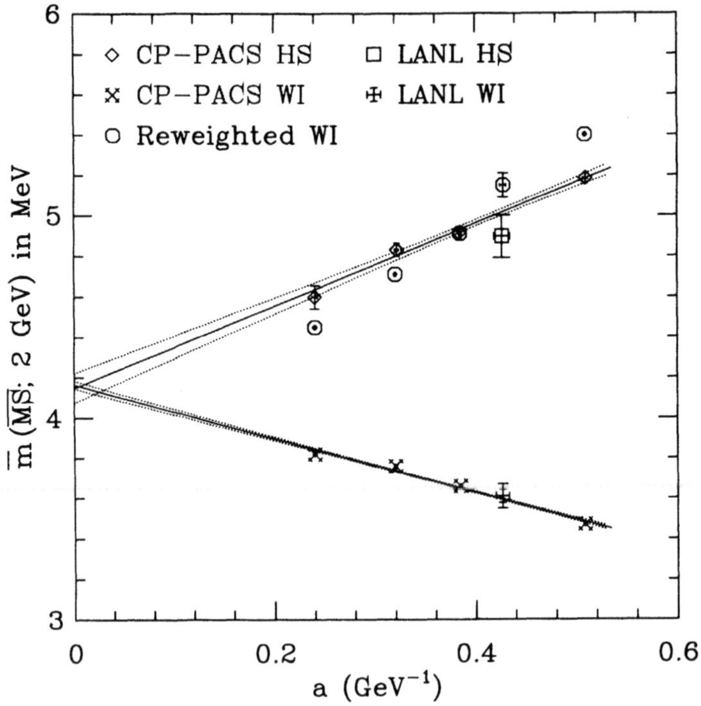

Figure 2. Linear extrapolation of $\overline{m}$ versus $a(M_\rho)$ for Wilson fermions using HS and WI methods. The WI data corrected by using non-perturbative estimates for the matching constants are also shown.

	Wilson	TI Clover	Staggered
$\overline{m}(M_\pi)$	4.1(1)	3.8(1)	3.5(1)
$m_s(M_K)$	107(2)	99(3)	91(2)
$m_s(M_\phi)$	139(11)	117(8)	109(5) .

used to extract it, M_K versus M_{K^*} (or equivalently M_ϕ) could be due to the quenched approximation or again an artifact of keeping only the lowest order correction term in the extrapolations. To disentangle these discretization and quenching errors we again need precise unquenched data.

Thus, for our best estimate of quenched results we average the data and use the spread as the error. To these, we add a second uncertainty of 10% as due to the determination of the scale $1/a$ (another estimate of quenching errors). The final results, in $\overline{\text{MS}}$ scheme evaluated at 2 GeV, are [32]

$$\begin{aligned} \overline{m} &= 3.8(4)(4)\ \text{MeV} \\ m_s &= 110(20)(11)\ \text{MeV}. \end{aligned} \tag{11}$$

The important question is how do these estimates change on unquenching. The 1996 analyses suggested that unquenching could lower the quark masses by $\approx 20\%$ [34, 35], however, as discussed in [32] I no longer feel confident making an assessment of the

Table 1. Values and bounds on $\overline{m}$ and m_s, in $\overline{\text{MS}}$ scheme at 2 GeV, from sumrule analyses.

reference	$\overline{m}$ (MeV)	m_s (MeV)
[39] 1989	$= 6.2(0.4)$	$= 138(8)$
[36] 1995	$= 4.7(1.0)$	
[38] 1995	$= 5.1(0.7)$	$= 144(21)$
[37] 1995		$= 137(23)$
[40] 1996		$= 148(15)$
[41] 1997		$= 91 - 116$
[42] 1997		$= 115(22)$
[43] 1997	$= 4.9(1.9)$	
[45] 1997	$\geq 3.8 - 6$	$\geq 118 - 189$
[46] 1997	≥ 3.4	$\geq 88(9)$
[47] 1997	$\geq 4.1 - 4.4$	$\geq 104 - 116$

magnitude of the effect. The data does still indicate that the sign of the effect is negative, *i.e.* that unquenching lowers the masses. An estimate of the size requires more unquenched data.

To end this section let me comment on a comparison of the quenched estimates with values extracted from sum-rules as there seems to be a general feeling that the two estimates are vastly different. In fact the recent analyses indicate that the quenched lattice results and the sum-rule estimates are actually consistent. A large part of the apparent difference is due to the use of different scales at which results are presented. Lattice QCD results are usually stated at $\mu = 2$ GeV, while the sum-rules community uses $\mu = 1$ GeV, and the running of the masses between these two scales is an $\approx 30\%$ effect in full QCD. This issue is important enough that I would like to briefly review the status of sum-rule estimates.

6 Sum rule determinations of $\overline{m}$ and m_s

A summary of light quark masses from sum-rules is given in Table 1. Sum rule calculations proceed in one of two ways. (i) Using axial or vector current Ward identities one writes a relation between two 2-point correlation functions. One of these is evaluated perturbatively after using the operator product expansion, and the other by saturating with intermediate hadronic states [36, 37]. The quark masses are the constant of proportionality between these two correlation functions. (ii) Evaluating a given correlation function both by saturating with known hadronic states and by evaluating it perturbatively [38]. The perturbative expression depends on quark masses, and defines the renormalization scheme in which they are measured. The main sources of systematic errors arise from using (i) finite order calculation of the perturbative expressions, and (ii) the ansatz for the hadronic spectral function. Of these the most severe is the second as there does not exist enough experimental data to constrain the spectral function even for $\mu < 2$ GeV. Since there are narrow resonances in this region, one cannot match the two expressions point by point in energy scale. The two common approaches are to match the moments integrated up to some sufficiently high scale (finite energy sum rules) or to match the Borel transforms. The hope then is that the result is independent of this scale or of the Borel parameter.

Progress in sum-rules analyses has also been incremental as in LQCD. The perturbative expressions have now been calculated to $O(\alpha_s^3)$ [40], and the value of $\Lambda_{QCD}^{(3)}$ has settled at $\approx$ 380MeV. A detailed analysis of the convergence of the perturbation expansion suggests that the error associated with the truncation at $O(\alpha_s^3)$ is $\approx 10\%$ for $\mu \geq 2$ GeV [40].

Improving the spectral function has proven to be much harder. For example, Colangelo et al. [41] have extended the analysis of m_s in [37, 40] by constructing the hadronic spectral function up to the first resonance ($K^*(1430)$) from known $K\pi$ phase shift data. Similarly, Jamin [42] has used a different parametrization of the Omnes representation of the scalar form factor using the same phase shift data. In both cases the reanalysis lowers the estimate of the strange quark mass significantly. The new estimates, listed in Table 1, are consistent with the quenched estimates discussed in Section 5.

One can circumvent the uncertainties in the ansatz for the spectral function by deriving rigorous lower bounds using just the positivity of the spectral function [44, 45, 46, 47]. Of these the most stringent are in [47] which rule out $\overline{m} < 3$ and $m_s < 80$ MeV for $\mu \lesssim 2.5$ GeV. The bounds, however, have a significant dependence on the scale μ as evident by comparing the above values to those in the last row in Table 1, and the open question is how to fix μ, *i.e.* the upper limit of integration in the finite energy sum rules at which duality between PQCD and hadronic physics becomes valid? Unfortunately, this question cannot be answered ab initio.

7 Implications for ϵ'/ϵ

The Standard Model (SM) prediction of ϵ'/ϵ can be written as [48]

$$\epsilon'/\epsilon = A\Big\{c_0 + [c_6 B_6^{1/2} + c_8 B_8^{3/2}]M_r\Big\} , \tag{12}$$

where $M_r = (158\text{MeV}/(m_s + m_d))^2$ and all quantities are to be evaluated at the scale $m_c = 1.3$GeV. Eq. 12 highlights the dependence on the light quark masses and the bag parameters $B_6^{1/2}$ and $B_8^{3/2}$. For the other SM parameters that are needed in obtaining this expression we use the central values quoted by Buras *et al.* [48]. Then, we get $A = 1.29 \times 10^{-4}$, $c_0 = -1.4$, $c_6 = 7.9$, $c_8 = -4.0$. Thus, to a good approximation $\epsilon'/\epsilon \propto M_r$.

Conventional analysis, with $m_s + m_d = 158$MeV and $B_6^{1/2} = B_8^{3/2} = 1$, gives $\epsilon'/\epsilon \approx 3.2 \times 10^{-4}$. The uncertainties in the remaining SM parameters used to determine A, c_0, c_6, and c_8 in Eq. 12 are large enough that, in fact, any value between -1×10^{-4} and 16×10^{-4} is acceptable[48]. Current experimental estimates are $7.4(5.9) \times 10^{-4}$ from Fermilab E731 [49] and $23(7) \times 10^{-4}$ from CERN NA31 [50]. So at present there is no resolution of the issue whether the CKM ansatz explains all observed CP violation.

The new generation of experiments, Fermilab E832, CERN NA48, and DAΦNE KLOE, will reduce the uncertainty to $\approx 1 \times 10^{-4}$. First results from these experiments should be available in the next couple of years. Thus, it is very important to tighten the theoretical prediction.

As is clear from Eq. 12, both the values of quark masses and the interplay between $B_6^{1/2}$ and $B_8^{3/2}$ will have a significant impact on ϵ'/ϵ. The lower values of quark masses suggested by lattice QCD analyses would increase the estimate. The status of results for the various B-parameters relevant to the study of CP violation are discussed in the next section.

8 B-parameters, B_K, B_6, $B_7^{3/2}$, $B_8^{3/2}$

Considerable effort has been devoted by the lattice community to calculate the various B-parameters needed in the standard model expressions describing CP violation. A summary of the results and the existing sources of uncertainties are as follows.

8.1 B_K

The standard model expression for the parameter ϵ, which characterizes the strength of the mixing of CP odd and even states in K_L and K_S, is of the form [51]

$$|\epsilon| \sim \text{Im}(V_{td}V_{ts}^*) B_K(\mu)\ \Phi(\frac{m_c}{M_W}, \frac{m_t}{M_W}, \mu) \ , \tag{13}$$

where Φ is a known function involving Inami-Lim functions and CKM elements. This relation provides a crucial constraint in the effort to pin down the ρ and η parameters in the Wolfenstein parameterization of the CKM matrix [52]. The quantity B_K parameterizes the QCD corrections to the basic box diagram responsible for $K^0 - \overline{K^0}$ mixing. This transition matrix element is what we calculate on the lattice.

The calculation of B_K is one of the highlights of LQCD simulations. It was one of the first quantities for which theoretical estimates were made, using quenched chiral perturbation theory, of the lattice size dependence, dependence on quark masses, and on the effects of quenching [53, 54, 26, 55]. Numerical data in the staggered formulation (which has the advantage of retaining a chiral symmetry which preserves the continuum like behavior of the matrix elements) is consistent with these estimates in both the sign and the magnitude. Since all these corrections have turned out to be small, results for B_K with staggered fermions have remained stable over the last five years.

Three collaborations have pursued calculations of B_K using staggered fermions. The results are $B_K(NDR, 2\ \text{GeV}) = 0.62(2)(2)$ by Kilcup, Gupta, and Sharpe [55], 0.552(7) by Pekurovsky and Kilcup [56], and 0.628(42) by the JLQCD collaboration[57]. Of these, the results by the JLQCD collaboration are based on a far more extensive analysis. The quality of their data are precise enough to include both the leading $O(a^2)$ discretization corrections, and the $O(\alpha_s^2)$ corrections in the 1-loop matching factors. Their data, along with the extrapolation to $a = 0$ limit including both factors, are shown in Fig. 3. I consider theirs the current best estimate of the quenched value.

The two remaining uncertainties in the above estimate of B_K are quenching errors and SU(3) breaking effects (all current results have been obtained using degenerate quarks $m_u = m_d = m_s$, with kaons composed of two quarks of mass $\sim m_s/2$ instead of m_s and m_d). There exists only preliminary unquenched data [58] which suggest that the effect of sea quarks is to increase the estimate by $\approx 5\%$. Lastly, Sharpe has used χPT to estimate that the SU(3) breaking effects could also increase B_K by another $4 - 8\%$ [26]. Confirmation of these corrections requires precise unquenched data which is still some years away.

In a number of phenomenological applications what one wants is the renormalization group invariant quantity $\hat{B}_K$ defined, at two-loops, as

$$\hat{B}_K = B_K(\mu)\ (\alpha_s(\mu))^{-\gamma_0/2\beta_0}(1 + \frac{\alpha_s(\mu)}{4\pi}[\frac{\beta_1\gamma_0 - \beta_0\gamma_0}{2\beta_0^2}]) \ . \tag{14}$$

Unfortunately, to convert the quenched JLQCD number, 0.628(42), one has to face the issue of the choice of the value of α_s and the number of flavors n_f. It turns out that the two-loop evolution of B_K is such that one gets essentially the same number for the

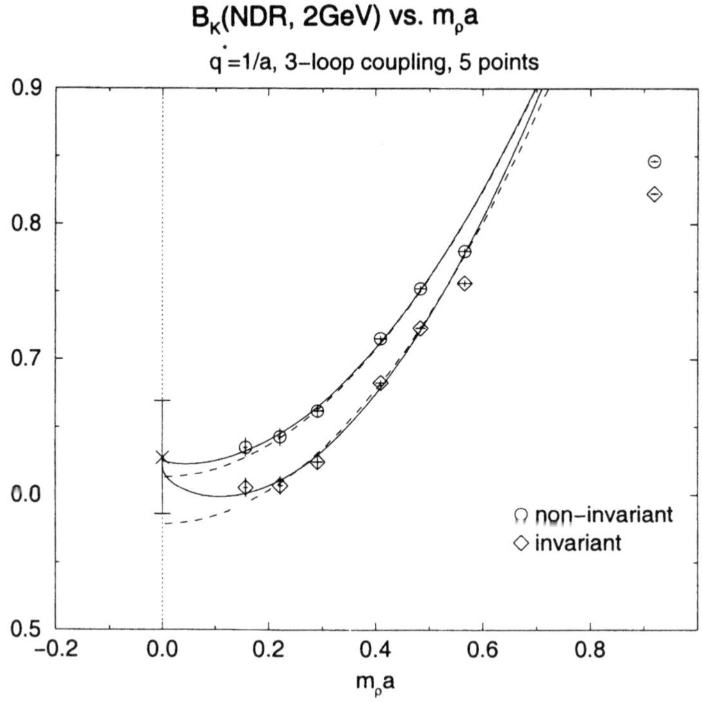

Figure 3. JLQCD data for $B_K(NDR, 2$ GeV), and their extrapolation to the continuum limit keeping both the $O(a^2)$ discretization corrections, and the $O(\alpha_s^2)$ in the 1-loop matching factors for two different discretizations of the weak operator. The extrapolation of the data without including the $O(\alpha_s^2)$ corrections are shown by the dashed lines.

quenched theory, 0.87(6) for $n_f = 0$ and $\alpha_s(2GeV) = 0.192$, and 0.84(6) for the physical case of $n_f = 3$ and $\alpha_s(M_\tau) = 0.354$. One might interpret this near equality to imply that the quenching errors are small. Such an argument is based on the assumption that there exists a perturbative scale at which the full and quenched theories match. Since there is no *a priori* reason to believe that this is true, my preference is to double the difference and assign a second error of 0.06, an estimate also suggested by the preliminary unquenched data and the χPT analysis. With this caveat I arrive at the lattice prediction

$$\hat{B}_K = 0.86(6)(6). \tag{15}$$

8.2 $\langle \pi^+\pi^0|\mathcal{O}_4|K^+\rangle$ and its relation to B_K

The $\Delta S = 2$ operator $\overline{s}L_\mu d\overline{s}L_\mu d$ responsible for the $K^0 - \overline{K^0}$ transition belongs to the same **27** representation of SU(3) as the $\Delta S = 1, \Delta I = 3/2$ operator $\mathcal{O}_4 = \overline{s}L_\mu d(\overline{u}L_\mu u - \overline{d}L_\mu d) + \overline{s}L_\mu u\overline{u}L_\mu d$. At tree level in χPT, one gets the relation [59, 60]

$$\langle K^0|\mathcal{O}_{\Delta S=2}|\overline{K^0}\rangle = \frac{\sqrt{2}f_\pi}{3i}\frac{2M_K^2}{M_K^2 - M_\pi^2}\langle \pi^+\pi^0|\mathcal{O}_4|K^+\rangle \tag{16}$$

So one way to calculate B_K is to measure $\langle \pi^+\pi^0|\mathcal{O}_4|K^+\rangle$ on the lattice and then use Eq. 16. The motivation for doing this is that, for Wilson-like lattice actions, $\mathcal{O}_4$ is only multiplicatively renormalized due to CPS symmetry [61], whereas the $\Delta S = 2$ operator mixes with all other chirality operators of dimension six. Using 1-loop values for these mixing coefficients has proven inadequate, though the recent implementation of non-perturbative estimates has made this situation much better [62].

The first lattice calculations of the $\Delta I = 3/2$ part of the $K^+ \to \pi^+\pi^0$ amplitude [63, 65] gave roughly twice the experimental value, even though the B_K extracted from this "wrong" amplitude "agrees" with modern lattice estimates. The important question therefore is how reliable are the χPT relations between $\langle\pi^+\pi^0|\mathcal{O}_4|K^+\rangle|_{lattice}$ and $\langle\pi^+\pi^0|\mathcal{O}_4|K^+\rangle|_{physical}$, and between $\langle\pi^+\pi^0|\mathcal{O}_4|K^+\rangle|_{lattice}$ and $\langle K^0|\mathcal{O}_{\Delta S=2}|\overline{K^0}\rangle$, *i.e.* does one or both fail?

There are three sources of systematic errors in lattice calculations of $\langle\pi^+\pi^0|\mathcal{O}_4|K^+\rangle$ that could explain the contradiction. The calculation is done in the quenched approximation, on finite size lattices, and with unphysical kinematics (in the lattice calculation the final state pions are degenerate with the kaon since $m_u = m_d = m_s$, and are at rest). Of the three possibilities, the last is the most serious as it gives a factor of two even at the tree-level

$$\frac{2M_K^2}{M_K^2 - M_\pi^2}\langle\pi^+\pi^0|\mathcal{O}_4|K^+\rangle|_{physical} = \langle\pi^+\pi^0|\mathcal{O}_4|K^+\rangle|_{unphysical} \,. \qquad (17)$$

This tree-level correction was taken onto account in [63, 65]. The source of the remaining discrepancy by a factor of two was anticipated by Bernard as due to the failure of the tree-level expression [64]. Recently, Golterman and Leung [59] have calculated the 1-loop corrections to Eq. 17. This calculation involves a number of unknown $O(p^4)$ chiral coefficients of the weak interactions and thus has a number of caveats. However, under reasonable assumptions about the value of these $O(p^4)$ constants, the corrections due to finite volume, quenching, and unphysical kinematics all go in the right direction, and the total 1-loop correction can modify Eq. 17 by roughly a factor of two. On the other hand the modification to the connection with B_K at the physical point is small.

The JLQCD Collaboration [66] has recently updated the calculations in [63, 65]. By improving the statistical errors they are able to validate the trends predicted by 1-loop χPT expressions. Thus one has a plausible resolution of the problem. I say plausible because the calculation involves a number of unknown chiral couplings in both the full and quenched theory and also because of the size of the 1-loop correction. The conclusive statement is the failure of χPT for the relation Eq. 17.

The other relevant question is what bearing does the analysis of Golterman and Leung [59] have on the calculations of other B-parameters. In the calculation of B_K, based on measuring the $\langle K^0|\mathcal{O}_{\Delta S=2}|\overline{K^0}\rangle$ transition matrix element as discussed in section 8.1, χPT has been used only to understand effects of finite volume and chiral logs. These are found to be small and the data show the predicted behavior. Based on this success of χPT we estimate that the two remaining errors – quenching and the use of degenerate quarks – are each $\approx 5\%$ as suggested by 1-loop χPT. On the other hand, in present calculations of B_6, $B_7^{3/2}$, and $B_8^{3/2}$ χPT is used in an essential way, *i.e.* to relate $\langle\pi^+\pi^0|\mathcal{O}|K^+\rangle$ to $\langle\pi^+|\mathcal{O}|K^+\rangle$. Second, the calculations are done for unphysical kinematics, *i.e.* the final state pion is degenerate with the kaon. It would be interesting to know the size of the one-loop corrections to these relations.

8.3 $B_7^{3/2}$ and $B_8^{3/2}$

Assuming the reduction of $\langle\pi^+\pi^0|\mathcal{O}_{7,8}^{\Delta I=3/2}|K^+\rangle$ to $\langle\pi^+|\mathcal{O}_{7,8}^{\Delta I=3/2}|K^+\rangle$ using tree-level χPT is reliable, the lattice calculations of $B_7^{3/2}$ and $B_8^{3/2}$ are as straightforward as

Fermion Action	Z	β	$1/a$ GeV	$B_7^{3/2}$	$B_8^{3/2}$
(A) Staggered [55]	1-loop	6.0, 6.2	$a \to 0$	0.62(3)(6)	0.77(4)(4)
(B) Wilson [67]	1-loop	6.0	2.3	0.58(2)(7)	0.81(3)(3)
(C) Tree-level Clover [62]	1-loop	6.0	2.0	0.58(2)	0.83(2)
(D) Tree-level Clover [62]	Non-pert.	6.0	2.0	0.72(5)	1.03(3) .

those for B_K. There are three "modern" quenched estimates of $B_7^{3/2}$ and $B_8^{3/2}$ which supercede all previous reported values. These, in the NDR-$\overline{\text{MS}}$ scheme at 2 GeV, are where I have also given the type of lattice action used, the β's at which the calculation was done, the lattice scale $1/a$ at which the results were extracted, and how the 1-loop matching constants Z were determined.

The difference between estimates (C) and (D) is the use of non-perturbative versus perturbative Z's. Thus (D) is the more reliable of the APE numbers. The agreement between (B) and (C), in spite of the difference in the action, is a check that the numerics are stable. All three of these results suffer from the fact that these calculations were done at $\beta = 6.0$ ($1/a \approx 2$ GeV) and there does not yet exist data at other β needed to do the extrapolation to the continuum limit.

The result (A) does incorporate an extrapolation to a=0, but with only two beta values. For example, $B_8^{3/2} = 1.24(1)$ and $1.03(2)$ at $\beta = 6.0$ and 6.2 respectively. Due to the large slope in a^2, such an extrapolation based on two points should be considered preliminary. Lastly, one needs to demonstrate that corrections to 1-loop $Z's$ are under control.

8.4 B_6

The recent work of Pekurovsky and Kilcup [56] provides the best lattice estimate for B_6. Their results $B_6 = 0.67(4)(5)$ for quenched and $0.76(3)(5)$ for two flavors have the following systematics that are not under control. The calculation is done for degenerate quarks, $m_u = m_d = m_s$ and uses the lowest order χPT to relate $\langle \pi\pi|\mathcal{O}|K^+\rangle$ to $\langle \pi|\mathcal{O}|K^+\rangle$. There is no reason to believe that higher order corrections may not be as or more significant as discussed above for $\mathcal{O}_4$. The second issue is that the 1-loop perturbative corrections in the matching coefficients are large. Lastly, there is no estimate for the discretization errors as the calculation has been done at only one value of β. Thus, at this point there is no solid prediction from the lattice.

9 Conclusions

In view of the new generation of ongoing experiments to measure ϵ'/ϵ with the proposed accuracy of 1×10^{-4}, it is very important to firm up the theoretical prediction. The standard model estimate depends very sensitively on the sum $m_s + m_d$ and on the interplay between the strong and electromagnetic penguin operators, *i.e.* $B_6^{1/2}$ and $B_8^{3/2}$. Quenched lattice results for $(m_s + m_d)(2 \text{ GeV})$ are settling down at 115(25) MeV, and preliminary evidence is that unquenching further lowers these estimates. The calculations of $B_6^{1/2}$ and $B_8^{3/2}$ are less advanced. Hopefully we can provide reliable quenched estimates for these parameters in the next year or so. Thereafter, we shall start to chip away at realistic full QCD simulations.

Acknowledgements

I would like to thank the organizers of ORBIS SCIENTIAE 1997-II for a very stimulating conference, and my collaborators Tanmoy Bhattacharya, Greg Kilcup, Kim Maltman, and Steve Sharpe for many discussions. I also thank the DOE and the Advanced Computing Laboratory for support of our work.

References

[1] M. Creutz, *Quarks, Gluons, and Lattices*, Cambridge University Press, 1983
[2] *Quantum Fields on the Computer*, Ed. M. Creutz, World Scientific, 1992.
[3] *Quantum Fields on a Lattice*, I. Montvay and G. Münster, Cambridge Monographs on
[4] H. J. Rothe, *Quantum Gauge Theories: An Introduction*, World Scientific, 1997.
[5] H.B.Nielsen and M. Ninomiya, Nucl. Phys. **B185** (1981) 20.
[6] Y. Shamir, Nucl. Phys. (Proc. Suppl.) **B53** (1997) 664; Nucl. Phys. (Proc. Suppl.) **B47** (1996) 212.
[7] R. Narayanan, hep-lat/9707035.
[8] M. Testa, hep-lat/9707007.
[9] M. Alford, T. Klassen, and P. Lepage, Nucl. Phys. **B496** (1997) 377; and hep-lat/9712005
[10] M. Lüscher, *et al.*, Nucl. Phys. **B478** (1996) 365.
[11] P. Hasenfratz, hep-lat/9709110.
[12] K. G. Wilson, Phys. Rev. **D10** (1974) 2445; in *New Phenomena in Subnuclear Physics* (Erice 1975), Ed. A. Zichichi, Plenum New York, 1977.
[13] B. Sheikholesalami and R. Wohlert, Nucl. Phys. **B259** (1985) 572.
[14] J. Kogut and L. Susskind, Phys. Rev. **D11** (1975) 395; T. Banks, J. Kogut, and L. Susskind, Phys. Rev. **D13** (1996) 1043; L. Susskind, Phys. Rev. **D16** (1977) 3031.
[15] M. Lüscher, *Selected Topics In Lattice Field Theory*, Lectures at the 1988 Les Houches summer school, North Holland, 1990.
[16] K. Symanzik, in *Mathematical Problems in Theoretical Physics*, Springer Lecture Notes in Physics, bf 153 (1982) 47; M. Lüscher and P. Weisz, it Commun. Math. Phys. 97 1985 59; G. Heatlie, *et al.*, Nucl. Phys. **B352** (1991) 266.
[17] H. Wittig, hep-lat/9710013; M. Peardon, hep-lat/9710029
[18] P. Lepage and P. Mackenzie, Phys. Rev. **D48** (1993) 2250.
[19] P. Lepage and B. Thacker, Phys. Rev. **D43** (1991) 196; G. Lepage, L. Magnea, C. Nakhleh, U. Magnea and K. Hornbostel, Phys. Rev. **D46** (1992) 4052.
[20] A. El-Khadra, A. Kronfeld, and P. Mackenzie, Phys. Rev. **D55** (1997) 3933.
[21] M. Neubert, Phys. Rev. **245** (1994) 259; G. Martinelli and C. Sachrajda, Nucl. Phys. **B478** (1996) 660; J. Mandula and M. Ogilvie, hep-lat/9703020.
[22] G.C. Rossi, Nucl. Phys. (Proc. Suppl.) **B53** (1977) 3; M. Lüscher, *et al.*, Nucl. Phys. **B478** (1996) 365; G. Martinelli, *et al.*, Nucl. Phys. **B445** (1995) 81.
[23] S. Sharpe, Nucl. Phys. (Proc. Suppl.) **B17** (1990) 146, Phys. Rev. **D41** (1990) 1990, Phys. Rev. **D46** (1992) 3146, Nucl. Phys. (Proc. Suppl.) **B30** (1993) 213.
[24] C. Bernard and M. Golterman, Phys. Rev. **D46** (1992) 853; Nucl. Phys. (Proc. Suppl.) **B30** (1993) 217; Nucl. Phys. (Proc. Suppl.) **B37** (1994) 334; M. Golterman, hep-lat/9405002 and hep-lat/9411005.
[25] R. Gupta, Nucl. Phys. (Proc. Suppl.) **B42** (1995) 85.
[26] S. Sharpe, Nucl. Phys. (Proc. Suppl.) **B53** (1997) 181.

[27] J. Gasser and H. Leutwyler, Phys. Rep. **C87** (1982) 77.
[28] H. Leutwyler, Nucl. Phys. **B337** (1990) 108.
[29] H. Leutwyler, hep-ph/9609467.
[30] D. Kaplan and A. Manohar, Phys. Rev. Lett. **56** (1986) 2004.
[31] T. Banks, Y. Nir, and N. Sieberg, hep-ph/9403203.
[32] T. Bhattacharya, and R. Gupta, hep-lat/9710095.
[33] T. Bhattacharya, R. Gupta, G. Kilcup, and S. Sharpe, Phys. Rev. **D53** (1996) 6486.
[34] R. Gupta and T. Bhattacharya *et al.*, Phys. Rev. **D55** (1997) 7203.
[35] B. Gough *et al.*, Phys. Rev. Lett. **79** (1997) 1622.
[36] J. Bijnens, J. Prades, and E. de Rafael, Phys. Lett. **348B** (1995) 226.
[37] M. Jamin and M. Münz, Z. Phys. **C66** (1995) 633.
[38] S. Narison, Phys. Lett. **358B** (1995) 113.
[39] S. Narison, Phys. Lett. **216B** (1989) 335.
[40] K.G. Chetyrkin, D. Pirjol, and K. Schilcher, Phys. Lett. **404B** (1997) 337.
[41] P. Colangelo, F. De Fazio, G. Nardulli, and N. Paver, Phys. Lett. **408B** (1997) 340.
[42] M. Jamin, hep-ph/9709484.
[43] J. Prades, hep-ph/9708395.
[44] T. Bhattacharya, R. Gupta, and K. Maltman, hep-ph/9703455.
[45] F.J. Ynduráin, hep-ph/9708300.
[46] H.G. Dosch and S. Narison, hep-ph/9709215.
[47] L. Lellouch, E. de Rafael, and J. Taron, Phys. Lett. **414B** (1997) 195.
[48] A. Buras, M. Jamin, and M. E. Lautenbacher, Phys. Lett. **389B** (1996) 749; hep-ph/9608365.
[49] L. K. Gibbons, *et al.*, Phys. Rev. Lett. **70** (1993) 1203.
[50] G. D. Barr, *et al.*, Phys. Lett. **317B** (1993) 233.
[51] G. Buchalla, A. Buras, and M. E. Lautenbacher, Rev. Mod. Phys. **68** (1996) 1125.
[52] A. Buras, hep-ph/9711217.
[53] G. Kilcup, S. Sharpe, R. Gupta, and A. Patel, Phys. Rev. Lett. **64** (1990) 25.
[54] S. Sharpe, Nucl. Phys. (Proc. Suppl.) **B34** (1994) 403.
[55] G. Kilcup, R. Gupta, and S. Sharpe, Phys. Rev. **D57** (1998) 1654, hep-lat/9707006.
[56] D. Pekurovsky and G. Kilcup, hep-lat/9709146.
[57] S. Aoki, *et al.*, JLQCD collaboration, hep-lat/9710073.
[58] G. Kilcup, D. Pekurovsky, and L. Venkatraman, Nucl. Phys. (Proc. Suppl.) **B53** (1997) 345.
[59] M. Golterman and K. C. Leung, Phys. Rev. **D56** (1997) 2950.
[60] J. Donoghue, E. Golowich, and B. Holstein, Phys. Lett. **119B** (1982) 412.
[61] C. Bernard, *et al.*, Phys. Rev. **D32** (1985) 2343.
[62] L. Conti, A. Donini, V. Gimenez, G. Martinelli, M. Talevi, A. Vladikas, hep-lat/9711053.
[63] C. Bernard and A. Soni, Nucl. Phys. (Proc. Suppl.) **B9** (1990) 155; Nucl. Phys. (Proc. Suppl.) **B17** (1990) 495.
[64] C. Bernard, in *From Actions to Answers*, Proceedings of the 1989 TASI School, edited by T. DeGrand and D. Toussaint, (World Scientific, 1990).
[65] M.B.Gavela, *et al.*, Phys. Lett. **211B** (1988) 139; Nucl. Phys. **B306** (1988) 677.
[66] S. Aoki, *et al.*, JLQCD Collaboration, hep-lat/9711046.
[67] R. Gupta, T. Bhattacharya, and S. Sharpe, Phys. Rev. **D55** (1997) 4036.
[68] G. Kilcup, Nucl. Phys. (Proc. Suppl.) **B20** (1991) 417.

SECTION V
Progress on New and Old Ideas II

PARTICLES AS BOUND STATES IN THEIR OWN POTENTIALS

R. P. Woodard

Department of Physics
University of Florida
Gainesville, FL 32611

1. INTRODUCTION: A PARABLE OF POLITICAL CORRECTNESS

I wish to speak out against a form of bigotry. The prejudice in question might be termed, *integro-centrism*, and it consists of the belief that asymptotic series may contain only non-negative, integer powers of the expansion coefficient. Not only is this exclusionary against non-integer powers and logarithms, it even discrimates against sign-challenged integers! I shall also argue that integro-centrism may be imposing a kind of cultural genocide on quantum gravity and on the problem of mass.

Imagine that you are the asymptotic expansion $\widetilde{f}(g)$ of some quantum field theoretic quantity $f(g)$. Without succumbing to negative stereotypes we can assume you have the following form:

$$\widetilde{f}(g) = \sum_{n=0}^{\infty} f_n \phi_n(g) \; . \tag{1.1}$$

We can also assume that the $\phi_n(g)$ are elementary functions which have been arranged in a (value-neutral) order such that:

$$\lim_{g \to 0} \frac{\phi_{n+1}(g)}{\phi_n(g)} = 0 \; . \tag{1.2}$$

Finally, the fact that you are asymptotic means that the difference between $f(g)$ and the sum of your first N terms must vanish faster than your N-th term as g goes to zero:

$$\lim_{g \to 0} \Big(f(g) - \sum_{n=0}^{N} f_n \phi_n(g) \Big) \frac{1}{\phi_N(g)} = 0 \; . \tag{1.3}$$

In the Ward and June Cleaver world of conventional perturbation theory the coefficient functions would be integer powers — $\phi_n(g) = g^n$ — and their coefficients could be obtained by taking derivatives of the original function $f(g)$:

$$f_n = \frac{1}{n!} \frac{\partial^n f(g)}{\partial g^n} \Big|_{g=0} \; . \tag{1.4}$$

Suppose, however, that you are leading an alternate lifestyle which includes logarithms or fractional powers. For example, you might have the form:

$$\widetilde{f}(g) = 1 + 3g \ln(g) + O(g^2) \; . \tag{1.5}$$

Physics of Mass
Edited by Kursunoglu *et al.*, Kluwer Academic / Plenum Publishers, New York, 1999

Although your actual first order correction is small for small g, an integro-centric bigot would claim it is logarithmically divergent:

$$f_1 = \lim_{g \to 0} \Big(3\ln(g) + 3 + O(g) \Big) \ . \tag{1.6}$$

And he would compute the higher terms to consist of an oscillating tower of increasingly virulent divergences:

$$f_n = \lim_{g \to 0} \Big(-\tfrac{3}{n}(-g)^{1-n} + \dots \Big) \quad , \qquad n \geq 2 \ . \tag{1.7}$$

His frustration with your non-conformism might provoke him to abandon the quantum field theory behind $f(g)$ in favor of some yet-to-be-specified model in a peculiar dimension. He might even take to making optimistic pronouncements about our ability to exactly solve this model, and hence its correspondence limits of Yang-Mills and General Relativity, in 5-10 years (3-8 years from now, and counting).

Aside from poking fun at the political and scientific prejudices of my colleagues this paper does have some serious points to make. The first of these is that there is no reason why the perturbative non-renormalizability[1–5] of General Relativity necessarily implies the need for an alternate theory of quantum gravity. It has long been realized that the problem could derive instead from the appearance of logarithms or fractional powers of Newton's constant in the correct asymptotic expansion of quantum gravity.[6–9] To underscore that this would not be without precedent I devote Section 2 to a discussion of the analogous phenomenon in two simple systems from statistical mechanics.

The second point I wish to make is that the breakdown of conventional perturbation theory in quantum gravity is likely to be associated with ultraviolet divergences. The idea is that gravity screens effects which tend to make the stress tensor divergent. If so, it must be that the divergence returns when Newton's constant goes to zero, which means the correct asymptotic series must contain logarithms or negative powers. There is no doubt that this does occur on the classical level. Arnowit, Deser and Misner found an explicit example in the finite self-energy of point charged particles.[10] Section 3 is devoted to a brief review of their result.

So far I have been discussing old stuff. Although many people have suspected that quantum gravity regulates ultraviolet divergences[6–9] no one has been able to make anything of the idea for want of a non-perturbative calculational technique. Divergences *do* evoke an infinite response from gravitation, but only at the next order in perturbation theory. What is needed is a way of reorganizing perturbation theory so that the gravitational response has a chance of keeping up with divergences. The main point of this paper is that I have found such a reorganization, at least for the special case of certain types of matter self-energies.

I begin the derivation in Section 4 by writing down an exact functional integral representation for the mass of a charged, gravitating scalar in quantum field theory. In Section 5 I show that this expression reduces, in the classical limit, to the point particle system studied by ADM,[10] with an extra term representing the negative pressure needed to hold the point charge together. In Section 6 I return to the original, exact expression, and show how it can be rearranged to give an expansion in the number of closed loops which do not include a least some part of the incoming and outgoing matter line. Further, the 0-th order term in this new expansion has the simple interpretation of computing the binding energy of a quantum mechanical particle which moves in the gravitational and electromagnetic potentials induced by its own probability current. This is the origin of the title.

Gravitational attraction must overcome electrostatic repulsion in order for a particle to bind to its own potentials. In Section 7 I obtain the unsurprising result that this can only happen for a scalar which has a Planck scale mass. In the final section I argue that substantially lighter masses may be obtainable for particles with spin.

2. TWO EXAMPLES FROM STATISTICAL MECHANICS

Exotic terms occur in many familiar asymptotic expansions. Consider the logarithm of the grand canonical partition function for non-interacting, non-relativistic bosons in a three dimensional volume V:

$$\ln\Big(\Xi\Big) = V\, n_Q \sum_{k=1}^{\infty} k^{-\frac{5}{2}} \exp(k\beta\mu) . \tag{2.1}$$

Here n_Q is the quantum concentration, μ is the chemical potential, and $\beta = (k_B T)^{-1}$. Near condensation one has $0 < -\beta\mu \ll 1$ so it should make sense to expand $\ln(\Xi)$ for small $\beta\mu$. Straightforward perturbation theory corresponds to the following expansion:

$$\ln\Big(\Xi\Big) = V\, n_Q \sum_{k=1}^{\infty} k^{-\frac{5}{2}} \sum_{\ell=0}^{\infty} (k\beta\mu)^{\ell} \tag{2.2a}$$

$$\rightarrow V\, n_Q \sum_{\ell=0}^{\infty} (\beta\mu)^{\ell} \sum_{k=1}^{\infty} k^{\ell-\frac{5}{2}} . \tag{2.2b}$$

Although the $\ell = 0$ and $\ell = 1$ terms are finite, the sum over k diverges for $\ell \geq 2$.

The divergences we have encountered do not mean that higher corrections are large, just that they are not as small as $(\beta\mu)^2$. One sees this by expanding the second derivative around its integral approximation:

$$\frac{\partial^2 \ln\Big(\Xi\Big)}{\partial(\beta\mu)^2} = V\, n_Q \sum_{k=1}^{\infty} k^{-\frac{1}{2}} \exp(k\beta\mu) \tag{2.3a}$$

$$= V\, n_Q \Bigg\{ \int_0^{\infty} dy\, y^{-\frac{1}{2}} \exp(y\beta\mu) + \sum_{k=1}^{\infty} \Big[k^{-\frac{1}{2}} \exp(k\beta\mu) - \int_{k-1}^{k} dy\, y^{-\frac{1}{2}} \exp(y\beta\mu) \Big] \Bigg\} \tag{2.3b}$$

$$= V\, n_Q \Bigg\{ \Big(\frac{-\pi}{\beta\mu}\Big)^{\frac{1}{2}} + \sum_{k=1}^{\infty} \Big[k^{-\frac{1}{2}} - 2\, k^{\frac{1}{2}} + 2\,(k-1)^{\frac{1}{2}} \Big] + O\Big(\beta\mu\Big) \Bigg\} . \tag{2.3c}$$

Integration reveals the true asymptotic expansion:

$$\ln\Big(\Xi\Big) = V\, n_Q \Big\{ \zeta\Big(\tfrac{5}{2}\Big) + \zeta\Big(\tfrac{3}{2}\Big)\, \beta\mu + \tfrac{4}{3}\sqrt{\pi}(-\beta\mu)^{\frac{3}{2}} + O\Big(\beta^2\mu^2\Big) \Big\} . \tag{2.4}$$

The oscillating series of ever-increasing divergences in the perturbative expansion (2.2b) has resolved itself into a perfectly finite, fractional power.

Logarithms can also invalidate perturbation theory. Consider the canonical partition function for a non-interacting particle of mass m in a three dimensional volume V:

$$Z = \frac{V}{2\pi^2\hbar^3} \int_0^{\infty} dp\, p^2 \exp\Big[-\beta\sqrt{p^2c^2 + m^2c^4} + \beta mc^2\Big] \tag{2.5a}$$

$$= \frac{V}{2\pi^2\hbar^3 c^3} \int_0^{\infty} dK\, (K + mc^2)\, \sqrt{K^2 + 2\,Kmc^2}\, \exp(-\beta K) . \tag{2.5b}$$

When the rest mass energy is small compared to the thermal energy it ought to make sense to expand in the small parameter $x \equiv \beta mc^2$. But straightforward perturbation theory fails again:

$$Z = \frac{V}{2\pi^2} \left(\frac{k_B T}{\hbar c}\right)^3 \int_0^\infty dt\, t^2 e^{-t} \left(1 + \tfrac{x}{t}\right) \sqrt{1 + 2\tfrac{x}{t}} \tag{2.6a}$$

$$= \frac{V}{2\pi^2} \left(\frac{k_B T}{\hbar c}\right)^3 \int_0^\infty dt\, t^2 e^{-t} \left\{1 + 2\tfrac{x}{t} + \tfrac{1}{2}\left(\tfrac{x}{t}\right)^2 - \sum_{n=3}^{\infty} \frac{(n-3)\,(2n-5)!!}{n!} \left(-\tfrac{x}{t}\right)^n \right\}. \tag{2.6b}$$

It seems as though the term of order x^3 vanishes, and that the higher terms have increasingly divergent coefficients with oscillating signs. In fact the x^3 term is non-zero, and the apparent divergences merely signal contamination with logarithms:

$$Z = \frac{V}{2\pi^2} \left(\frac{k_B T}{\hbar c}\right)^3 \left\{2 + 2\,x + \tfrac{1}{2}\,x^2 - \tfrac{1}{6}\,x^3 - \tfrac{1}{48}\,x^4 \ln(x) + O\!\left(x^4\right)\right\}. \tag{2.7}$$

3. THE ADM MECHANISM

Arnowitt, Deser and Misner showed that perturbation theory also breaks down in computing the self-energy of a classical, charged, gravitating point particle.[10] It is simplest to model the particle as a stationary spherical shell of radius ϵ, charge e and bare mass m_0. In Newtonian gravity its energy would be:

$$E = m_0 + \frac{e^2}{8\pi\epsilon} - \frac{G m_0^2}{2\epsilon}. \tag{3.1}$$

It turns out that all the effects of general relativity are accounted for by replacing E and m_0 with the full mass:*

$$m_\epsilon = m_0 + \frac{e^2}{8\pi\epsilon} - \frac{G m_\epsilon^2}{2\epsilon} \tag{3.2a}$$

$$= \frac{\epsilon}{G}\left[-1 + \sqrt{1 + \frac{2G}{\epsilon}\left(m_0 + \frac{e^2}{8\pi\epsilon}\right)}\,\right]. \tag{3.2b}$$

The perturbative result is obtained by expanding the square root:

$$m_{\text{pert}} = m_0 + \frac{e^2}{8\pi\epsilon} + \sum_{n=2}^{\infty} \frac{(2n-3)!!}{n!} \left(-\frac{G}{\epsilon}\right)^{n-1} \left(m_0 + \frac{e^2}{8\pi\epsilon}\right)^n, \tag{3.3}$$

and shows the oscillating series of increasingly singular terms characteristic of the previous examples. The alternating sign derives from the fact that gravity is attractive. The positive divergence of order e^2/ϵ evokes a negative divergence or order Ge^4/ϵ^3, which results in a positive divergence of order $G^2 e^6/\epsilon^5$, and so on. The reason these terms are increasingly singular is that the gravitational response to an effect at one order is delayed to a higher order in perturbation theory.

The correct result is obtained by taking ϵ to zero before expanding in the coupling constants e^2 and G:

$$\lim_{\epsilon \to 0} m_\epsilon = \left(\frac{e^2}{4\pi G}\right)^{\frac{1}{2}}. \tag{3.4}$$

* It should be noted that Arnowitt, Deser and Misner rigorously solved the constraint equations of general relativity and electrodynamics, and then used the asymptotic metric to compute the ADM mass.[10] They also developed the simple model I am presenting.

Like the examples of Section 2 it is finite but not analytic in the coupling constants e^2 and G. Unlike the previous examples, it diverges for small G. This is because gravity has regulated the linear self-energy divergence which results for a non-gravitating charged particle.

One can understand the process from the fact that gravity has a built-in tendency to oppose divergences. A charge shell does not want to contract in pure electromagnetism; the act of compressing it calls forth a huge energy density concentrated in the nearby electric field. Gravity, on the other hand, tends to make things collapse, especially large concentrations of energy density. The dynamical signature of this tendency is the large negative energy density concentrated in the Newtonian gravitational potential. In the limit the two effects balance and a finite total mass results.

Said this way, there seems no reason why gravitational interactions should not act to cancel divergences in quantum field theory. It is especially significant, in this context, that the divergences of some quantum field theories — such as QED — are weaker than the linear ones which ADM have shown that classical gravity regulates. The frustrating thing is that one cannot hope to see the cancellation perturbatively. In perturbation theory the gravitational response to an effect at any order must be delayed to a higher order. This is why the perturbative result (3.3) consists of an oscillating series of ever higher divergences. What is needed is an approximation technique in which the gravitational response is able to keep pace with what is going on in other sectors.

A final point of interest is that any finite bare mass drops out of the exact result (3.4) in the limit $\epsilon \to 0$. This makes for an interesting contrast with the usual program of renormalization. Without gravity one would pick the desired physical mass, m_p, and then adjust the bare mass to be whatever divergent quantity was necessary to give it:

$$m_0 = m_p - \frac{e^2}{8\pi\epsilon} \,. \tag{3.5}$$

Of course the same procedure would work with gravity as well:

$$m_0 = m_p - \frac{e^2}{8\pi\epsilon} + \frac{Gm_p^2}{2\epsilon} \,. \tag{3.6}$$

The difference with gravity is that we have an alternative: keep m_0 finite and let the dynamical cancellation of divergences produce a unique result for the physical mass. The ADM mechanism is in fact the classical realization of the old dream of computing a particle's mass from its self-interactions.

4. MASS OF A CHARGED GRAVITATING SCALAR IN QFT

The purpose of this section is to obtain a conveneient functional integral representation for the standard quantum field theoretic definition of a particle's mass as the pole of its propagator. For simplicity I will consider a charged, gravitating scalar, the Lagrangian for which is:

$$\begin{aligned}\mathcal{L} = \frac{1}{16\pi G}\Big(R\sqrt{-g} - \text{S.T.}\Big) - \tfrac{1}{4}F_{\mu\nu}F_{\rho\sigma}g^{\mu\rho}g^{\nu\sigma}\sqrt{-g} \\ - \Big(\partial_\mu - iqA_\mu\Big)\phi^*\Big(\partial_\nu + iqA_\nu\Big)\phi\, g^{\mu\nu}\sqrt{-g} - m_0^2\,\phi^*\phi\sqrt{-g} \,. \end{aligned} \tag{4.1a}$$

The symbol "S.T." denotes the gravitational surface term needed to purge the Lagrangian of second derivatives:

$$\text{S.T.} \equiv \partial_\mu\Big[(g_{\nu\rho,\sigma} - g_{\rho\sigma,\nu})g^{\mu\nu}g^{\rho\sigma}\sqrt{-g}\Big] \,. \tag{4.1b}$$

If we temporarily regulate infrared divergences and agree to understand operator relations in the weak sense then it is possible to write the operators which annihilate outgoing particles and create incoming ones as simple limits:*

$$a_k^{\rm out} = \lim_{t_+\to\infty} \frac{ie^{i\omega t_+}}{\sqrt{2\omega Z}} \overleftrightarrow{W}_+ \widetilde{\phi}(t_+,\vec{k}) \; , \tag{4.2a}$$

$$\left(a_k^{\rm in}\right)^+ = \lim_{t_-\to-\infty} \frac{ie^{-i\omega t_-}}{\sqrt{2\omega Z}} \overleftrightarrow{W}_- \widetilde{\phi}^*(t_-,\vec{k}) \; , \tag{4.2b}$$

where the energy is:

$$\omega \equiv \sqrt{k^2+m^2} \; . \tag{4.3}$$

Consider single particle states whose wave functions in the infinite past and future are $\psi_\mp$, respectively. The inner product between two such states can be given the following expression:

$$\left\langle \psi_+^{\rm out} \middle| \psi_-^{\rm in} \right\rangle = \int \frac{d^3k}{(2\pi)^3} \frac{\psi_+^*(\vec{k})\psi_-(\vec{k})}{2\omega Z} \lim_{t_\pm\to\pm\infty} e^{i\omega(t_+-t_-)} \overleftrightarrow{W}_+ \overleftrightarrow{W}_- \times \int d^3x \, e^{-i\vec{k}\cdot\vec{x}} \left\langle \Omega^{\rm out} \middle| \phi(t_+,\vec{x})\phi^*(t_-,\vec{0}) \middle| \Omega^{\rm in} \right\rangle . \tag{4.4}$$

One way of computing the mass is to tune the parameter m in the energy (4.3) to the precise value for which expression (4.4) assumes the form:

$$\left\langle \psi_+^{\rm out} \middle| \psi_-^{\rm in} \right\rangle = \int \frac{d^3k}{(2\pi)^3} \frac{1}{2\omega} \psi_+^*(\vec{k})\psi_-(\vec{k}) \; . \tag{4.5}$$

This agrees with the usual definition of the mass as the pole of the propagator.

A somewhat more direct way of computing the mass is to focus on the second line of (4.4) which we can write as a phase:

$$e^{-i\xi(t_+,t_-,k)} \equiv \int d^3x \, e^{-i\vec{k}\cdot\vec{x}} \left\langle \Omega^{\rm out} \middle| \phi(t_+,\vec{x})\phi^*(t_-,\vec{0}) \middle| \Omega^{\rm in} \right\rangle . \tag{4.6}$$

Dividing by the time interval and then taking it to infinity we obtain the energy:

$$\lim_{t_\pm\to\pm\infty} \left(\frac{\xi(t_+,t_-,k)}{t_+-t_-} \right) = \sqrt{k^2+m^2} \; . \tag{4.7}$$

Note that by using this method we avoid the problem of infrared divergences. These affect only the field strength renormalization, not the mass.

It is straightforward to write the phase as a functional integral:

$$e^{-i\xi} = \int d^3x \, e^{-i\vec{k}\cdot\vec{x}} \int [dg][dA][d\phi] \, \phi(t_+,\vec{x}) \, \phi^*(t_-,\vec{0}) \, e^{iS_{\rm GR}[g]+iS_{\rm EM}[g,A]+iS_\phi[g,A,\phi]} \; . \tag{4.8}$$

The next step is to integrate out the scalar. In the presence of an arbitrary metric and electromagnetic background its kinetic operator is:

$$\mathcal{D}[g,A] \equiv \frac{1}{\sqrt{-g}} \Big(\partial_\mu + iqA_\mu\Big) \sqrt{-g} g^{\mu\nu} \Big(\partial_\nu + iqA_\nu\Big) \; . \tag{4.9}$$

We can use this operator to express the scalar-induced effective action:

$$\Gamma_\phi[g,A] \equiv -i \ln\Big(\det[-\mathcal{D} + m_0^2 - i\epsilon]\Big) \; , \tag{4.10a}$$

---

* The notation employed in these formulae is standard: Z is the field strength renormalization, the Wronskian is $\overleftrightarrow{W} \equiv \overrightarrow{\partial}_0 - \overleftarrow{\partial}_0$, and a tilde over the scalar field denotes its spatial Fourier transform.

and the scalar propagator in the presence of an general background:

$$D[g,A]\Big(t_+,\vec{x};t_-,\vec{0}\Big) \equiv \Big\langle t_+,\vec{x}\Big|\frac{i}{\mathcal{D}-m_0^2+i\epsilon}\Big|t_-,\vec{0}\Big\rangle \,. \tag{4.10b}$$

With these objects the phase can be reduced to a functional integral over only metrics and vector potentials:

$$e^{-i\xi} = \int d^3x\; e^{-i\vec{k}\cdot\vec{x}}\int [dg][dA]\; D[g,A]\Big(t_+,\vec{x};t_-,\vec{0}\Big)\; e^{iS_{\rm GR}[g]+iS_{\rm EM}[g,A]+i\Gamma_\phi[g,A]} \,, \tag{4.11}$$

Contact is made with particle dynamics by writing the general propagator in Schwinger form:

$$\Big\langle t_+,\vec{x}\Big|\frac{i}{\mathcal{D}-m_0^2+i\epsilon}\Big|t_-,\vec{0}\Big\rangle = \int_0^\infty ds\;\Big\langle t_+,\vec{x}\Big|\exp\Big[is(\mathcal{D}-m_0^2+i\epsilon)\Big]\Big|t_-,\vec{0}\Big\rangle \,. \tag{4.12}$$

One then regards the exponent as the Hamiltonian of a first quantized particle and the expectation value is converted into a functional integral in the usual way. We can give this a reparameterization invariant form by regarding the proper time as the unfixed part of the einbein $e(\tau)$ in $\dot{e}=0$ gauge:

$$\int_0^\infty ds = \int [de]\;\delta[\dot{e}] \tag{4.13}$$

Integrating out the canonical momenta and absorbing any ordering terms into the measure gives:

$$\begin{aligned}\int d^3x\; e^{-i\vec{k}\cdot\vec{x}}&\Big\langle t_+,\vec{x}\Big|\frac{i}{\mathcal{D}-m_0^2+i\epsilon}\Big|t_-,\vec{0}\Big\rangle \\ &= \int [de][d^4\chi]\;\delta[\dot{e}[\;\delta\Big(\chi^0(0)-t_-\Big)\;\delta\Big(\chi^0(1)-t_+\Big)\;\delta^3\Big(\vec{\chi}(0)\Big) \\ &\quad\times \exp\Big\{-i\vec{k}\cdot\vec{\chi}(1)+i\int_0^1 d\tau\Big[\tfrac{1}{4e}g_{\mu\nu}\dot{\chi}^\mu\dot{\chi}^\nu - em_0^2 - q\dot{\chi}^\mu A_\mu\Big]\Big\}\end{aligned} \tag{4.14}$$

One now makes the change of variables defined by the reparameterization which changes the gauge condition to:

$$\chi^0(\tau) \equiv t_- + (t_+ - t_-)\tau \,. \tag{4.15}$$

The integral over the einbein is done using the functional equivalent of the identity:

$$\int_0^\infty \frac{dx}{\sqrt{x}}\exp\Big[-ax-\tfrac{b}{x}\Big] = \sqrt{\tfrac{\pi}{a}}\exp\Big[-\sqrt{4ab}\Big] \,. \tag{4.16}$$

The final form is:

$$\begin{aligned}\int d^3x\; e^{-i\vec{k}\cdot\vec{x}}&\Big\langle t_+,\vec{x}\Big|\frac{i}{\mathcal{D}-m_0^2+i\epsilon}\Big|t_-,\vec{0}\Big\rangle \\ &= \int [d^3\chi]\delta^3\Big(\vec{\chi}(0)\Big)\exp\Big\{-i\vec{k}\cdot\vec{\chi}(1) - i\int_0^1 d\tau\Big[m_0\sqrt{-g_{\mu\nu}\dot{\chi}^\mu\dot{\chi}^\nu} + q\dot{\chi}^\mu A_\mu\Big]\Big\}\end{aligned} \tag{4.17}$$

where $\chi^0(\tau)$ is understood to be defined by (4.15).

Substituting (4.17) into (4.11) gives the following expression for the phase:

$$\begin{aligned}e^{-i\xi} = &\int [dg][dA][d^3\chi]\;\delta^3\Big(\vec{\chi}(0)\Big)\exp\Big\{-i\vec{k}\cdot\vec{\chi}(1)\Big\} \\ &\times\exp\Big\{iS_{\rm GR}[g]+iS_{\rm EM}[g,A]+iS_{\rm part}[g,A,\chi]+i\Gamma_\phi[g,A]\Big\} \,,\end{aligned} \tag{4.18a}$$

where the particle action is:

$$S_{\text{part}}[g,A,\chi] \equiv -\int_0^1 d\tau \left[m_0 \sqrt{-g_{\mu\nu}\Big(\chi(\tau)\Big)\dot{\chi}^\mu(\tau)\dot{\chi}^\nu(\tau)} + q\dot{\chi}^\mu(\tau)A_\mu\Big(\chi(\tau)\Big)\right] . \tag{4.18b}$$

It should be noted again that various ordering corrections have be subsumed into the measure. Also note, again, that $\chi^0(\tau)$ is the non-dynamical function (4.15).

5. THE CLASSICAL LIMIT

The purpose of this section is to show how the classical limit of ξ relates to the ADM[10] mechanism discussed in Section 3. This is crucial to seeing that the reorganization of of perturbation theory I shall propose in the next section in fact manifests the gravitational regulation of ultraviolet divergences at lowest order.

We can forget about the scalar-induced effective action Γ_ϕ because it is a quantum effect. What is necessary for our purposes is to solve the classical field equations derived from the action:*

$$S_{\text{class}}[g,A,\chi] = S_{\text{GR}}[g] + S_{\text{EM}}[g,A] + S_{\text{part}}[g,A,\chi] . \tag{5.1}$$

The boundary conditions for the metric and the vector potential come from the asymptotic in and out vacua. Those for the particle are:

$$\chi^i(0) = 0 \quad , \quad \dot{\chi}^i(1) = k^i . \tag{5.2a}$$

We can save ourselves a small amount of effort by instead imposing:

$$\chi^i(0) = 0 \quad , \quad \dot{\chi}^i(1) = 0 , \tag{5.2b}$$

and then boosting up to (5.2a). If (5.2b) is used one finds the classical limit of the phase by evaluating the action at the solution:

$$\xi_{\text{class}}(t_+, t_-, 0) = -S_{\text{class}}[\widehat{g}, \widehat{A}, \widehat{\chi}] . \tag{5.3a}$$

One then divides out the time interval and takes the asymptotic limit:

$$m_{\text{class}} = \lim_{t_\pm \to \pm\infty} \left(\frac{\xi(t_+, t_-, 0)}{t_+ - t_-}\right) . \tag{5.3b}$$

Although gravity does regulate this problem, just as it did for that of ADM,[10] some of the intermediate expressions will be singular unless the point particle is smeared out. ADM resolved this issue by converting the particle into a spherical shell of radius ϵ in isotropic coordinates. I shall do the same, but I face the additional problem, which they did not, of keeping the system static for all time. I shall accordingly employ a perfect fluid regularization in which the point particle is converted into a swarm of particles labelled by an internal vector $\vec{\sigma}$:

$$\chi^i(\tau) \longrightarrow X^i(\tau, \vec{\sigma}) \tag{5.4}$$

The particle's action goes to that of a perfect fluid:

$$S_{\text{part}}[g,A,\chi] \longrightarrow -\int d\tau d^3\sigma \left\{ \sqrt{-g_{\alpha\beta}\dot{X}^\alpha \dot{X}^\beta}\left[\mu(\vec{\sigma}) + \frac{\Pi(\vec{\sigma})}{\sqrt{-g}}\right] + \frac{q}{m_0}\mu(\vec{\sigma})\dot{X}^\mu A_\mu \right\} , \tag{5.5}$$

---

* The same technique has been used, in the context of 2-body scattering, by Fabbrichesi, Pettorino, Veneziano and Vilkovisky[11].

with number density $n(x)$ given by $\mu(\vec{\sigma})$:

$$n(x) = \frac{1}{\sqrt{-g}} \int d\tau d^3\sigma \, \frac{\mu(\vec{\sigma})}{m_0} \delta^4\Big(x - X(\tau,\vec{\sigma})\Big) \sqrt{-g_{\alpha\beta}\dot{X}^\alpha \dot{X}^\beta} \, , \tag{5.6a}$$

and pressure $p(x)$ given by $\Pi(\vec{\sigma})$:

$$p(x) = -\frac{1}{g} \int d\tau d^3\sigma \, \Pi(\vec{\sigma}) \, \delta^4\Big(x - X(\tau,\vec{\sigma})\Big) \sqrt{-g_{\alpha\beta}\dot{X}^\alpha \dot{X}^\beta} \, , \tag{5.6b}$$

I shall follow ADM in taking the mass density to be that of a spherical shell:

$$\mu(\vec{\sigma}) \equiv \frac{m_0 \delta(\sigma - \epsilon)}{4\pi\epsilon^2} \, , \tag{5.7a}$$

however, I shall impose a negative internal pressure:

$$\Pi(\vec{\sigma}) = -f(\epsilon)\theta(\epsilon - \sigma) \, , \tag{5.7b}$$

to hold the shell together. Duff has shown that this does not affect the ADM mass.[12] The function $f(\epsilon)$ is a non-dynamical constant to be determined shortly.

The manifest spherical symmetry and the assumed time translation invariance of this problem suggest that we look for a solution of the form:

$$\widehat{g}_{\mu\nu} dx^\mu dx^\nu = -A^2(r)dt^2 + B^2(r) d\vec{x} \cdot d\vec{x} \, , \tag{5.8a}$$

$$\widehat{A}_\mu dx^\mu = A_0(r) dt \, , \tag{5.8b}$$

$$\widehat{X}^\mu(\tau,\vec{\sigma}) = \delta^\mu_0 \Big(t_- + (t_+ - t_-)\tau\Big) + \delta^\mu_i \sigma^i \, . \tag{5.8c}$$

The solution for $A(r)$ has the form $A(r) = \alpha(r)/B(r)$ with:

$$\alpha(r) = \theta(r-\epsilon)\Big[1 + \frac{Q^2 - M^2}{4r^2}\Big] + \theta(\epsilon - r)\Big[1 + \Big(\frac{Q^2 - M^2}{4\epsilon^2}\Big)\Big(2 - \frac{r^2}{\epsilon^2}\Big)\Big] \, . \tag{5.9a}$$

The two other functions work out to be:

$$B(r) = \theta(r-\epsilon)\Big[1 + \frac{M}{r} - \Big(\frac{Q^2 - M^2}{4r^2}\Big)\Big] + \theta(\epsilon - r)\Big[1 + \frac{M}{\epsilon} - \Big(\frac{Q^2 - M^2}{4\epsilon^2}\Big)\Big] \, , \tag{5.9b}$$

$$A_0(r) = \frac{q}{4\pi B(r)} \Big\{ \frac{\theta(r-\epsilon)}{r} + \frac{\theta(\epsilon - r)}{\epsilon} \Big\} \, . \tag{5.9c}$$

The parameters in relations (5.9a-c) have been represented as lengths according to the standard convention of geometrodynamics:

$$M_0 \equiv G m_0 \quad , \quad Q \equiv q\sqrt{\frac{G}{4\pi}} \, , \tag{5.10a}$$

$$M \equiv \epsilon \Big[-1 + \sqrt{1 + \frac{2M_0}{\epsilon} + \frac{Q^2}{\epsilon^2}}\Big] \, . \tag{5.10b}$$

The necessary internal pressure constant $f(\epsilon)$ works out to be:

$$f(\epsilon) = \frac{B^3(\epsilon)}{8\pi G} \Big(\frac{Q^2 - M^2}{\epsilon^4}\Big) \, . \tag{5.11}$$

Since the solution is static the action must be a constant multiplied by the time interval. However, this constant turns outs not to be minus the ADM mass but rather a sort of enthalpy reflecting the presence of the pressure in the perfect fluid regularization of the particle action:

$$\Big(S_{\rm GR} + S_{\rm EM} + S_{\rm part}\Big)\Big|_{\rm solution} = -\frac{M}{G}(t_+ - t_-) - \int d^4x\sqrt{-g}\,p\,, \qquad (5.12a)$$

$$\equiv -\Big(U + pV\Big)(t_+ - t_-)\,. \qquad (5.12b)$$

The energy U and the pV term can be evaluated for any ϵ. They have the following simple forms:

$$U = \frac{M}{G} \quad , \quad pV = -\frac{1}{3G}(M - M_0)\,. \qquad (5.13)$$

In the limit $\epsilon \to 0$ the energy just gives the ADM mass (3.4). The pV term remains finite in this limit, but neither does it vanish. Its physical interpretation seems to be that gravity is not sufficient to hold the charge together. This means the calculation is not really consistent. I shall do better shortly, but do not let this obscure two important facts:

(1) Contact has been established, modulo the pV term, between the standard quantum field theoretic definition of mass and the classical ADM calculation of the self-energy of a gravitating, point charged particle.

(2) Even with the pV term, gravity has suppressed what would otherwise be a divergent result.

6. QUANTUM MECHANICAL INTERPRETATION

The purpose of this section is to introduce the promised reorganization of conventional perturbation theory. The starting point is the expression (4.18a) obtained for the phase at the end of Section 4:

$$e^{-i\xi} = \int [dg][dA]\,\exp\Big\{iS_{\rm GR}[g] + iS_{\rm EM}[g,A] + i\Gamma_\phi[g,A]\Big\}$$
$$\times \int [d\chi]\,\delta^3\Big(\vec{\chi}(0)\Big)\,\exp\Big\{-i\vec{k}\cdot\vec{\chi}(1) + iS_{\rm part}[g,A,\chi]\Big\}\,. \qquad (6.1)$$

The second line of this expression can be interpreted as the amplitude for a quantum mechanical particle to go from a delta function at $t = t_-$:

$$\psi_-(\vec{x}) = \delta^3(\vec{x})\,, \qquad (6.2a)$$

to a plane wave at $t = t_+$:

$$\psi_+(\vec{x}) = e^{i\vec{k}\cdot\vec{x}}\,, \qquad (6.2b)$$

Let us denote the associated action as:

$$\exp\Big\{iS_{\rm prop}[g,A]\Big\} \equiv \int [d\chi]\,\delta^3\Big(\vec{\chi}(0)\Big)\,\exp\Big\{-i\vec{k}\cdot\vec{\chi}(1) + iS_{\rm part}[g,A,\chi]\Big\}\,. \qquad (6.3)$$

I define the 0-th order term in the reorganized perturbation theory to be the stationary phase approximation to the functional integral over metrics and vector potentials with the following action:

$$S_{\rm class}[g,A] = S_{\rm GR}[g] + S_{\rm EM}[g,A] + S_{\rm prop}[g,A]\,. \qquad (6.4)$$

To see what diagrams this term captures it is simplest to identify the ones it misses. No closed scalar loops are included since the scalar-induced effective action

Γ_ϕ was excluded from (6.4). This does not mean there are no scalar lines at all. Owing to the presence of S_{prop} there must be a single, continuous scalar line in all the included diagrams. Any number of graviton and gauge lines can be attached to this line. However, the restriction to stationary phase means that we include no closed gauge and/or graviton loops which do not include some portion of the single, continuous scalar line. *So the 0-th order term I have defined consists of all diagrams with a single, continuous scalar line and no closed loops which do not include this line.* One can imagine that the next order term consists of diagrams with one closed loop external to the in-out line, and so on.

It is obvious from the way I have defined it that the 0-th order term constitutes a gauge invariant resummation of an infinite subset of diagrams. It is also obvious this 0-th term contains the classical limit considered in the previous section. Further, it will be rendered *less* singular, not more, by the inevitable quantum spread of the particle. All of this implies that the 0-th term just defined must manifest the gravitational suppression of divergences.

The physical interpretation of the 0-th order approximation to ξ is the phase developed by a quantum mechanical particle moving in the potentials generated by its own probability current. Whether or not there is any chance of being able to compute it depends upon which of the following two possibilities is realized:

(1) The particle cannot form bound states in its own potentials; or

(2) The particle can form bound states in its own potentials.

In case (1) we are left with a complicated, time dependent scattering problem which seems to be intractable. However, many simplifications are possible in case (2).

If bound states form one can forget about the asymptotic wavefunctions (6.2a-b). In the limit of infinite time separation the phase will be dominated by the lowest energy state. Further, one need only compute $S_{\text{prop}}[g, A]$ for a class of metrics and vector potentials which is broad enough to include the eventual solution. In the scalar problem we could immediately reduce from nine functions of x^μ to a static, spherically symmetric system characterized by only three functions of a single variable:

$$g_{\mu\nu}dx^\mu dx^\nu = -A^2(r)dt^2 + B^2(r)d\vec{x}\cdot d\vec{x} \qquad (6.5a)$$

$$A_\mu dx^\mu = A_0(r)dt \qquad (6.5b)$$

Finally, variational methods can be usefully applied. If one simply guesses the wavefunction, assuming static potentials, and then minimizes the total energy, the result will be an upper bound on the true 0-th order mass. Note, in this context, that *any* finite result would be awe inspiring.

7. QUANTUM MECHANICS IN REISSNER-NORDSTROM

The purpose of this section is to ascertain which of the two cases pertains to a charged, gravitating scalar: can it bind to its own potentials or not? We can immediately specialize to the static, spherically symmetric potentials (6.5a-b). Modulo the effects of operator ordering, the Hamiltonian is:

$$H = \frac{A(r)}{B(r)}\sqrt{\|\vec{p}\|^2 + m_0^2 B^2(r)} + qA_0(r) \qquad (7.1)$$

Assuming the particle is bound, we can invoke Birkhoff's theorem to fix the potentials outside most of the particle's probability density:

$$A_{\text{ext}}(r) = \frac{1}{B_{\text{ext}}}\left[1 + \left(\frac{Q^2 - M^2}{4r^2}\right)\right], \qquad (7.2a)$$

$$B_{\text{ext}}(r) = 1 + \frac{M}{r} - \left(\frac{Q^2 - M^2}{4r^2}\right), \qquad (7.2b)$$

$$A_0^{\text{ext}}(r) = \frac{q}{4\pi}\frac{1}{r B_{\text{ext}}} \; . \tag{7.2c}$$

The various parameters have been expressed as lengths in the usual geometrodynamical convention:

$$Q \equiv q\sqrt{\frac{G}{4\pi}} \quad , \qquad M_0 \equiv m_0 G \; , \tag{7.3}$$

however, it should be noted that M is at this stage undetermined. We can also assume that the momentum is dominated by uncertainty pressure:

$$p \sim \frac{\hbar}{r} \qquad \Longrightarrow P \equiv G\, p \sim \frac{L_P^2}{r} \; , \tag{7.4}$$

where L_P is the Planck length.

If we geometrodynamicize the Hamitonian ($\mathcal{H} \equiv GH$), then its form beyond most of the probability density is:

$$\mathcal{H}_{\text{ext}} = \frac{A_{\text{ext}}}{B_{\text{ext}}}\sqrt{P^2 + M_0^2 B_{\text{ext}}^2} + \frac{Q^2}{r B_{\text{ext}}} \tag{7.5}$$

At large r this becomes:

$$\mathcal{H}_{\text{ext}} \sim M_0 + (Q^2 - M_0 M)\frac{1}{r} \qquad r \gg Q \tag{7.6}$$

One consequence of (7.3) is that the ratio Q/L_P goes like the square root of the fine structure constant, whereas the M_0/L_P is the ratio of m_0 to the Planck mass. It follows that any particle which is relevant to low energy physics must obey:

$$M_0, M \ll Q \tag{7.7}$$

In this case we see that the Hamiltonian falls off asymptotically, suggesting that no reasonably light bound state can form.

It is not really consistent to use the external potentials in the interior but doing so fails to reveal an inner region of binding. In isotropic coordinates the singularity occurs at $r_0 = (Q - M)/2$. Specializing to a point just slightly outside gives only another repulsive Hamiltonian:

$$\mathcal{H}_{\text{ext}} \sim \frac{(Q - M)^2 L_P^2}{4Q\epsilon^2} \qquad r = \tfrac{1}{2}(Q - M) + \epsilon \tag{7.8}$$

It seems fair to conclude that any charged scalar bound states would necessarily have Planck scale masses. On the other hand, setting $Q = 0$ in (7.6) seems to suggest that the chargeless scalar can form a bound state. Expression (7.8) suggests that quantum uncertainty pressure protects it from collapse, unlike the neutral scalar studied classically by ADM.[10]

8. DISCUSSION

I have proposed a gauge invariant reorganization of conventional perturbation theory in which gravitational regularization is a 0-th order effect. The existence of any new technique deserves comment because it might be thought that the possibilities for one have been pretty well exhausted by now. A new expansion must be in terms of some parameter, such as the dimension of spacetime[13] or the number of matter fields,[14] and there simply aren't any plausible parameters that have not been tried.

The secret of my expansion is that it does not conform to the usual rules which require the parameter to appear in the Lagrangian. I have instead exploited a parameter which depends, to some extent, on the thing being computed. This parameter is the number of closed loops which are external to continuous matter lines that come in from the asymptotic past and proceed out to the asymptotic future. Not all processes have such lines. However, the technique can be used on those that do, and any evidence for the non-perturbative viability of quantum General Relativity would be interesting.

Of particular interest are the self-energies of matter particles. The classical computation of ADM,[10] summarized in Section 3, suggests that this is a natural setting for conventional perturbation theory to break down. If quantum gravity regulates ultraviolet divergences as classical gravity certainly does then the asymptotic expansion must contain inverse powers or logarithms.

I was able to reexpress the standard definition of the pole of the propagator in terms of the new expansion. The 0-th order term has the physical interpretation of the phase developed by a quantum mechanical particle moving in the potentials induced by its own probability current. If these potentials cannot form bound states one has an intractable scattering problem. However, the 0-th term is eminently calculable if there are bound states. In this case the lowest energy state dominates. One can also assume that the potentials are static, and that they possess simplifying symmetries. If nothing else works, it is always possible to obtain an upper bound on the mass through variational techniques.

The explicit analysis of Section 7 indicates that there is probably not a charged bound state scalar of less than about the Planck mass. However, it does seem possible that light neutral scalars can form. Adding spin complicates the gravitational and electrodynamic potentials enormously. It also adds a new parameter in the form of the spin-to-mass ratio a. It may be very significant that, whereas the charge parameter completely dominates the mass, the spin-to-mass ratio is larger by almost the same ratio. For an electron one finds:

$$M \sim 10^{-55}\ \text{cm} \ll -Q \sim 10^{-34}\ \text{cm} \ll a \sim 10^{-11}\ \text{cm}\ , \tag{8.1}$$

so it is not unreasonable to expect large spin-dependent forces. The physical interpretation for this is that different portions of a rapidly spinning body see one another through enormous relative boosts. What is a minuscule matter density in our frame can therefore seem overwhelming from the instantaneous rest frame of a spinning observer.* So there is some hope for getting light fermionic bound states.

ACKNOWLEDGEMENTS

It is a pleasure to acknowledge almost twenty years of conversations on this subject with S. Deser and N. C. Tsamis. This work was partially supported by DOE contract DE-FG02-97ER41029 and by the Institute for Fundamental Theory.

REFERENCES

1. G. 't Hooft and M. Veltman, *Ann. Inst. Henri Poincaré* 20:69 (1974).
2. M. Goroff and A. Sagnotti, *Phys. Lett.* B106:81 (1985); *Nucl. Phys.* B266:709 (1986).
3. S. Deser and P. van Nieuwenhuizen, *Phys. Rev.* D10:401 (1974).
4. S. Deser and P. van Nieuwenhuizen, *Phys. Rev.* D10:411 (1974).
5. S. Deser, Hung Sheng Tsao and P. van Nieuwenhuizen, *Phys. Rev.* D10:3337 (1974).
6. S. Deser, *Rev. Mod. Phys.* 29:417 (1957).
7. B. S. DeWitt, *Phys. Rev. Lett.* 13:114 (1964).
8. I. B. Khriplovich, *Soviet J. Nucl. Phys.* 3:415 (1966).
9. C. J. Isham, A. Salam, and J. Strathdee, *Phys. Rev.* D3:1805 (1971).

* I wish to thank D. N. Page for elucidating this point.

10. R. Arnowit, S. Deser and C. W. Misner, *Phys. Rev. Lett.* 4:375 (1960);
Phys. Rev. 120:313 (1960);
Phys. Rev. 120:321 (1960);
Ann. Phys. 33:88 (1965).
11. M. Fabbrichesi, R. Pettorino, G. Veneziano and G. A. Vilkovisky, *Nucl. Phys.* B419:147 (1994).
12. M. J. Duff, *Phys. Rev.* D7:2317 (1973).
13. A. Strominger, *Phys. Rev.* D24:3082 (1982).
14. A. Strominger, A gauge invariant resummation of quantum gravity, in: *International Symposium on Gauge Theory and Gravitation, Nara, Japan, Aug. 20-24, 1982*, eds., K. Kikkawa, N. Nakanishi and H. Nariai, Springer-Verlag, (1983).

ENHANCED SYMMETRIES AND TENSOR THEORIES IN SIX DIMENSIONS

L. Dolan

Department of Physics, University of North Carolina
Chapel Hill, North Carolina 27599-3255, USA

INTRODUCTION

The IIB superstring compactified on $R^6 \times K3$ is a $D = 6$ theory with $N = (2,0)$ spacetime (local) supersymmetry, which at generic points in the $K3$ moduli space has the massless spectrum of a supergravity multiplet with Spin(4) content $(3,3) + 5(1,3) + 4(2,3)$ and 21 tensor supermultiplets $(3,1) + 5(1,1) + 4(2,1)$, see [1,2]. Recently there has been evidence of a new type of quantum theory in $D = 6$ that is probably not a field theory[3,4].

To construct this theory, again consider the type IIB superstring on $R^6 \times K3$, but this time not at a generic point but at a point in $K3$ moduli space where a 2-cycle becomes small and a θ parameter $\rightarrow 0$, for example at an A_1 enhanced symmetry point of the ADE possibilities.[5,6] This $D = 6$ theory with $N = (2,0)$ spacetime (global) supersymmetry, has a massless spectrum of 1 tensor multiplet. The spectrum also includes a not very heavy (tension $\sim \frac{\epsilon}{g_s}$) BPS string which couples to $B^-_{\mu\nu}$, the two-form with self-dual field strength, existing in the tensor multiplet. (The IIA superstring on $R^6 \times K3$ at an A_1 enhanced symmetry point changes a $D = 6, N = (1,1)$ vector supermultiplet $((2,2) + 4(1,1) + 2(2,1) + 2(1,2))$ to an $SU(2)$ multiplet. Instead, here the weakly coupled type IIB on $R^6 \times K3$ at an A_1 enhanced symmetry point develops a non-critical string, that is a string that propagates without gravity.)[7–9]

The same theory is constructed from the world-volume theory of two parallel almost coincident fivebranes in M theory (or two parallel almost coincident NS five-branes in IIA). So we see that a 5-brane action and a compactified-to-six-dimensions string theory both describe the same physics, ie the physics on the six-dimensional hyperplane. This observation ties together brane theories and and M-theory/string compactifications.

Since further compactification of this system to $D = 5$ is then T-dual to IIA on $R^6 \times K3$ at an A_1 singularity $\times S^1$, there must be a corresponding two parallel

Physics of Mass
Edited by Kursunoglu *et al.*, Kluwer Academic / Plenum Publishers, New York, 1999

fivebrane picture in IIB. In this case, rather than a tensor theory with a non-critical string in $D = 6$, we have an enhanced gauge symmetry, that is a $D = 6$ non-abelian vector multiplet. This is as follows. Two parallel D fivebranes in type IIB have a world-volume theory given by $D = 6, N = (1,1)$ with a $U(2)$ vector supermultiplet, and gauge coupling $\frac{1}{g_D^2} = \frac{M_s^2}{g_s}$. After an S-duality transformation this is two parallel NS fivebranes in IIB with the same world volume theory but now the gauge coupling is $\frac{1}{g_{NS}^2} = M_s^2, since(g_s \to \frac{1}{g_s}$, and a Weyl rescaling of the coordinates x^μ is performed in the Yang Mills action). Now the gauge coupling is independent of the string coupling g_s, and the theory becomes $D = 6, N = (1,1)$ with an $SU(2)$ vector supermultiplet in the limit $g_s = 0$.

We note that the world-volume theory of one M-theory fivebrane is a $D = 6$ theory with $N = (2,0)$ spacetime (global) supersymmetry and 1 tensor supermultiplet, where the five massless scalars are the fluctuations in the directions transverse to the fivebrane. There is no massless graviton on the fivebrane, since the graviton propagates in the bulk, ie in all eleven dimensions.

In order to compare the above constructions with features of conformal field theory (cft), we first consider the (perturbative) partition functions of the IIA and IIB superstrings compactified on $R^6 \times K3$, in the T^4/Z_2 orbifold limit. This results in the same massless spectrum as for a generic point in the $K3$ moduli space. We then note the similarity in form of the partition function to that which occurs when we replace the T^4/Z_2 conformal field theory with that of the tube metric conformal field theory[10–12], which has central charge $c = 6$ and corresponds to the tranverse degrees of freedom of the NS fivebrane. This latter partition function uses the A_{k+1} modular invariant of the ADE classification[13], built from level k, $SU(2)$ character formulae. Since $k = n_H - 2$, where n_H is the number of parallel NS fivebranes, we can compare this cft with the cft on the Higgs branch of a $D = 2$ system with $n_H(n_V)$ hyper (vector) supermultiplets with $c = 6(n_H - n_V)$. Although a cft description is believed to break down at a point of enhanced symmetry, ie as the n_H fivebranes coincide, it is interesting to look at the role of the tube metric cft, which describes some non-perturbative fivebrane phenomena.

PARTITION FUNCTION FOR TYPE II ON $R^6 \times T^4/Z_2$

We compute explicitly[14] type II strings on $\mathbf{R}^6 \times \mathbf{T}^4/\mathbf{Z}_2$, where $\mathbf{T}^4/\mathbf{Z}_2$ is the $\mathbf{Z}_2$ orbifold limit of $K3$. In the light-cone description, the left and right-moving modes are each taken to be described by 8 bosonic and 8 fermionic worldsheet (primary) fields: $\tilde{A}_n^{\hat{\imath}}, \tilde{A}_s^I, \tilde{\psi}_r^{\hat{\imath}}, \tilde{\psi}_r^I$; $A_n^{\hat{\imath}}, A_s^I, \psi_r^{\hat{\imath}}, \psi_r^I$; $1 \le \hat{\imath}, \le 4$ and $5 \le I, \le 8$, where the superscript $\hat{\imath}$ refers to the transverse spatial degrees of freedom, the superscript I to the internal ones, and the subscripts s, r each correspond to either integer or half-integer modding depending on the sector. The partition function or one loop contribution to the vacuum to vacuum amplitude in D space-time dimensions is

$$\Lambda = -\frac{1}{4\pi(\alpha')^{\frac{D}{2}}} \int_{\mathcal{F}} d^2\tau (\operatorname{Im}\tau)^{-2-(\frac{D-2}{2})} |f(\omega)|^{-2(D-2)} \Lambda_f \tag{1}$$

where Λ_f is the partition function for the fermionic and internal bosonic degrees of freedom,

$$\Lambda_f = \sum_{\alpha\in\Omega} \delta_\alpha \operatorname{tr}_\alpha \{\bar{\omega}^{\bar{L}_0 - \frac{1}{2}} \omega^{L_0 - \frac{1}{2}} \prod_{\beta\in\Omega} P_{\alpha,\beta}\}, \tag{2}$$

i.e. the spectrum of a theory will consist of a set of sectors Ω, characterized by the modding of the internal bosons ($s \in \mathbf{Z}$, untwisted), ($s \in \mathbf{Z}+\frac{1}{2}$, twisted), and of the fermions ($r \in \mathbf{Z}$, untwisted Ramond (R) or twisted Neveu-Schwarz (NS)), and ($r \in \mathbf{Z}+\frac{1}{2}$, untwisted NS or twisted R). The quantities δ_α and the projection operators $P_{\alpha,\beta}$ are discussed below. The integration region $\mathcal{F} = \{\tau : |\tau| > 1 \ |Re\tau| < \frac{1}{2}\}$ is a fundamental region for the modular group that is generated by $\tau \to \tau+1$, $\tau \to -\frac{1}{\tau}$; and $\tilde{L}_0$, L_0 refer to left, right movers.

To define the orbifold choose a complex basis for the internal fermions, for example for the left-movers: $f^1 = \frac{1}{\sqrt{2}}(h^5+ih^6)$, $\bar{f}^1 = \frac{1}{\sqrt{2}}(h^5-ih^6)$, $f^2 = \frac{1}{\sqrt{2}}(h^7+ih^8)$, $\bar{f}^2 = \frac{1}{\sqrt{2}}(h^7-ih^8)$. Then the $\mathbf{Z}_2$ transformation θ acting on the internal fermions in terms of the number operator $F = \sum_{i=1,2;r} : f^i_r \bar{f}^i_{-r} :$ (where $: f^i_0 \bar{f}^j_0 := - : \bar{f}^j_0 f^i_0 :$), so that $\theta = (-1)^F$; and similarly θ acts on the internal bosons by $\theta A^I_s \theta^{-1} = -A^I_s$. Oscillators with space-time indices are invariant under θ, and $D=6$.

A sector α is labelled by a twelve-dimensional vector whose components are 0 for NS and 1 for R:

$$\rho_a = (\tilde{\rho}_1, \ldots, \tilde{\rho}_4; \tilde{\rho}'_1, \tilde{\rho}'_2; \rho_1, \ldots, \rho_4; \rho'_1, \rho'_2) \tag{3}$$

This vector corresponds to boundary conditions of left- and right- modes separately described by 4 real and 2 complex fermions.

The set of states on which the theory is unitary is specified by states that survive projections defined by number operators which generalize the GSO projection. The projections are defined by requiring the parity of the number operators, N_β defined in (9), to take on definite values $\epsilon(\alpha,\beta)$ on any state in the sector α, *i.e.*

$$(-1)^{N_\beta}\big|_a = \epsilon(\alpha,\beta), \tag{4}$$

where each $\epsilon(\alpha,\beta)$ is either ± 1. The (perturbative) spectrum of a model is specified by a set of sectors Ω, together with a set $\{(-1)^{N_\beta} : \beta \in \Omega\}$ of parity operators, and their values $\epsilon(\alpha,\beta)$ on the sectors $\alpha \in \Omega$.

Eq. (2) can be expressed as a sum over spin strutures:

$$\Lambda_f = \frac{1}{2^{K+1}} \sum_{\alpha\in\Omega} \sum_{\beta\in\Omega} \delta_a \ \epsilon(a,\beta) \ \mathrm{tr}_\alpha \{\bar{\omega}^{\tilde{L}_0 - \frac{1}{2}} \omega^{L_0 - \frac{1}{2}} (-1)^{N_\beta}\} \tag{5}$$

where K is the number of basis vectors which generate Ω. We denote the trace in eq.(2.5) by $\{a,\beta\}$, so that, without the factor of 2^{-K-1}, the sum is

$$\sum_{a,\beta} \delta_a \ \epsilon(a,\beta) \ \{a,\beta\}, \tag{6}$$

where

$$\begin{aligned} \{a,\beta\} &= \mathrm{tr}_\alpha\{\bar{\omega}^{\tilde{L}_0-\frac{1}{2}}\omega^{L_0-\frac{1}{2}}(-1)^{N_\beta}\} \\ &= |\omega|^{-1}|f(\omega)|^{-20} \prod_{i=1}^{4} \left(\bar{\Theta}\begin{bmatrix}\tilde{\rho}^i_\alpha \\ \tilde{\mu}^i_\beta\end{bmatrix}(0|\tau)\right)^{\frac{1}{2}} \prod_{i=5}^{6}\left(\bar{\Theta}\begin{bmatrix}\tilde{\rho}^i_\alpha \\ \tilde{\mu}^i_\beta\end{bmatrix}(0|\tau)\right) \\ &\times \prod_{j=1}^{4}\left(\Theta\begin{bmatrix}\rho^j_\alpha \\ \mu^j_\beta\end{bmatrix}(0|\tau)\right)^{\frac{1}{2}} \prod_{j=5}^{6}\left(\Theta\begin{bmatrix}\rho^j_\alpha \\ \mu^j_\beta\end{bmatrix}(0|\tau)\right) \quad \times \text{internal bosons}, \end{aligned} \tag{7}$$

and $(\tilde{\rho}^i_\alpha, \rho^j_\alpha)$ and $(\tilde{\mu}^i_\beta, \mu^j_\beta)$ are the the twelve-component vectors describing the sectors α and β respectively, *i.e.* the components are 0 for NS and 1 for R [see for example (2.8)]. $\Theta\left[\begin{smallmatrix}\rho^i\\ \mu^i\end{smallmatrix}\right](0|\tau)$ and $f(\omega)$ are given by (2.14), and $\delta_a = \delta^L_a \delta^R_a$ where $\delta_a = 1$ if the states of the sector a are space-time bosons and $\delta_a = -1$ if the states are space-time fermions. A consistent (perturbative) string theory is such that under modular transformations the integrand of (1) is invariant.

The type II string on $\mathbf{R}^6 \times \mathbf{T}^4/\mathbf{Z}_2$ has eight sectors, whose fermion boundary condition vectors (3) are given by

$$\begin{aligned} \rho_{b_2} &= (1^4, 1^2; 0^4, 0^2) & \rho_{b_0 b_1 b_2} &= (1^4, 0^2; 0^4, 1^2) \\ \rho_{b_0 b_2} &= (0^4, 0^2; 1^4, 1^2) & \rho_{b_1 b_2} &= (0^4, 1^2; 1^4, 0^2) \\ \rho_{b_0 b_1} &= (0^4, 1^2; 0^4, 1^2) & \rho_{b_1} &= (1^4, 0^2; 1^4, 0^2) \\ \rho_{b_0} &= (1^4, 1^2; 1^4, 1^2) & \rho_{\phi} &= (0^4, 0^2; 0^4, 0^2)\,, \end{aligned} \tag{8}$$

where $\{\phi, b_0, b_0 b_2, b_2\}$ are the sectors that have untwisted bosons, and the $\mathbf{Z}_2$ twisted sectors are written as $\{b_0 b_1, b_1, b_1 b_2, b_0 b_1 b_2\}$. For this theory, the eigenvalues $\epsilon(\alpha,\beta)$ of the parity operators are given in Table 1, where λ, ρ, μ, ν take values ± 1, and different choices of λ, ρ, μ, ν do not change the theory. Table 1 is derived by requiring modular invariance for the part of Λ given by $\frac{1}{8}\sum_{\alpha,\beta \in \{\phi, b_0, b_0 b_2, b_2\}} \delta_a\, \epsilon(a,\beta)\, \mathrm{tr}_\alpha\{\bar{\omega}^{\bar{L}_0 - \frac{1}{2}} \omega^{L_0 - \frac{1}{2}} (-1)^{N_\beta}\}$, and then computing the remaining values of $\epsilon(a,\beta)$ using $\epsilon(a,\beta\gamma) = \epsilon(a,\beta)\,\epsilon(a,\gamma)$, which follows from (4).

In the fermionic picture, we define the parity of the number operator N_β acting on the sector α by

$$(-1)^{N_\beta}\big|_\alpha = (-1)^{\rho_\beta \cdot F}\big|_\alpha\,, \tag{9}$$

where F is a vector whose components are the operators $F_j = \sum_r : f^j_r \bar{f}^j_{-r} :$ for complex fermions and $\sum_{r\geq 0}^{\infty} \psi^j_{-r}\psi^j_r$ for real fermions, and r is modded according to the boundary condition of the "j^{th}" fermion in the sector α.

$$\rho_\beta \cdot F = \sum_{j=1}^{4} \tilde{\rho}_j \tilde{F}_j + \sum_{j=1}^{2} \tilde{\rho}'_j \tilde{F}'_j + \sum_{j=1}^{4} \rho_j F_j + \sum_{j=1}^{2} \rho'_j F'_j \tag{10}$$

and the sums $\tilde{F}_j$ and F_j distinquish left and right movers, but the pair $\bar{f}$ and f denotes a complex fermion which is either wholly left moving or right moving. In

Table 1

$\beta \longrightarrow$

$\epsilon(a,\beta)$	Ø	b_0	b_1	b_2	b_0b_1	b_0b_2	b_1b_2	$b_0b_1b_2$
Ø	1	1	1	-1	1	-1	-1	-1
b_0	1	λ	ρ	μ	$\lambda\rho$	$\lambda\mu$	$\rho\mu$	$\lambda\rho\mu$
a $\downarrow$ b_1	1	ρ	ρ	ν	1	$\rho\nu$	$\rho\nu$	ν
b_2	1	$-\mu$	ν	μ	$-\mu\nu$	-1	$\mu\nu$	$-\nu$
b_0b_1	1	$\lambda\rho$	1	$-\mu\nu$	$\lambda\rho$	$-\lambda\rho\mu\nu$	$-\mu\nu$	$-\lambda\rho\mu\nu$
b_0b_2	1	$-\lambda\mu$	$\rho\nu$	-1	$-\lambda\rho\mu\nu$	$\lambda\mu$	$-\rho\nu$	$\lambda\rho\mu\nu$
b_1b_2	1	$-\rho\mu$	$\rho\nu$	$-\mu\nu$	$-\mu\nu$	$\rho\nu$	$-\mu\rho$	1
$b_0b_1b_2$	1	$-\lambda\rho\mu$	ν	ν	$-\lambda\rho\mu\nu$	$-\lambda\rho\mu\nu$	1	$-\lambda\rho\mu$

general, the projection operators are defined by

$$P_{a,\beta} = \frac{1}{2}\{1 + \epsilon(a,\beta)(-1)^{N_\beta}\}, \qquad \beta \in \Omega. \tag{11}$$

The functions in (7) have $\omega = e^{2\pi i\tau}$ and

$$\Theta\left[\begin{smallmatrix}\rho\\ \mu\end{smallmatrix}\right](\nu|\tau) = \sum_{n\in\mathbf{Z}} e^{i\pi\tau(n+\frac{\rho}{2})^2} e^{i2\pi(n+\frac{\rho}{2})(\nu+\frac{\mu}{2})-i\frac{\pi}{2}\rho\mu} \tag{12a}$$

$$\bar\Theta\left[\begin{smallmatrix}\tilde\rho\\ \tilde\mu\end{smallmatrix}\right](\nu|\tau) = \sum_{n\in\mathbf{Z}} e^{-i\pi\bar\tau(n+\frac{\tilde\rho}{2})^2} e^{-i2\pi(n+\frac{\tilde\rho}{2})(\nu+\frac{\tilde\mu}{2})+i\frac{\pi}{2}\tilde\rho\tilde\mu} \tag{12b}$$

$$\eta(\tau) = \omega^{\frac{1}{24}} f(\omega) = \omega^{\frac{1}{24}} \prod_{n=1}^{\infty}(1-\omega^n). \tag{12c}$$

Collecting the contributions from the different (non-zero) spin structures, we have for type II strings on $\mathbf{R}^6 \times \mathbf{T}^4/\mathbf{Z}_2$ that

$$\Lambda = -\frac{1}{4\pi(\alpha')^3}\int_{\mathcal{F}} d^2\tau (\mathrm{Im}\,\tau)^{-4}\,|\eta(\tau)|^{-8}\,\Lambda'_f \tag{13a}$$

$$\begin{aligned}\Lambda'_f = \frac{1}{8}\Big[&\frac{\theta_{\Gamma_{4,4}}(\bar\tau,\tau)}{|\eta(\tau)|^8}\quad |\eta(\tau)|^{-8}(\bar\theta_3^4-\bar\theta_4^4-\bar\theta_2^4)(\theta_3^4-\theta_4^4-\theta_2^4)\\ &+\frac{2^4|\eta(\tau)|^4}{\bar\theta_2^2\theta_2^2}\quad |\eta(\tau)|^{-8}\,\bar\theta_3^2\bar\theta_4^2\theta_3^2\theta_4^2\,(1-1-1+1)\\ &+\frac{2^4|\eta(\tau)|^4}{\bar\theta_4^2\theta_4^2}\quad |\eta(\tau)|^{-8}\,\bar\theta_3^2\bar\theta_2^2\theta_3^2\theta_2^2\,(2-2)\\ &+\frac{2^4|\eta(\tau)|^4}{\bar\theta_3^2\theta_3^2}\quad |\eta(\tau)|^{-8}\,\bar\theta_4^2\bar\theta_2^2\theta_4^2\theta_2^2\,(2-2)\ \Big].\end{aligned} \tag{13b}$$

In (13b), the factors $(1-1-1+1)$ are all space-time boson contributions, while in lines 3 and 4 each the factors $(2-2)$ contribute 2 from space-time bosons and (-2) from space-time fermions. In line 1, the lattice theta function $\theta_{\Gamma_{4,4}}(\bar\tau,\tau) = \sum_{p_L,p_R\in\Gamma_{4,4}} \bar\omega^{\frac{1}{2}p_L^2}\omega^{\frac{1}{2}p_R^2}$ is defined for any even, self-dual eight-dimensional Lorentzian lattice $\Gamma_{4,4}$, and is modular invariant.

MASSLESS SPECTRA

In this section we give the massless spectra in these theories, using the projection operators (11), rather than directly using facts about the cohomology and moduli space of K3. Our procedure insures that the spectrum at a given mass level agrees with the coefficient of a suitable power of $\bar\omega\omega$ in (13). The partition function (13) describes a precise set of states on which the theory is known to satisfy (perturbative) unitarity.

The massless spectrum of the Type IIA superstring on $\mathbf{R}^6 \times \mathbf{T}^4/\mathbf{Z}_2$ is given in terms of representations of the $D=6$ lightcone little group $\mathrm{Spin}(4) \cong SU(2)\times SU(2)$ which form $D=6$, $N=(1,1)$ spacetime supersymmetry multiplets. The supergravity multiplet is

$$(3,3)+(3,1)+(1,3)+4(2,2)+(1,1)+2(3,2)+2(2,3)+2(2,1)+2(1,2). \tag{14a}$$

It couples to 20 vector supermultiplets with spin content:

$$(2,2)+4(1,1)+2(2,1)+2(1,2)\,. \qquad (14b)$$

The Ramond ground states in $D=6$ for the type IIA superstring on $\mathbf{R}^6\times\mathbf{T}^4$ or $\mathbf{R}^6\times\mathbf{T}^4/\mathbf{Z}_2$ corresponds to the spin content:

$$\begin{aligned}\bar{F}_1&=\text{even} & &|(2,1)\rangle_{\text{Left}}\\ \bar{F}_1&=\text{odd} & &|(1,2)\rangle_{\text{Left}}\\ \bar{F}_3&=\text{even} & &|(1,2)\rangle_{\text{Right}}\\ \bar{F}_3&=\text{odd} & &|(2,1)\rangle_{\text{Right}}\,.\end{aligned} \qquad (15)$$

Using (5), we find the massless states in the untwisted RR sector have spin content 8(2,2). These eight vectors are from $2|(2,1)\rangle_{\text{Left}}\times 2|(1,2)\rangle_{\text{Right}}$ and $2|(1,2)\rangle_{\text{Left}}\times 2|(2,1)\rangle_{\text{Right}}$. Similar arguments show the R-NS sector contains massless states $2(2,3)+10(2,1)$ given by $2|(2,1)\rangle_{\text{Left}}\times\psi^{J}_{-\frac{1}{2}}|0\rangle$ and $2|(1,2)\rangle_{\text{Left}}\times\psi^{\hat{J}}_{-\frac{1}{2}}|0\rangle$; the NS-R sector contains $2(3,2)+10(1,2)$; and the four twisted sectors contain 16 massless vector supermultiplets given by $4\,|16\rangle$ and $|(2,1)\rangle_{\text{Left}}\times|(1,2)\rangle_{\text{Right}}\times|16\rangle$, $2\times|(1,2)\rangle_{\text{Right}}\times|16\rangle$, $|(2,1)\rangle_{\text{Left}}\times 2\times|16\rangle$, where $|16\rangle$ is the degeneracy of the ground state of the twisted bosonic operators $\tilde{A}^I_s, A^J_s$[15].

For the type IIB superstring, the analog of (15) is

$$\begin{aligned}\bar{F}_1&=\text{even} & &|(1,2)\rangle_{\text{Left}}\\ \bar{F}_1&=\text{odd} & &|(2,1)\rangle_{\text{Left}}\\ \bar{F}_3&=\text{even} & &|(1,2)\rangle_{\text{Right}}\\ \bar{F}_3&=\text{odd} & &|(2,1)\rangle_{\text{Right}}\,.\end{aligned} \qquad (16)$$

So for the Type IIB superstring on $\mathbf{R}^6\times\mathbf{T}^4/\mathbf{Z}_2$ the partition function is the same (2.15) as for IIA on $\mathbf{R}^6\times\mathbf{T}^4/\mathbf{Z}_2$, the number of massless states is the same, but the Spin (4) representations are now form $D=6$, $N=(0,2)$ supermultiplet: the supergravity multiplet

$$(3,3)+5(3,1)+4(3,2) \qquad (17a)$$

is coupled to 21 tensor supermultiplets:

$$(1,3)+5(1,1)+4(1,2)\,. \qquad (17b)$$

From the projections (2.13), the boundary values (2.8), Table 1, and (16), it follows that the massless states in the untwisted sectors are: $(3,3)+(1,3)+(3,1)+17(1,1)$ from NS-NS, $2|(1,2)\rangle_{\text{Left}}\times 2|(1,2)\rangle_{\text{Right}}+2|(2,1)\rangle_{\text{Left}}\times 2|(2,1)\rangle_{\text{Right}}=4(1,3)+4(3,1)+8(1,1)$ from RR, and $4(3,2)+20(1,2)$ from R-NS and NS-R. In the twisted sectors, the massless spectrum is $64(1,1)$ from b_0b_1, $16(1,1)+16(1,3)$ from b_1, $64(1,2)$ from b_1b_2 and $b_0b_1b_2$. The supergravity multiplet and 5 of the tensor multiplets come from the untwisted sector, and 16 tensor multiplets come from the twisted sector.

4. Partition function for tube metric conformal field theory

The partition function for the type II superstring on $\mathbf{R}^6\times W4$, for $W4$ constructed from a level k, $SU(2)$ Wess-Zumino-Witten model, a Liouville boson with background charge $Q=-\frac{\sqrt{2}}{\sqrt{k+2}}$, and four free fermions, is given by

$$\Lambda = -\frac{1}{4\pi(\alpha')^3}\int_{\mathcal{F}} d^2\tau (\mathrm{Im}\,\tau)^{-2}\,|\eta(\tau)|^{-8}\,\Lambda'_f$$

$$\Lambda'_f = \frac{1}{4}\frac{(\mathrm{Im}\,\tau)^{-\frac{1}{2}}}{|\eta(\tau)|^2} Z_k(\bar\tau,\tau) \quad |\eta(\tau)|^{-8}(\bar\theta_3^4 - \bar\theta_4^4 - \bar\theta_2^4)(\theta_3^4 - \theta_4^4 - \theta_2^4)\,, \tag{18}$$

where the diagonal modular invariant

$$Z_k(\bar\tau,\tau) = \sum_{\lambda=1}^{k+1} \bar\chi_{k,\lambda}(\bar\tau)\chi_{k,\lambda}(\tau) \tag{19}$$

is in correspondence with $su(2+k)$, i.e. A_{k+1} in the ADE classification[13] of modular invariant combinations of level k affine $SU(2)$ characters. An irreducible highest weight representation of an affine algebra $\hat g$ is an infinite-dimensional tower of irreducible representations of the finite-dimensional algebra g, and is classified by its highest weight. Allowed highest weights for level k affine $SU(2)$ are $\lambda = 2\ell+1$, where ℓ is the spin of the $SU(2)$ representation at the top of the tower, and $0 < \ell < \frac{k}{2}$.

The character formula, which counts the states in a given irreducible representation of the level k affine $SU(2)$ is

$$\chi_{k,\lambda}(\tau) = \frac{1}{\eta^3(\tau)}\sum_{n\in Z}(n2(k+2)+\lambda)\,\omega^{\frac{(n2(k+2)+\lambda)^2}{4(k+2)}}\,. \tag{20}$$

The invariant $Z_k(\bar\tau,\tau) = \sum_{\lambda=1}^{k+1}\bar\chi_{k,\lambda}(\bar\tau)\chi_{k,\lambda}(\tau)$ is defined for $k \geq 0$. For $k = 0$, $Z_0(\bar\tau,\tau) = \bar\chi_{0,1}(\bar\tau)\chi_{0,1}(\tau) = 1$ since $\chi_{0,1}(\tau) = \frac{1}{\eta^3(\tau)}\omega^{\frac{1}{8}}\sum_{n\in Z}(4n+1)\,\omega^{2n^2+n} = 1$ using the Jacobi triple product identity for η^3.

The other ADE modular invariants, for example those corresponding to $D_{\frac{k}{2}+2}$, k even, are discussed in [13].

A modular invariant partition function for the type II superstring on $\mathbf{R}^6\times W4/\mathbf{Z}_2$ is

$$\Lambda = -\frac{1}{4\pi(\alpha')^3}\int_{\mathcal{F}} d^2\tau (\mathrm{Im}\,\tau)^{-4}\,|\eta(\tau)|^{-8}\,\Lambda'_f$$

$$\begin{aligned}\Lambda'_f = \frac{1}{8}\frac{(\mathrm{Im}\,\tau)^{-\frac{1}{2}}}{|\eta(\tau)|^2}\,[\,& Z_k\left[\begin{smallmatrix}0\\0\end{smallmatrix}\right](\bar\tau,\tau) \quad |\eta(\tau)|^{-8}(\bar\theta_3^4 - \bar\theta_4^4 - \bar\theta_2^4)(\theta_3^4 - \theta_4^4 - \theta_2^4)\\ &+ Z_k\left[\begin{smallmatrix}0\\1\end{smallmatrix}\right](\bar\tau,\tau) \quad |\eta(\tau)|^{-8}\,\bar\theta_3^2\bar\theta_4^2\theta_3^2\theta_4^2\,(1-1-1+1)\\ &+ Z_k\left[\begin{smallmatrix}1\\0\end{smallmatrix}\right](\bar\tau,\tau) \quad |\eta(\tau)|^{-8}\,\bar\theta_3^2\bar\theta_2^2\theta_3^2\theta_2^2\,(2-2)\\ &+ Z_k\left[\begin{smallmatrix}1\\1\end{smallmatrix}\right](\bar\tau,\tau) \quad |\eta(\tau)|^{-8}\,\bar\theta_4^2\bar\theta_2^2\theta_4^2\theta_2^2\,(2-2)\,]\end{aligned} \tag{21}$$

where

$$Z_k\left[\begin{smallmatrix}\alpha\\\beta\end{smallmatrix}\right](\bar\tau,\tau) = \sum_{\lambda=1}^{k+1} e^{i\pi\beta(\lambda-1)}\bar\chi_{k,\lambda}(\bar\tau)\ \chi_{k,\lambda+\alpha(k+2-2\lambda)}(\tau) \tag{22}$$

transforms under modular transformations as

$$\begin{aligned} Z_k\left[\begin{smallmatrix}\alpha\\\beta\end{smallmatrix}\right] &\to e^{-i\pi(\frac{k}{2})\alpha^2} Z_k\left[\begin{smallmatrix}\alpha\\\beta+\alpha\end{smallmatrix}\right] && \text{for} \quad \tau\to\tau+1\\ Z_k\left[\begin{smallmatrix}\alpha\\\beta\end{smallmatrix}\right] &\to e^{i\pi k\alpha\beta} Z_k\left[\begin{smallmatrix}\beta\\\alpha\end{smallmatrix}\right] && \text{for} \quad \tau\to -\frac{1}{\tau}\end{aligned} \tag{23}$$

and $Z_k \left[{0 \atop 0} \right](\bar{\tau}, \tau) = Z_k(\bar{\tau}, \tau)$ in (19). We note that (21) is identical to (13) when the internal bosons A^I are replaced with Liouville and WZW modes J^0, J^i.

CONCLUSIONS

The partition functions (18) and (21) are constructed from the A_{k+1} modular invariant $Z_k(\bar{\tau}, \tau)$ and related twisted expressions (22). These correspond to excitations of a type II fundamental string in a background described by degrees of freedom transverse to the NS fivebrane. The incorporation of exact results on Liouville cft may modify which states survive in this theory, and hence their gauge symmetry properties. We note the occurrence of the A_{k+1} modular invariants, and contrast this theory with a type II compactification[7–9] described by a 2D supersymmetric gauge theory leading to a $D = 6$, $N = (1, 1)$ theory with massless spectrum of 1 SG multiplet, 19 $U(1)$ and 1 $SU(2)$ vector supermultiplets. A deeper understanding of how the conventional type II cft breaks down in this case, due to massless solitons, may guide us to a more economical description of how string theory picks the vacuum. It is believed that the appearance of these massless non-perturbative BPS states may be an important mechanism for the way in which nature incorporates gauge symmetry in string theory.

REFERENCES

1. J. Schwarz, "Lectures On Superstring And M Theory Dualities", 1996 TASI Lectures, Nucl. Phys. Proc. Suppl. **55B**: 1-32 (1997), hep-th/9607201.
2. M.B. Green, J. Schwarz and E. Witten, *Superstring Theory*(Cambridge University Press, 1987).
3. N. Seiberg, "Matrix Description of M theory on T^5 and T^5/Z_2, hep-th/9795221.
4. E. Witten, "New Gauge Theories in Six Dimensions", hep-th/9710065.
5. P. Aspinwall, "Enhanced Gauge Symmetries And K3 Surfaces", hep-th/9507012, Phys. Lett. **B357** (1995) 329.
6. P. Aspinwall. "K3 Surfaces And String Duality", hep-th/9611137.
7. E. Witten, "Some Comments On String Dynamics", *Strings '95* (World Scientific, 1996), ed. I. Bars et. al., 501, hep-th/9507021.
8. E. Witten, "On The Conformal Field Theory Of The Higgs Branch", hep-th/9707093.
9. D. Diaconescu and N. Seiberg, "The Coulomb Branch of (4,4) Supersymmetric Field Theories In Two Dimensions", hep-th/9707158.
10. A. Strominger, "Heterotic Solitons", Nucl. Phys. **B343** (1990) 167.
11. C. Callan, J. Harvey, and A. Strominger, "Worldsheet approach to heterotic instantons and solitons", Nucl. Phys. **B359** (1991) 611; and "Worldbrane actions for string solitons", Nucl. Phys. **B 367** (1991) 60.
12. C. Callan, "Instantons and Solitons in Heterotic String Theory", Swieca Summer School, June 1991; hep-th/9109052. C. Callan, J. Harvey, and A. Strominger, "Supersymmetric String Solitions", 1991 Trieste Spring School on String Theory

and Quantum Gravity, hep-th/9111030.

13. A. Cappelli, C. Itzykson, J.B.Zuber, Nucl. Phys. **B280** (1987) 445; Comm. Math. Phys. **113** (1987) 1.

14. L. Dolan and M. Langham, "Partition Functions, Duality, and the Tube Metric", hep-th/9711114.

15. R. Bluhm, L. Dolan and P. Goddard, "Unitarity and Modular Invariance as Constraints on Four-dimensional Superstrings", Nucl. Phys. **B309** (1988) 330.

QUARK MASSES IN DUAL THEORIES *

K. Tanaka

Department of Physics
The Ohio State University
174 West 18th Avenue
Columbus, OH 43210, USA

ABSTRACT

We consider the $N = 2$, $N_f = 2$ Seiberg-Witten dual theory with two bare quark masses. We find that it is possible to have an arbitrary large mass ratio.

Recently, a great deal of progress was made in dual theories fostered by Seiberg and Witten who provided the exact vacuum structure and spectrum of four dimensional $N = 2$ Supersymmetric SU(2) QCD with matter [1,2].

We are interested in the hierarchy of quark masses in an $N = 2$, $N_f = 2$ (f= flavor) Seiberg Witten toy model with two bare masses. In order to motivate the equation of family of curves for $N_f = 2$, that parametrize the modular space of the quantum vacau, we briefly outline its derivation.

The low energy effective Lagrangian involves the $N = 2$ vector multiplet A that contains gauge fields A_μ, two Weyl fermions λ, ψ and a scalar ϕ. In terms of $N = 1$ supersymmetry, the fields are vector multiplet $W_\alpha(A_\mu, \lambda)$ and chiral multiplet $\Phi(\phi, \psi)$. The Lagrangian and Kahler potential can be written in terms of a prepotential $\mathcal{F}(A)$ that is an arbitrary holomorphic function of complex variables. The scalar component of the $N = 2$ vector multiplet A is a. The metric on the moduli space parametrized by a is

$$(ds)^2 = Im\tau(a)dad\bar{a}$$

The $\tau(a)$ is a holomorphic function, and $\frac{\partial^2\mathcal{F}}{\partial a^2} = \tau(a)$. Define $a_D = \frac{\partial\mathcal{F}}{\partial a}$ then $(ds)^2 = Imda_Dd\bar{a}$. The tree and one loop contribution to $\mathcal{F}$ is

$$\mathcal{F}_{\text{oneloop}} = i\frac{2}{\pi}A^2 \ln\frac{A}{\Lambda} \qquad \Lambda = scale$$

$\mathcal{F}$ follows from singularity structure and monodromy.

From a_D and $\mathcal{F}_{\text{oneloop}}$ for large a, one has

*Talk given at International Conference on Orbis Scientiae 1997-II, December, 1997, Miami Beach, Florida

Physics of Mass
Edited by Kursunoglu *et al.*, Kluwer Academic / Plenum Publishers, New York, 1998

$$a_D = \frac{4ia}{\pi}\ln(a/\Lambda) + 2ia/\pi$$

that is not a single valued function. Put $u = 2a^2$, and the monodromy is determined by drawing a circle or closed loop around a point u on the u plane at large u,

$$\begin{array}{ll} u \to e^{2\pi i}u, \text{ or } & \ln u \to \ln u + 2\pi i, \\ a \to e^{\pi i}a \to -a & (\ln a \to \ln a + \pi i). \end{array}$$

Then,

$$\begin{aligned} a_D &\to -a_D + 4a, \\ a &\to -a, \end{aligned}$$

or

$$\begin{pmatrix} a_D \\ a \end{pmatrix} \to M_\infty \begin{pmatrix} a_D \\ a \end{pmatrix}, \text{where } M_\infty = \begin{pmatrix} -1 & 4 \\ 0 & -1 \end{pmatrix}.$$

Consider magnetic coupling [3, 4]

$$\beta_D(g_D) = \mu\frac{dg_D}{d\mu} = \frac{g_D^3}{8\pi^2} \text{ or } \mu\frac{d\tau_D}{d\mu} = -\frac{i}{\pi},$$

where $\tau_D = 4\pi i/g_D^2(a_D)$ when $\theta_D = 0$. Since the scale μ here is a_D

$$d\tau_D = -\frac{i}{\pi}\frac{da_D}{a_D},$$

or

$$\tau_D = -\frac{i}{\pi}\ln a_D.$$

Define $h_D(A) = \frac{\partial\mathcal{F}}{\partial A_D}$, A_D = dual vector multiplet, and scalar component of A_D is a_D, then

$$\frac{dh_D(a)}{da_D} = \frac{\partial^2\mathcal{F}}{\partial a_D \partial a_D} = \tau_D$$

hence

$$h_D(a) = \int \tau_D da_D = -\frac{i}{\pi}\int \ln a_D da_D$$

In the $N = 2$ gauge theory, the central charge Z is

$$Z = a(n_e + \tau_{cl}n_m) = (an_e + a_D n_m),$$

where $\tau_{cl} \sim \frac{\theta}{2\pi} + \frac{4\pi i}{g^2}$, n_e and n_m are integers. The BPS bound as the mass of the dyon is $M \geq \sqrt{2}|Z|$.

For later convenience,

$$Z = (n_m, n_e)\begin{pmatrix} a_D \\ a \end{pmatrix} \equiv qa^\alpha.$$

Note $\sqrt{2}a_D$ is the mass of the monopole and duality symmetry is $a \leftrightarrow a_D$, $n_e \leftrightarrow n_m$, $\tau \leftrightarrow -\frac{1}{\tau} = \tau_D$. The contribution of monopole to BPS bound is $\sqrt{2}a_D n_m$. The full duality group is SL(2, Z) under which a, a_D transform as a doublet.

Take a massless monopole at u_0

$$u_D(u_0) = 0.$$

Near $u_0 (= \Lambda^2)$,

$$a_D(u) = c_0(u - u_0),$$

$$a(u) = -h_D = \frac{i}{2\pi} c_0 (u - u_0) \ln(u - u_0) + a_0,$$

where

$$a_0 = a(u_0), \qquad dh_D = \tau_D da_D.$$

Let u circle around u_0 in the u plane,

$$\ln(u - u_0) \to \ln(u - u_0) + 2\pi i$$

or

$$a(u) \to a \quad a_D,$$

$$a_D(u) \to a_D,$$

hence

$$M_{u_0} = \begin{pmatrix} 1 & 0 \\ -1 & 1 \end{pmatrix}$$

When the three monodromies are taken in counter clockwise direction where from a base point a large loop that includes those that include u_0 and $-u_0$, the monodromies obey

$$M_{u_0} M_{-u_0} = M_\infty,$$

so

$$M_{-u_0} = \begin{pmatrix} -1 & 4 \\ -1 & 3 \end{pmatrix}.$$

The pair of singularies $\pm u_0$ is required by the Z_2 symmetry $u \to -u$. The transformation of functions around a complex singularity are monodromies and the matrices are called monodromy matrices. Note

$$h(A) = \frac{\partial \mathcal{F}}{\partial A} \quad \tau(A) = \frac{\partial^2 \mathcal{F}}{\partial A^2} = \frac{\partial h(A)}{\partial A} = h',$$

$$A_D = \frac{\partial \mathcal{F}}{\partial A} = h(A) \text{ or } A = -h_D(A_D).$$

Duality transformation has freedom to shift A_D by a constant, and

$$-\frac{1}{\tau(A)} = -\frac{1}{h'(A)} = \tau_D(A_D) = h'_D(A_D),$$

$$S \begin{pmatrix} h_D \\ h \end{pmatrix} = \begin{pmatrix} 0 & 1 \\ -1 & 0 \end{pmatrix} \begin{pmatrix} h_D \\ h \end{pmatrix} = \begin{pmatrix} h \\ -h_D \end{pmatrix},$$

duality implements SL(2, Z) generator S.

A natural physical interpretation of singularities in u plane is that some additional massless particles are appearing at a particular value of u [5]. What particle becomes massless to generate the singularity? The massless particle that produces a monodromy M satisfies,

$$
\begin{aligned}
qM &= q, & q &= (n_m, n_e), & \\
q &= (1,0), & qM_{u_0} &= (1,0)\begin{pmatrix} 1 & 0 \\ -1 & 1 \end{pmatrix} = (1,0), & \text{monopole.} \\
q &= (1,-1), & qM_{u_0} &= (1,-2)\begin{pmatrix} -1 & 4 \\ -1 & 3 \end{pmatrix} = (1,-2) & \text{dyon.} \\
q &= (n_m, n_e) & qM_{\infty} &= (n_m, n_e)\begin{pmatrix} -1 & 4 \\ 0 & -1 \end{pmatrix} = (-n_m, 4n_m - n_e) &
\end{aligned}
$$

There is no simple solution.
The moduli space M of quantum vacua is the u plane with singularities. We described the u plane punctured at the singularities $-\Lambda^2$, Λ^2 and ∞ with their monodromy matrices.

The family of curves of the moduli space can be parametrized by the cubic equation

$$y^2 = (x + \Lambda^2)(x - \Lambda^2)(x - u).$$

For every u, there is a genus one Riemann surface-torus determined by the equation above with discriminant

$$\Delta = \prod_{i<j}(e_i - e_j)^2 = (e_1 - e_2)^2(e_1 - e_3)^2(e_2 - e_3)^2 = 4\Lambda^4(\Lambda^4 - u^2)^2.$$

The previously found $M_{-\Lambda}, M_{\Lambda}, M_0$ are obtained in Refs. [1] and [2]. The x plane has 4 branch points at $x = -\Lambda^2, \Lambda^2, u, \infty$. The branch points are joined pair wise by two cuts along $-\Lambda^2$ to Λ^2 and u to ∞ in x plane.

Consider the toy model massive $N_f = 2$, with masses m_1 and m_2. The most general structure of the curve is

$$y^2 = (x^2 - t\Lambda_2^4)(x - u) + \mu^2\Lambda_2^3(ax + bu) + 2cm^2\Lambda_2^4,$$

where

$$\mu^2 = m_1 m_2, \qquad m^2 = \frac{1}{2}(m_1^2 + m_2^2).$$

and t, a, b, c are constants to be determined.

The $U(1)_R$ charge (global phase rotation) is conserved, where the assignment of charges for the parameters and x and y are

$$
\begin{matrix} u & x & \Lambda & m & y \\ 4 & 4 & 2 & 2 & 6. \end{matrix}
$$

Define λ_2 such that when m_2 is large low energy $N_f = 1$ is finite $\Lambda_1^3 = m_2\lambda_2^2$ $m_2 \to \infty, \lambda_2^2 \to 0$.

Then, we obtain $a = \frac{1}{4}, b = 0$ and $c = -\frac{1}{64}$ and

$$y^2 = (x^2 - t\Lambda_2^4)(x - u) + \frac{1}{4}m_1 m_2 x - \frac{1}{64}(m_1^2 + m_2^2)\Lambda_2^4.$$

To find t, put $m_2 = 0$, $u = m_1^2$ in discriminant Δ, put coefficients of the leading term in m_1^8 in Δ equal to zero, and get $t = \frac{1}{64}$. Then the equation with $s = \Lambda^2/8$ is

$$y^2 = (x^2 - s^2)(x - u) + 2\mu^2 sx - 2m^2 s^2.$$

We are interested in examining the possible mass heirarchy in this $N_f = 2$ model. Define $m_1 = M + D$, $m_2 = M - D$ so $\mu^2 = m_1 m_2 = M^2 - D^2$. We determined the Δ with masses μ^2 and m^2. Then, m^2 is replaced by $m^2 = \mu^2 + (m^2 - \mu^2)$. The μ^2 part leads to the previous equal mass case and the coefficient of $(m^2 - \mu^2) = 2D^2$ leads to the constraint. We find

$$\Delta = 4s^2[(u+s)^2 - 8\mu^2 s](u - s - \mu^2)^2$$

with a constraint

$$D^2[(3s^2 - su)M^2 + suD^2 - us^2 + \frac{u^3}{9}] = 0.$$

If $D = 0$, $\mu^2 = m^2$ and the equal mass result is obtained [2, 6].

If $D \neq 0$, we get

$$(3s^2 - su)M^2 + suD^2 - us^2 + \frac{u^3}{9} = 0.$$

Consider u near the double zero of $\Delta = 0$,

$$\mu^2 = u - s = M^2 - D^2,$$

then we find

$$M^2 = \frac{u^2}{3s} - \frac{u^3}{27s^2} = u\left(\frac{u}{3s} - \frac{u^2}{27s^2}\right)$$

$$D^2 = u\left[\frac{s}{u} - 1 + \left(\frac{u}{3s} - \frac{u^2}{27s^2}\right)\right]$$

When $m_1 > 0, m_2 > 0$, we require

$$M^2 > D^2 \quad M^2 > 0 \quad D^2 > 0$$

and note

$$\frac{m_1}{m_2} = \frac{M+D}{M-D} = \frac{(M+D)^2}{M^2 - D^2}.$$

For $\frac{u}{s} = 1 + \epsilon, \quad \epsilon > 0, \quad s = \Lambda^2/8$

$$\frac{m_1}{m_2} = \frac{32}{27}\frac{1}{\epsilon}.$$

Large mass ratio is possible. For $\frac{u}{s} = 2$

$$\frac{m_1}{m_2} = \frac{29}{27} + \frac{4}{27}\sqrt{7} \approx 1.47.$$

Consider next the two other zeros of Δ

$$\mu = \pm(u + s)/\Lambda,$$

then

$$M^2 = u\left(3 + \frac{16}{3}\frac{u}{s} + \frac{71}{27}\frac{u^2}{s^2}\right),$$

$$D^2 = u\left(-13 - \frac{8}{3}\frac{u}{s} - 8\frac{s}{u} + \frac{71}{27}\frac{u^2}{s^2}\right).$$

For $\frac{u}{s} = 3 \quad M^2 = 42.6 \quad D^2 = 0 \quad m_1 = m_2 = \Lambda/2$ and a large mass ratio is not possible. For $\frac{u}{s} = 4, \quad m_1/m_2 \approx 3$.

We verify that the relation u/s is in the strong coupling regime for all of the preceding values [7]. The weak coupling regime satisfies $\Lambda^2 \ll u$. For the case of double zeros, $u/s = 1+\epsilon$ or $u/\Lambda^2 \approx 1/8$ or $\Lambda^2 \sim 8u$ so $\Lambda^2 \ll u$ is not satisfied. The case of $\mu = 1/2(u+s)/\Lambda, \quad \Lambda^2 = 8u/3$ so again $\Lambda^2 \ll u$ is not satisfied.

The problem with $N_f = 3$, 3 masses m_a, m_b, and m_c is being studied with G. Cleaver.

The author wishes to thank B. Kursunoglu and A. Perlmutter for their hospitality in Florida and G. Veneziano for his hospitality at Cern, and G. Cleaver for valuable discussions.

References

1. N. Seiberg and E. Witten, Nucl. Phys. **B** 426 (1994) 19.
2. N. Seiberg and E. Witten, Nucl. Phys. **B** 431 (1994) 484.
3. L. Alvarez-Gaumé and S. F. Hassan hep-th/9701069.
4. A. Bilal, Proceedings of the "61 Recontre entre Physiciens Théoriciens et Mathématiciens," Strasbourg, France, December, 1995. hep-th/9601007.
5. N. Seiberg, Phys. Rev. D **49**, 6857 (1994).
6. A. Hanany and Y. Oz, Nucl. Phys. B **452** (1995) 283.
7. M. E. Peskin, Proceedings of the 1996 Theoretical Advanced Study Institute, Fields, Strings, and Duality. Boulder, Colorado, June, 1996. hep-th/9702094.

THE BaBar EXPERIMENT AT SLAC

B.T. Meadows

University of Cincinnati, Cincinnati, OH 45221, USA
Representing the BaBar collaboration

INTRODUCTION

CP violation has been known to exist for 35 years and understanding its origin is an intrinsically important goal [1]. Experimental evidence comes from two sources - mixing and decay of K° mesons and the observed baryon anti-baryon asymmetry of the universe. Experimentally most results come from K° decays and, after many years of careful measurement, we know those ratios expected to be zero if CP were conserved, of amplitudes for K_L° and K_S° decay to $\pi^+\pi^-$ (η_{+-}) and to $\pi^\circ\pi^\circ$ ($\eta_{\circ\circ}$) to be similar ($\sim 2.3 \times 10^{-3} e^{44^\circ}$), and the CP violating asymmetry δ for semi-leptonic decays to lepton (anti-lepton) to be of the same order (3.3×10^{-3}.)

No clear evidence for direct CP violation exists. Measurements of the ratio of parameters ϵ' and ϵ which describe the CP violation in K° mixing and decay cannot rule out the possibility that ϵ' is zero (a non zero ϵ' would be evidence for direct CP violation.) This could mean that the violation occurs entirely in the mixing through a $\Delta Y = 2$ "superweak" interaction [2]

If CPT is conserved, cosmological matter/anti matter asymmetry requires, amongst other things, a CP violation mechanism [3]. The standard model provides such a mechanism, but it is unlikely that this will account for a large enough effect [4]. It is certainly a possibility therefore that in the B decay experiments discussed here which will confront the standard model predictions, large discrepancies will be found.

The B meson is another system which should provide information on the CP phenomenon. The B° in particular exhibits mixing just as the K° does, but predominantly through intermediate t quark states. The standard model predicts a large, time dependent CP violation in rare B° decays to CP self conjugate states arising from interference between decay and mixing. Up to the present, it has not been possible to observe this effect. The BaBar experiment at SLAC, is one which plans to use a dedicated asymmetric B factory principally for that purpose.

In the next section, the standard model of CP violation and how measurements of B meson decays may be used to confront it are briefly discussed. Next the main strategy to be used at the B factories, where decays of substantial samples of B° mesons

Physics of Mass
Edited by Kursunoglu *et al.*, Kluwer Academic / Plenum Publishers, New York, 1999

are to be used to determine the CP phases of the model, is discussed. The rationale for asymmetric machines and the experimental method is then described and in section 5 the experimental facilities are outlined. Finally, a summary of the CP reach of the BaBar experiment and what it may reveal is given.

TESTING THE STANDARD MODEL OF CP VIOLATION

CP conservation would require that transitions $B \to f$ *and* $\overline{B} \to \overline{f}$ occur at equal rates. If two weak processes with amplitudes A, A', weak phases δ_w, δ'_w and strong phases δ_s, δ'_s respectively contribute to the transition, then since, under CP, the weak Hamiltonian becomes Hermitian conjugated reversing the sign of the weak phases, but not the strong phases the rates for the transition and its CP conjugate become respectively

$$\begin{aligned}\Gamma(B \to f) &= A^2 + A'^2 + 2AA'\cos(\phi_s + \phi_w)\\ \overline{\Gamma}(\overline{B} \to \overline{f}) &= A^2 + A'^2 + 2AA'\cos(\phi_s - \phi_w)\end{aligned}$$

where $\phi_s = \delta_s - \delta'_s$ and $\phi_w = \delta_w - \delta'_w$. We concluded that CP violation occurs ($\Gamma \neq \overline{\Gamma}$) in a transition if there are at least two processes contributing to it; they have different weak phases ($\phi_w \neq 0$) AND different strong phases ($\phi_s \neq 0$). The violation is maximum when $A = A'$.

In the standard model, the CKM quark mixing matrix $V_{qq'}$, which connects electroweak and mass eigenstates [5], provides a simple and elegant prediction for CP violation in both decay and mixing amplitudes as well as in their interference for K° and B° mesons. Weak vertices are described at the quark level by charged currents

$$J_\mu = \sum_{q,q'} V_{qq'}\overline{q}'\gamma_\mu(1+\gamma_5)q \qquad (q = d, s, b\ ;\ q' = u, c, t)$$

As $V_{qq'}$ is a unitary, 3×3 matrix, it is defined by three real quantities and a complex phase. This phase is observable - it cannot be eliminated from the weak currents making up a Lagrangian by simply re-phasing the quark fields q. Factors of V at Feynman vertices for weak processes therefore impart different "CKM phases" to the amplitude appropriate for each process and so lead to CP violation.

A test of this model is to check that V is unitary, and that the CKM phases which it introduces correctly account for any observed CP asymmetries. One of the six unitarity conditions is that

$$V_{ud}V_{ub}^* \ +\ V_{cd}V_{cb}^* \ +\ V_{td}V_{tb}^* \ =\ 0$$

It is common to write the V matrix in the Wolfenstein variables [6] λ, η and ρ which come from an expansion of the matrix elements in powers of λ - the cosine of the Cabbibo angle. In this representation the phase of $V_{cd}^*V_{cb}$ is nearly zero and the unitarity condition may be represented graphically as in fig 1

The angles of this triangle are identified as

$$\alpha \equiv \arg(-V_{td}V_{tb}^* \,/\, V_{ud}V_{ub}^*)\ ;\ \beta \equiv \arg(-V_{cd}V_{cb}^* \,/\, V_{td}V_{tb}^*)\ ;\ \gamma \equiv \arg(-V_{ud}V_{ub}^* \,/\, V_{cd}V_{cb}^*).$$

In the standard model ρ and η are highly correlated. Data from K° decay (ϵ), B and charm decays which measure $|V_{ub}|$ and $|V_{ub}|/|V_{cb}|$, and from $B_d^\circ - \overline{B_d^\circ}$ mixing and calculations of hadronic matrix elements have been combined to determine their

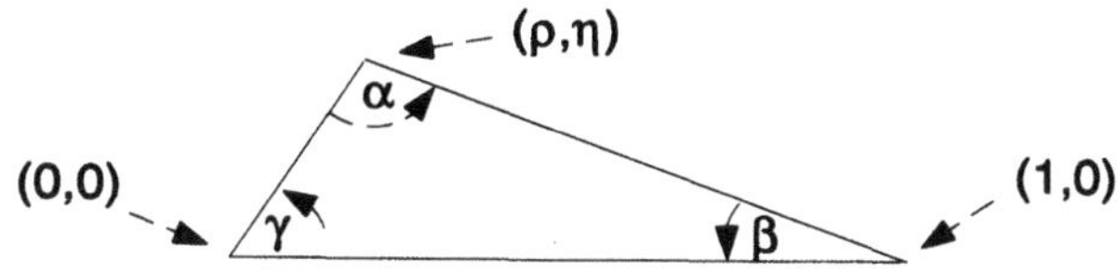

Figure 1. The Unitarity Triangle.

values. From the best estimates of ρ and η we infer that $\alpha \sim 90°$, $\beta \sim 17°$ and $\gamma \sim 73°$ [7, 8]. Direct measurement of the angles of the triangle have so far been inaccessible to experiment.

Asymmetric B Factory tests of the standard model account of CP violation centre upon attempts to overconstrain the triangle, by measuring α, β, γ in addition to improving measurement of CKM matrix elements, to see if the triangle closes.

THE MAIN STRATEGY - CP AND DECAY OF NEUTRAL B MESONS.

In decays of B° mesons to final states f accessible both to B° and to $\overline{B^\circ}$, mixing plays an important role. Interference between the two paths

$$B^\circ \to f \quad \text{and} \quad B^\circ \to \overline{B^\circ} \to f$$

leads to "mixing induced" CP violation. When the decays are dominated by a single amplitude D then $D(B^\circ \to f)$ and $\overline{D}(\overline{B^\circ} \to f)$ have opposite weak phases and identical strong phases so that measurement of the CP violation yields a measurement of the weak phases independent of strong interactions. When f is a CP self conjugate state, then D and $\overline{D}$ have the same magnitude and CP violation is maximised.

The mixing has a well defined time dependence, and a readily calculated CKM phase so that a known, time dependent CP asymmetry between B° and $\overline{B^\circ}$ should be observable and provide a means to measure the phase of D.

Mixing Induced Time Dependent Asymmetry

$B^\circ\overline{B^\circ}$ mixing is dominated by the process in figure 2 whose weak phase is δ_m. Mesons born at time $t = 0$ as pure states $|B^\circ>$ or $|\overline{B^\circ}>$ evolve as particle-antiparticle mixtures:

$$\begin{aligned} |B^\circ(t)> &= a(t)|B^\circ> + ie^{-2i\delta_m}b(t)|\overline{B^\circ}> \\ |\overline{B^\circ}(t)> &= b(t)|\overline{B^\circ}> + ie^{+2i\delta_m}a(t)|B^\circ> \end{aligned}$$

$$\text{where} \quad a(t) = e^{-i(m-i\frac{\Gamma}{2})t}\sin(\tfrac{\Delta m}{2}t) \;\; ; \;\; b(t) = e^{-i(m-i\frac{\Gamma}{2})t}\cos(\tfrac{\Delta m}{2}t)$$

$$m = (m_1 + m_2)/2 \;\; ; \;\; \Delta m = m_1 - m_2 \;\; ; \;\; \Gamma \approx \Gamma_1 \approx \Gamma_2$$

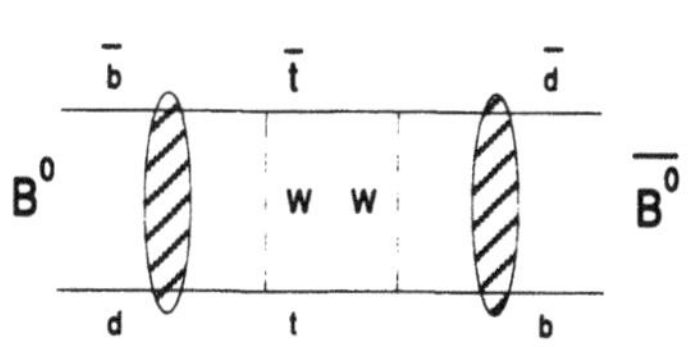

$$e^{2i\delta_m} = \frac{V_{td}V_{tb}^*}{V_{td}^*V_{tb}}$$

Figure 2. Dominant Process in $B^\circ - \overline{B^\circ}$ Mixing

and suffices 1 and 2 refer to CP odd and even eigenstates $|B_1>$, $|B_2>$.

It is then possible to compute the decay rates Γ_f^+ $(B^\circ \to f)$ and Γ_f^- $(\overline{B^\circ} \to f)$ and the time dependent asymmetry:

$$\Gamma_f^\pm(t) = \frac{1}{2}|M|^2 e^{-\Gamma t}\left[(1+|r_f|^2) \pm (1-|r_f|^2)\cos\Delta mt \mp 2\mathrm{Im}(r_f)\sin\Delta mt\right] \quad (1)$$

$$A_{CP}^f(t) = \frac{\Gamma_f^+(t) - \Gamma_f^-(t)}{\Gamma_f^+(t) + \Gamma_f^-(t)} = \frac{(1-|r_f|^2)\cos\Delta mt - \mathrm{Im}(r_f)\sin\Delta mt}{(1+|r_f|^2)} \quad (2)$$

$$\text{with} \quad r_f = e^{-2i\delta_m}\frac{< f|T|\overline{B^\circ} >}{< f|T|B^\circ >} \quad (3)$$

It is important to note that only the time dependence of the asymmetry is observable. If integrated over a large time period, the expression in (2) vanishes.

Decay dominated by single amplitude to CP self conjugate state

Decay amplitudes for B° and $\overline{B^\circ}$ mesons to a final state f which has CP quantum number $\eta_f = \pm 1$ are simply related -

$$CP < f|T|B^\circ > = \eta_f < f|T|\overline{B^\circ} >$$

If in addition, a SINGLE amplitude with weak phase δ_d and strong phase δ_s dominates the decay then the ratio $r_f = \eta_f e^{-2i(\delta_m + \delta_d)}$. The asymmetry then takes the simple form

$$A_{CP}^f(t) = \eta_f \sin\Delta mt \sin 2(\delta_m + \delta_d) \equiv \eta_f \sin\Delta mt \sin 2\phi_w \quad (4)$$

Such decays therefore provide the best measurement of weak phases since the time dependent asymmetry (4) is independent of strong interaction effects. It is only necessary to measure the amplitude of the $\sin\Delta mt$ term to measure ϕ_w and so δ_d.

In such decays, there is no $\cos\Delta mt$ term. Experimentally, detection of such a term would indicate that other decay amplitudes were significant.

Determination of β

This situation is closely realised in the decays of type $B^\circ \to J/\psi K_S^\circ$ in which the dominant tree level diagram (see figure 3) and the relatively weak penguin process have the same CKM phase. The weak phase ϕ_w in (4) is made up of the CKM phase δ_d^{JK} for this diagram, the phase δ_m^B for B° mixing and a further phase δ_{CP}^K which allows for the fact that K_S° is not a CP eigenstate (we assume that K° mixing of CP eigensytates occurs predominantly through a box diagram with intermediate c quarks). These phases combine to give

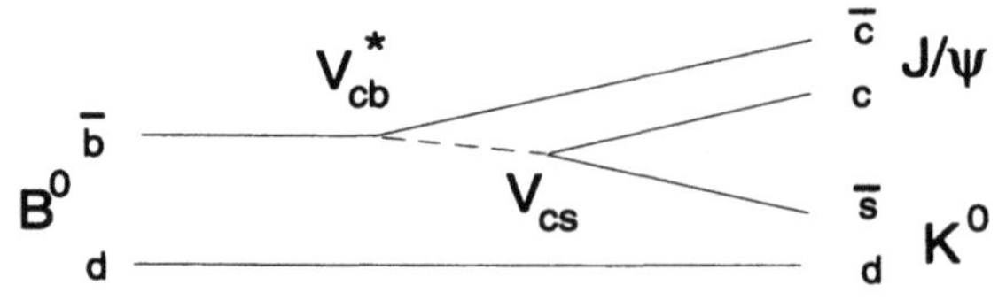

Figure 3. Tree diagram for $B^\circ \to J/\psi K^\circ$.

$$e^{2i\phi_w} = e^{2i(\delta_m^B+\delta_d^{JK}+\delta_{CP}^K)} = \left(\frac{V_{td}V_{tb}^*}{V_{td}^*V_{tb}}\right)\left(\frac{V_{cb}V_{cs}^*}{V_{cb}^*V_{cs}}\right)\left(\frac{V_{cs}V_{cd}^*}{V_{cs}^*V_{cd}}\right)$$

With the approximation $V_{tb} \approx 1$ and that $V_{cb}V_{cd}^*$ is almost real, this leads to $\phi_w = -\beta$. Therefore, measurement of the time dependent asymmetry (4) for this decay mode can provide a direct determination of $\beta = -\phi_w$.

Determination of α

Decays $B^\circ \to \pi^+\pi^-$ provide information on α. The tree and penguin diagrams however (figure 4) have different CKM phases. If we could neglect the penguin process then this and B° mixing would combine to introduce a phase factor ϕ_w where

$$e^{2i\phi_w} = e^{2i(\delta_m^B+\delta_d^{\pi\pi})} = \left(\frac{V_{td}V_{tb}^*}{V_{td}^*V_{tb}}\right)\left(\frac{V_{ub}V_{ud}^*}{V_{ub}^*V_{ud}}\right)$$

Then, since V_{tb} and V_{ud} are both ≈ 1, $\phi_w \approx \alpha$.
So the time dependent asymmetry for this decay provides a direct measurement of $\phi_w = \alpha$ *but only if we assume the tree process is the dominant one.*

A recent result from CLEO suggests that the penguin process may be quite important [9]. It is observed that decays of this type ($\pi\pi$, $\rho\pi$, etc.) are suppressed relative to the decays to $K\pi$ (or $K^*\pi$) which would not be expected if the tree process were dominant. These authors also note that the upper limit on the $B^\circ \to \pi^+\pi^-$ branching ratio is also rather small ($< 2 \times 10^{-5}$).

In this case $< f|T|B^\circ >$ is the sum of tree and penguin terms, each with its own strong and weak phase, and the ratio r_f in the asymmetry (2) is no longer independent of the strong phases. Experimentally, it would be observed that the asymmetry has both $\sin \Delta mt$ and $\cos \Delta mt$ terms, their coefficients giving information on the magnitude and phase of r_f.

To unambiguously determine α requires knowledge of the relative strengths and strong and weak phase differences between the tree and penguin processes which have different isospin content. Methods which exploit the isospin relations between decay rates in other charge states:

$$B^\circ(\overline{B^\circ}) \to \pi^\circ\pi^\circ \quad B^\pm \to \pi^\pm\pi^\circ$$

have been discussed as a means to bound the value of α. [10].

Determination of γ

Measurement of this angle should present the greatest experimental challenge. Decays which provide information on it include

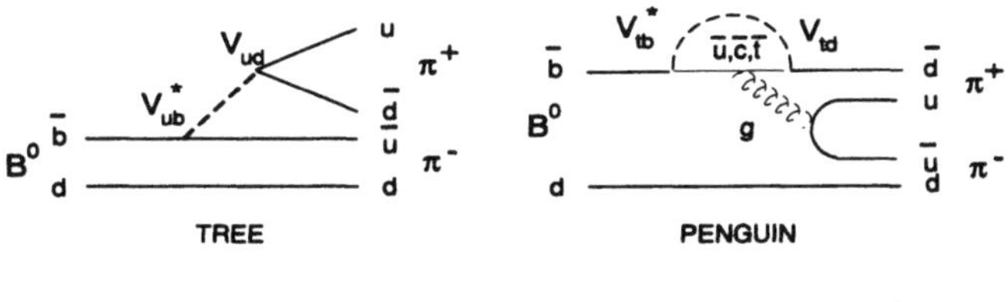

Figure 4. Main tree and penguin diagrams for $B^\circ \to \pi^+\pi^-$.

$$B_s \rightarrow \rho^\circ \overline{K^\circ}$$

in which the time dependent asymmetry (4) $\propto \sin\phi_w$ where $\phi_W = -\gamma$. This would require operation of the B factories at higher energy (at the $\Upsilon(5S)$) than for the other modes, and is not planned in the first years of operation.

Other information could be obtained from precise measurements of the rates for:

$$\begin{array}{lllll} B^\pm & \rightarrow & D^\circ K^\pm & \propto V_{ub}V_{cs}^* & (\text{Br. ratio} \sim 10^{-4}) \\ B^\pm & \rightarrow & \overline{D^\circ} K^\pm & \propto V_{us}V_{cb}^* & (\text{Br. ratio} \sim 10^{-5}) \end{array}$$

With sufficient accuracy, these could be used to extract γ from triangular relations between their amplitudes [11].

Use of other decay modes.

Decays to CP eigenstates proceed through Cabbibo suppressed channels and have small branching ratios. It is possible to use additional related channels [12] such as

$$\begin{array}{llll} B^\circ & \rightarrow & \pi\rho,\ \rho\rho,\ a_1\pi,\ etc.\ . & \text{for } \alpha \\ B^\circ & \rightarrow & J/\psi K_L^\circ,\ J/\psi K^{*\circ},\ \psi^* K,\ etc. & \text{for } \beta \end{array}$$

Use of the states with excited mesons will require the added complication of angular analysis to project out the helicity components which are CP eigenstates. Nevertheless, it is likely that the effective branching ratios available could increase by a factor of about five using these modes.

THE ASYMMETRY EXPERIMENT

To make these asymmetry measurements, it is necessary to produce large samples of B and $\overline{B}$ mesons, to be able to "tag" them as particle or anti particle and to be able to measure the time evolution of their rare decays to CP eigenstates.

The asymmetric production of B's through the reaction

$$e^+e^- \rightarrow \Upsilon(4S) \rightarrow B\overline{B}$$

was suggested [13] as a way to accomplish these goals. At SLAC, the asymmetry was chosen such that the e^+e^- collisions produce $\Upsilon(4S)$ with a $\beta\gamma = 0.56$, and these then decay to $B^\circ\overline{B^\circ}$ approximately 12% of the time. [1] The $B\overline{B}$ continue to move as a coherent $p-$wave system in the laboratory z direction at $\beta\gamma = 0.56$. Bose symmetry [14] prevents either B° from mixing into $\overline{B^\circ}$ until one of the B's decays (at time $t_{Tag} = 0$) and is tagged using either K or lepton to identify it as a B° or $\overline{B^\circ}$.

The other B gets its identity as $\overline{B^\circ}$ or B° at this time *but may already have decayed.* The evolution time $t \propto \Delta z$ (which may be < 0 as in the example illustrated in figure 5) is recorded if the decay is to a CP eigenstate and so the $B^\circ - \overline{B^\circ}$ asymmetry is measured as a function of t and using (4), a value for ϕ_w is determined.

[1] e^- momentum is larger than the e^+. At the $\Upsilon(4S)$ mass the continuum to signal ratio is ≈ 3 and a clean sample of B's is identifiable. B^+B^- and $B^\circ\overline{B^\circ}$ decays occur at approximately equal rates.

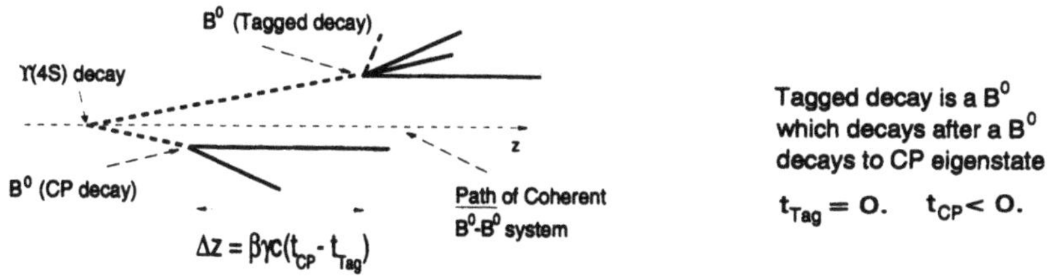

Figure 5. Tagging and timing $B°$ decays.

In decays to $J/\psi K°$ final states, this angle is expected to be equal to $-\beta$, and no $\cos \Delta mt$ term as in (2) is expected. This provides a valuable constraint on both the data and on the underlying theory. Useful checks can be made by comparing four classifications of events. Those with a $B°$ tag and $t > 0$ should have an A_{CP} with identical time dependence as those with a $\overline{B°}$ tag and $t < 0$ and the $B°$ tags with $t < 0$ should be identical to $\overline{B°}$ tags with $t > 0$.

A further check on the data can be made. The time averaged asymmetry should be zero.

A Few Experimental Technicalities

A limitation on the experimental precision arises from the uncertainty of the measurement of Δz crucial to the t dependence of A_{CP}. Also, for the $B°$ decay modes considered, the measured asymmetries are diluted by background B under each signal S by a factor $\sim \sqrt{S/(S+B)}$. Further, tagging (using charge of $e^{\pm}$, $\mu^{\pm}$ or $K^{\pm}$) is not possible in all cases, and mis-tagging is inevitable too due to both detector imperfections and to contrariness of nature itself:

a) Cascade decays $b \rightarrow c \rightarrow l$ give lepton l opposite charge from $b \rightarrow l$;
b) Double charm decay $B \rightarrow DD_sX$ give wrong charges
c) Cabbibo suppressed D decay gives wrong sign K

Estimates of all these effects have been made using a simulation of the detector and the reconstruction software [15].

B FACTORIES AND BaB̄ar

SLAC's new asymmetric B factory PEP II uses the old PEP accelerator as a high energy ring (HER) for storage of 9 GeV/c e^- from the linac. A new low energy ring (LER) will be built on top of this to provide (also from the linac) storage of 3.1 GeV/c e^+. The design luminosity of the machine - 3×10^{33} $cm^{-2}s^{-1}$ should provide ~ 30 fb^{-1} per year of operation accomplished in part by use of a large number of e^+ and e^- bunches, 1 cm long and 126 cm apart, crossing each other at 0°.

At this luminosity, the machine can produce 3×10^7 $B°\overline{B°}$ pairs (with $\beta\gamma = 0.56$ in the laboratory) per year, and is projected to start operation in Spring 1999 when the integrated luminosity at CLEO is expected to be ~ 10 fb^{-1}. Upgrades to operate at the $\Upsilon(5S)$ and at luminosity 10^{34} $cm^{-2}s^{-1}$ are envisaged. PEP II will produce collisions in May 1998, about a year before physics begins, and it is hoped that a great deal will be learned about how to reduce the backgrounds which such a high luminosity may produce before BaB̄ar is rolled in.

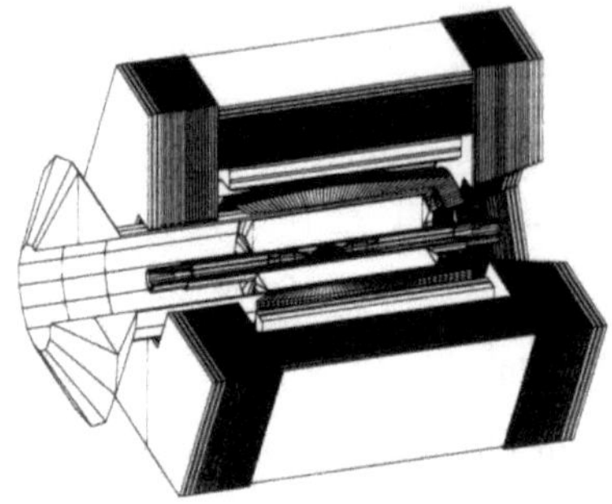

Figure 6. The BaB̄ar detector.

At KEK, the accelerator is being built with a slightly smaller asymmetry ($\beta\gamma = 0.47$) and a luminosity of $10^{34}\ cm^{-2}s^{-1}$. This machine is scheduled for completion for Spring 1999 and detector BELLE will be ready at that time.

The Detector, BaB̄ar

BaB̄ar , shown in figure 6 has been designed for the asymmetry experiment outlined. Important considerations were the ability to measure Δt, *i.e.*, the separation between B° decay vertices with precision small compared to $\beta\gamma c\tau_{B^\circ}$ ($\approx 250\ \mu m$); to identify and fully reconstruct all exclusive CP eigenstates to which a B° decays - including those with neutral particles; and the ability to perform good tagging.

The vertex detector is made from five layers of double sided Si strip detectors and is placed within 3.2 cm of the interaction region. The $B \to \pi\pi(\rho\pi)$ decays place particularly difficult constraints on particle identification. Distinction between $\pi^\pm$ and $K^\pm$ at momenta up to 4.5 GeV/c led to the use of a new kind of ring imaging device - the DIRC [17] - which provides 4σ separation at the highest momenta. Identification and measurement of π° and γ at a wide range of momenta is supplied by a CsI calorimeter and of K_L° and $\mu^\pm$ by the iron flux return (IFR) instrumented with resistive plate chambers.

Good tagging is provided by the DIRC, CsI calorimeter and IFR which provide lepton/hadron identification in the lower momentum region where tagging is most im-

Table I BaB̄ar **'s Anticipated** CP **Reach (per 30** fb^{-1}**)**

ϕ_w	CP State	Br. Ratio	ϵ	# reconstructed	$\Delta \sin\phi_w$
	$J/\psi K_S^\circ$	0.5×10^{-3}	0.41	1106	0.098
β	$J/\psi K_L^\circ$	0.5×10^{-3}	0.33	712	0.16
	$J/\psi K^{*\circ}$	1.6×10^{-3}	0.39	307	0.19
	D^+D^-	6×10^{-4}	0.25	248	0.21
β	$D^{*+}D^{*-}$	7×10^{-4}	0.15	485	0.15
	$J/\psi K^{*\circ}$	8×10^{-4}	?	?	~ 0.15
β	Combined			0.059	

portant, and a 40 layer, state of the art drift chamber provides momentum resolution of less than 1% at all momenta, and dE/dx information for separation of charged particle species in the tagging momentum range.

PEP and Ba$\overline{\text{B}}$ar Schedule

The HER at PEP has already provided the e^- beam current required and much has been learned about dealing with the backgrounds. The LER is built and is on schedule for collisions in May 1998.

The magnet and coil have been delivered to SLAC and the IFR is almost complete. All subsystems are on schedule for installation by year end 1998, roll in to PEP in February 1999.

SUMMARY

Estimates for the uncertainties in β which might be achieved in a year of running at the nominal luminosity (*i.e.*, an exposure of 30 fb^{-1}) have been made using a simulation of the Ba$\overline{\text{B}}$ar detector and of our ability to reconstruct and identify the relevant events. The estimates obtained were reoirted in the Ba$\overline{\text{B}}$ar technical design report [15] and are summarised for the most important channels in Table I These measurements of β should begin to emerge in the year 2000 and should provide an important test for the unitarity triangle. If $\beta \approx 17°$ as expected [8] then this measurement will have an uncertainty of $\sim 3 - 4°$. Of course, a measured value far from this would be of considerable importance, and would cast doubt on the origin of CP violation in the standard model, at least without the need for extensions to it.

Independent results on β may emerge on the same time scale from BELLE at KEK, and perhaps also from CDF at Fermilab when their Run II data are analysed. Uncertainties in both cases are expected to be similar. Another fixed target experiment at HERA is also planned for this time period and could yield important data on β.

Estimates for uncertainty in $\sin 2\alpha$ were also made in reference [15] at about 0.1. At the time, the expected branching ratio was $\sim 1.2 \times 10^{-5}$ and the penguin contribution thought to be small - both assumptions which may now appear optimistic [9]. These difficulties affect all experiments alike, and it appears that definitive information on α or for that matter γ will probably require several years longer given the small cross sections and the precision required.

In any event, it does appear that after 35 years, more experimental evidence for CP violation may be at hand.

References

[1] For recent reviews see: B. Winstein and L. Wolfenstein, The search for direct CP violation, *Rev. Mod. Phys.* **65**:1113 (1993) and Y. Nir and H. Quinn, CP violation in B physics, *Ann. Rev. Nucl. Part. Sci.* **42**:211 (1992), and references therein.

[2] L. Wolfenstein, Violation of CP invariance and the possibility of very weak interactions, *Phys. Rev. Lett.* **13**:562-564 (1964).

[3] A. D. Sakharov, Violation of CP invariance, C asymmetry, and baryon asymmetry of the universe, *JETP Lett.* 5:24-27 (1967).

[4] G. R. Farrar, CP violation and the baryonic asymmetry of the universe, *Nucl. Phys. Proc. Suppl.* **43**:312-332 (1995). G. R. Farrar and M.E. Shaposhnikov, Baryon asymmetry of the universe in the standard electroweak theory, *Phys. Rev.* **D50**:774 (1994). G. R. Farrar and M.E. Shaposhnikov, Baryon asymmetry of the universe in the minimal standard model, *Phys. Rev. Lett.* **70**:2833 (1993).

[5] M. Kobayashi and K. Maskawa, CP violation in the renormalizable theory of weak interaction, *Prog. Theor. Phys.* **49**:652-657 (1973).

[6] L. Wolfenstein, Parametrization of the Kobayashi-Maskawa matrix, *Phys. Rev. Lett.* **51**:1945 (1983).

[7] J.D. Richman and P.R. Burchat, Lepton and Semileptonic Decays of Charm and Bottom Hadrons, *Rev. Mod. Phys.* 67:893-976 (1995).

[8] H. Fritzsch, The kinematics and dynamics of flavor mixing, *Acta Physica Polonica* **B28**:2259-2277 (1997). Also see H. Fritzsch, Quark masses and the description and dynamics of flavor mixing, *Presented at this conference.* (1997).

[9] D. M. Asner, *et al.* , Search for exclusive charmless hadronic B decays, *Phys. Rev.* **D53**:1039-1050 (1996). See also M. Battle, *et al.* , Observation of B0 decay to two charmless mesons, *Phys. Rev. Lett.* **71**:3922-3926 (1993).

[10] Y. Grossman and H. Quinn, Bounding the effect of penguin diagrams in A(CP) ($B^\circ \to \pi^+\pi^-$), *hep-ph/9712306, SLAC-PUB-7713* (1997). M. Gronau and D. London, Isospin analysis of CP asymmetries in B decays, *Phys. Rev. Lett.* **65**:3381-3384 (1990).

[11] M. Gronau and D. Wyler, On determining a weak phase from CP asymmetries in charged B decays, *Phys. Lett.* **B265**:172-176 (1991).

[12] B. Kayser, Testing the standard model of CP violation in all manner of B decay modes, *PASCOS* 1991:837-862 QCD161:I69 (1991). A. Snyder and H. Quinn, Measuring CP asymmetry in $B \to \rho\pi$ decays without ambiguities, *Phys. Rev.* **D48**:2139-2144 (1993).

[13] P. Oddone, Detector Considerations, *Proc. UCLA workshop on Linear Collider B Factory Conceptual Design*, Ed. D. Stork, World Scientific (1987).

[14] This situation is discussed critically in B. Kayser, CP Violation, mixing, and quantum mechanics, *ICHEP* 96:1135-1138, QCD161:H51 (1996).

[15] Ba$\overline{\text{B}}$ar Collaboration, Technical Design Report, *SLAC-R* 95-457 (1995).

[16] BELLE Collaboration, Technical Design Report, *KEK Report* 95-1 (1995).

[17] H. Staengle, *et al.* , Test of a large scale prototype of the DIRC, a Cerenkov imaging detector based on total internal reflection for Ba$\overline{\text{B}}$ar at PEP-II, *Nucl. Instrum. Meth.* **A397**:261-282 (1997).

ON THE MASS OF THE KUIPER BELT

V. L. Teplitz, D.C. Rosenbaum, and R. J. Scalise

Physics Department, Southern Methodist University, Dallas, TX 75275

S. A. Stern

Southwest Research Institute, Boulder, CO 80302

J. D. Anderson

Jet Propulsion Laboratory, California Institute of Technology, Pasadena, CA 91109

Abstract

Evidence for the Kuiper Belt (of cometary material just past Neptune) and the processes that shape it are briefly reviewed. A summary of selected estimates of its mass is given. A two-sector model for the belt is summarized. A limit is placed on the amount of mass that could be present in the Kuiper Belt in the form of objects with sizes around a centimeter, from survival of the Pioneer 10 spacecraft's propellant tank during a dozen years in the belt. Work in progress on the belt's IR signal is reviewed: A useful formula is given for the IR signal from dust particles as a function of albedo, radius and heliocentric distance. Preliminary results are given for limits on the sector masses, in the two-sector model, as a function of time since last passage of the Sun through a giant molecular cloud. Possible indication for the time since such passage and possible support for the existence of a more massive outer sector are found in preliminary results for the ratios of the IR signal in the four different COBE DIRBE bands.

INTRODUCTION

One may see a similarity between the current growing realization that an additional component of the Solar System (SS) has been discovered and the discovery of the cosmic background radiation (CBR) a generation or so ago. Today we are in a similar position of trying to understanding the growing collection of data on the Kuiper Belt (KB). Just as the CBR provides a snapshot of the universe at recombination, the KB can show us the planetesimals from which the planets were formed. Figure 1 shows the major features of the Solar System. Note that the unit of distance is the astronomical unit (AU), the distance from the Sun to Earth. The spherically symmetric Oort cloud, which is the home of most comets, stretches between about $10^{3.3}$ and 10^5 AU from the Sun. We know the Oort cloud is spherically symmetric because the distribution of long period comets

Physics of Mass
Edited by Kursunoglu *et al.*, Kluwer Academic / Plenum Publishers, New York, 1999

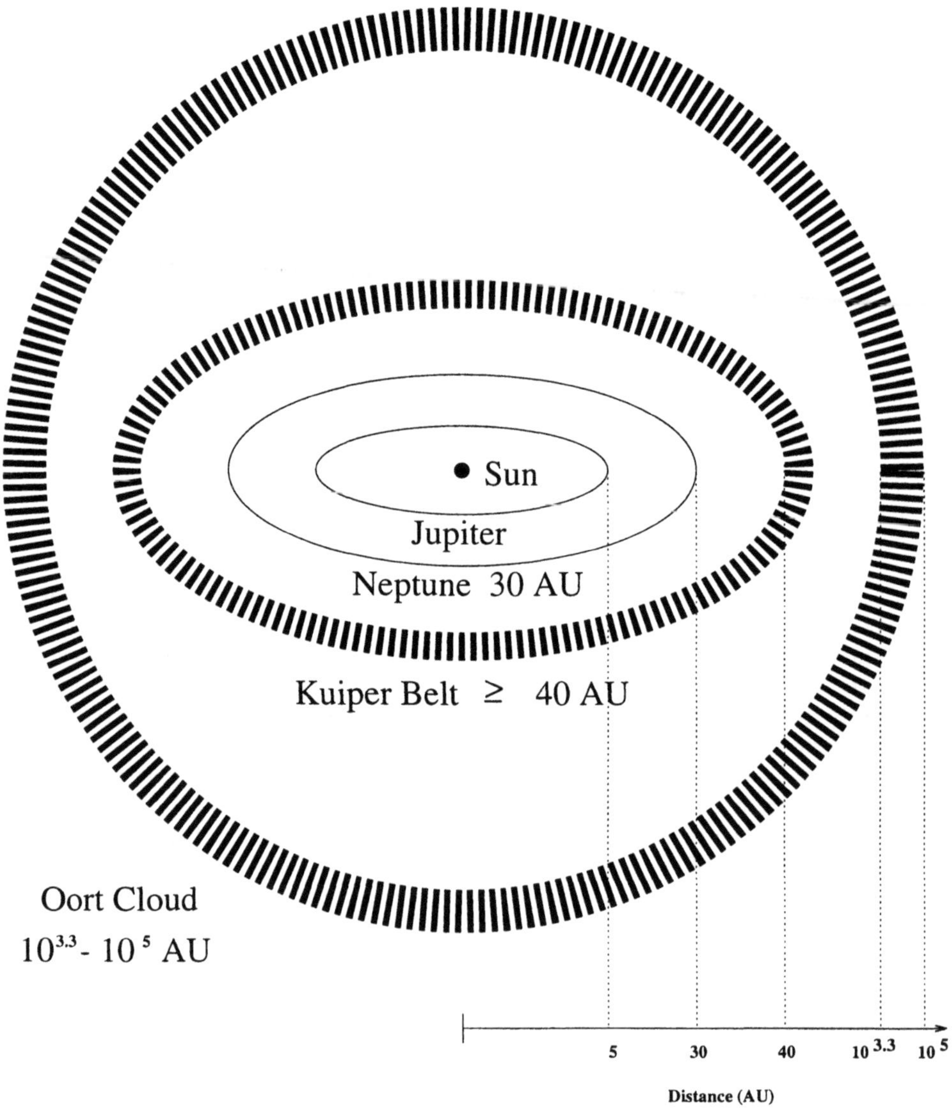

Figure 1. Selected Solar System components.

(over 200 years) is. The KB is roughly a wedge with full opening angle on the order of 30 – 45 degrees. It is the reservoir for comets traveling close to the plane of the ecliptic with periods under about 200 years. Comets are of around 1 to 10 km in diameter. The order of magnitude of the number of Oort cloud comets that have come to the inner Solar System and been seen and recorded is 10^3 and that of Kuiper comets come to the inner SS, 10^2 (the total number that have come to the inner SS over 4 Gyr is probably $10^7 - 10^8$ time these estimates). Comet composition is roughly that of "dirty snowballs," ice (water, CO_2, N_2, etc.) plus dust from supernovae. Over time, complicated organic molecules are made in comets as cosmic rays knock out electrons causing chemical reactions.

The two minute summary of the origin of the Oort cloud and the Kuiper Belt (Levy and Lunine, 1993) is as follows: Stars are formed from collapse of denser regions (cores) of giant molecular clouds (GMCs) which can have masses on the order of 10^5 $M_\odot$ (solar masses), sizes of 100 parsec or more, and densities of 10^3 protons/cm^3 or more. Newly forming stars develop dusty disks for about 10^7 years. Dust aggregates into centimeter-sized bodies, perhaps by Van der Waal forces, and then into kilometer-sized bodies called planetesimals. Planetesimals coalesce by gravitational forces and computer simulations show that, generally, one of the larger planetesimal in a zone tends to rapidly accumulate a large fraction of the others, thereby making a massive core. If the core is massive enough (10 Earth masses, $M_\oplus$, or more) hydrogen and other gases are gravitationally captured thereby making a giant gaseous planet, providing the core mass is achieved before the star's T-Tauri phase (a few 10^6 years) when a stellar wind blows away the stellar nebula. Growth of the planetesimals and the core requires low eccentricities so that collisions will have low relative velocities and result in adhesion rather than fragmentation. Development of the core also requires sufficient mass to make collisions frequent enough.

After formation of the massive gaseous planets, nearby orbits of planetesimals have been shown by computer modeling to become unstable with relatively short time scales and most planetesimals in such orbits are ejected from the SS. In some cases the perturbation caused by the planets falls short of ejecting the body. It is these failed ejections that populate the Oort cloud. The KB, on the other hand, is composed in part of planetesimals formed in a region insufficiently dense for aggregation to planets and sufficiently beyond the last planet for orbits to be stable. It also contains planetesimals scattered by the giant planets into stable KB orbits.

KUIPER BELT EVIDENCE, EVOLUTION AND PROPERTIES

Edgeworth (1943, 1949) and Kuiper (1951) independently postulated a belt of planetary material past the orbit of Neptune. Both argued that there was no reason to expect that the solar nebula should end abruptly at Neptune rather than continuing to decrease in density of solid material – with, perhaps, surface density falling as

$$\Sigma(R) = \Sigma_0 R^{-2} \tag{1}$$

with R being the heliocentric distance. Such a variation could mean as much as 20 $M_\oplus$ in the region 30 – 50 AU. There does not seem to have been a great deal of attention paid to their work until the 1980's when computer based statistical studies of planetary perturbations of comet orbits showed that planetary perturbations could not explain the so-called Jupiter family of comets – short period and low inclination – on the basis of an initially spherically symmetric distribution of Oort cloud comets. It was realized that the KB was the natural reservoir for such comets.

It was not until 1992, however, that a KB object was found (Jewitt and Luu, 1993) with the Keck telescope in Hawaii. At present, a few dozen such objects, with sizes in the range over 100 km, have been found by Jewitt and Luu and other observers. The size estimate is based on an assumption of an albedo on the order of $\alpha = 0.04$. In about two-thirds of the cases there is some follow-up orbit information. Space based observation in 1995 with the Hubble Space Telescope yielded "statistical detection" of twenty-five 10 km objects (Cochran et al., 1995). The method employed would be natural in high energy physics. A Kuiper Belt object (KBO) is found if it appears in several pictures taken several hours apart in pixels separated by the proper distance for an object moving with the velocity appropriate to an orbit in the KB. The problem is that there is background of cosmic rays which, although random, is sufficiently high that one expects coincidences appropriate to mimicking KBO velocities. Cochran et al. (1995) cleverly determined the background by analyzing their data for retrograde KBOs, of which of course none are expected. Since they found about 25 retrograde KBOs, but about 50 prograde ones and since cosmic ray coincidences should be the same in the two cases (in which case the probability is only 10^{-3} for the prograde and retrograde populations to be the same), they were able to report discovery of about 25 real KBOs – but without knowing which of the 50 are the real ones and without much orbit information. This is similar to the way COBE detected a statistical signal of CMB fluctuations but no given CMB fluctuation they saw could be assumed to be real. Such "observations" are different than the norm in astronomy, since the time of the cave person, which is based on assigning definite coordinates to astronomical object ("that one, over there just to the left of the branch"). Cochran et al. took further data in August 1997, under analysis at the time of this talk, which may confirm their earlier result. At issue is not so much the presence of 10 km KBOs but zeroing in on the relative number of 10 and 100 km KBOs which is important for understanding the distribution with size in the belt and its total mass.

Although the material in the KB dates from the formation of the Solar System, it is not pristine in the sense that the belt turns out to be a hotbed, relatively speaking, of activity. There are at least six relevant processes:

(i) Dynamical depletion. Computer modeling of solar orbits in the presence of planetary perturbations, beginning with the pioneering work of Holman and Wisdom (1993) and Levison and Duncan (1993) have now reached the stage at which the stability of a test body in a specific orbit over the age of the Solar System can be tested. Results of Duncan et al. (1995) show that, because of the perturbing effects of the planets, there are no stable (test mass) orbits in the inner Solar System and that it is necessary to go (in semimajor axis) to 42 AU before the first appear. Further in the region out to around 50 many orbits are unstable. Thus we expect the near portion of the KB to be rather depleted. The orbits in this region that are stable tend to be resonant ones. For example in the popular 2:3 mean motion resonance, the planetesimal makes 2 circuits while Neptune makes 3 and has perigees 180° away from Neptune on the average, thereby minimizing the planet's perturbing effect. Pluto's orbit is in the 2:3 resonance.

(ii) Collisional Erosion. One of us (Stern, 1995) pointed out the Jewitt and Luu observations imply relatively large inclinations for KBOs and hence erosive collisions, not adhesive ones, so that we know that KBOs have been eroding for most of the history of the Solar System.

(iii) Interstellar Medium (ISM) Erosion. The KB or at least its outer region is constantly subjected to particles with relatively high velocity (15 km/s) from the ISM. When the Sun passes through a molecular cloud, perhaps every $few \times 10^8$ years or so, a much higher flux impacts on the whole KB. One of us (Stern, 1990) has calculated that such passage results in loss of up to 10 meters of material from each comet (perhaps 10^{-3} the total KB mass). The debris and other dust is driven out of the Solar System. Thus we can say that the outer Solar System is cleaned and (polished) periodically.

Table 1. Some limits and estimates of Kuiper Belt mass since 1964

Reference	KB Mass (in $M_{\oplus}$)	Location (in AU)	Method
Whipple (1964)	$< 10 - 20$	40-50	Effect on Uranus and Neptune orbits
Hamid et al. (1968)	< 0.5	out to 40	Motions of periodic comets
Kuiper (1974)	3	35-50	Remainder from planet formation
Fernandez (1980)	1	35-50	Sufficient to solve short period comet problem
Anderson and Standish (1986)	< 5	inner edge at 35	Pioneer 10 acceleration anomalies
Duncan et al. (1988, 89)	0.06, each	Uranus-Nep -tune; 35-50	Mapping approach, scattering by Uranus and Neptune
Ip and Fernandez (1991)	> 0.002 ; ~ 0.2 if no planetoids	30-200	Earth sized "planetoid scattering" model
Dyson (1994)	0.5	35-50	"Optimistic" update of Kuiper (1951)
Anderson et al. (1995)	< few	35-50	Neptune orbit determinaton as improved by Voyager 2
Weissman (1995)	> 0.004 1.1	34-45 45-1000	Review
Jewitt and Luu (1995)	> 0.003 in 100 km objects	30-50	Direct observation
Cochran et al. (1995)	0.03 in comet sized objects	< 40	Statistical detection
Backman (1995); Stern (1996)	$< 10^{-5}$ in dust	35-100	COBE-DIRBE/ IRAS data
Anderson et al (1998b)	< 0.1 in cm sized objects	35-65	Pioneer 10 survival

(iv) Poynting-Robertson Effect. Because sunlight, in the rest frame of a body in solar orbit, is preferentially incident from the forward direction but is reradiated isotropically there is constant loss of angular momentum causing objects to spiral inward toward the Sun (see Burns, et al., 1979). The time for such loss increases linearly with object radius but is less than the age of the SS for KB objects less than about a centimeter

$$t_{PR} = \frac{4\,\pi\, c^2\, \rho\, a R^2}{3 L_{\odot}\,(1-\alpha)} = 6.75 \times 10^6\, \frac{\rho\, a R^2}{1-\alpha} \tag{2}$$

where ρ is the density and a the radius in cgs, R the heliocentric distance in AU, and α the albedo, of the object. $L_{\odot}$ is the Sun's luminosity, and t_{PR} is in years.

(v) Radiation Pressure. For particle radius a less than a few microns, radiation pressure, which drives particles out of the Solar System is more important than the Poynting-Robertson effect.

(vi) Cosmic Rays. Cosmic rays and solar UV play a major role in belt characteristics. The chemical reactions they cause darken Kuiper materials (after the polishing GMC passage). It takes on the order of 10^8 years to darken a meter thickness of material to albedo $\alpha = 0.04$ (Johnson, 1990). While more detailed study is needed, for the purpose of the present work we assume that the KB is cleared of objects of radius less than

10 meters by GMCs, and that larger objects and collision fragments are darkened from albedo $\alpha = 0.5$ to $\alpha = 0.04$ in 10^8 years.

Other than Poynting-Robertson and radiation pressure, these processes are not well understood. The work below is a step toward such understanding. It also illustrates how improved knowledge of the size distribution and IR characteristics of the small body population can shed light on those processes. For example improved IR data and collision modeling might tell us how long ago the Sun last passed through a GMC.

THE MASS OF THE KUIPER BELT

In Table 1 we present a sampling of KB mass estimates with apologies to all those whose estimates have been omitted. We note that the estimate of some of us (Anderson et al., 1995) was based on improved ephemerides for outer planets from Voyager flybys but was only a rough estimate of an upper bound because the precise work addressed only bounds on spherically symmetric distributions of non luminous matter interior to Neptune's orbit. The technique used could, however, be readily extended to a disk or wedge exterior to Neptune. One feature of the table is the wide range of the estimates. Another is the fact that some refer to total mass and others to mass in objects in some size range.

Note that if we assume a number distribution in object size of the form $n(a) \sim a^{-\gamma}$. the dividing line between mass dominance by large objects and by small is a^{-4}. $a^{-\gamma}$ with $\gamma = 3.5$ is the equilibrium distribution that is maintained under collisions in which either the cross section is independent of velocity or all bodies have the same velocity (Dohnanyi, 1969; Tanaka, 1995) The equilibrium value $\gamma = 3.5$ is insensitive to whether collisions are adhesive or fragmenting and the nature of the fragment size distribution function. As data improves we will learn the value of γ in various size ranges and thereby be able to test collision models.

The estimates of Backman et al. (1995) and Stern (1996a) of dust mass are based on the IR signal in the plane of the ecliptic. Their work includes models for dust production and removal over the age of the Solar System which permit estimates on the order of a third of an Earth mass for total KB mass. Preliminary results reported below build on these works but extend them by considering (i) the dependence of the IR signal on the time since last passage through a GMC and (ii) constraints for the IR signal on the two sector KB model addressed just below.

One of us (Stern, 1996a) and Stern and Colwell (1997) have devised a two sector model of the KB. In brief, these works note that the current large inclinations observed for KBOs imply erosive collisions. They compute that for the adhesive collisions needed to build up observed 100 km KBOs the KB must have had much smaller inclinations and significantly more mass (on the order of 20 $M_\oplus$ in the $30 - 50$ AU region) at the time of KBO formation. Assuming KBO formation at the time of planet formation the inner sector of the KB would subsequently have lost considerable mass from dynamical and collisional erosion. It likely would have had significant inclination (and eccentricity) increase if, as is believed (Malhotra, 1995), Neptune migrated 5–10 AU outward (from angular momentum considerations in the course of planetesimal ejection). The migration would pump up eccentricities and inclinations in part by capture into "mean motion resonances" with Neptune —orbits like that of Pluto designed to avoid perturbing close approaches to Neptune. The outer sector would not have suffered the dynamical depletion from Neptune and would have suffered less collisional erosion because of reduced density and lower velocities. Hence it could have much larger mass, approaching that of Eq. (1). Below we take the dividing line between the inner and outer sectors as 70 AU for the purpose of this paper but it is surely not sharp and could be located as much as 20 AU closer or farther. For the numerical computation we take the outer bound of the outer sector to be at 120 AU. In fact it should be located at a distance such that the density of early Solar System dust was too low to build up kilometer-sized planetesimals. Better estimates of this distance are needed. Before addressing the IR signal, however, we turn to an eyewitness account of the nature of the KB.

PIONEER 10 BOUND ON KUIPER BELT MASS

Pioneer 10, launched in 1972 (and contact with it turned off in 1997), is unique among human-made objects in having spent nearly a dozen years operating in the KB. Pioneer 11 and Voyagers 1 and 2 also explored the outer Solar System but were diverted out of the plane of the ecliptic before reaching the KB. We limit the mass of the KB from the simple fact that Pioneer 10 was not observed to be damaged. We apply the well known collision formula.

$$N_C = \sigma\, n\, v\, t \tag{3}$$

where: we take $N_C = 4$ so as to have a 2 sigma effect from no collision; t is the decade in the KB; v is the relative velocity between the object and Pioneer which, because orbital velocities are low, is essentially Pioneer's 12 km/s (2.5 AU/yr) velocity; and σ is the area of Pioneer in which damage would be noticed. Much of Pioneer's cross section was designed not to be affected by small body collisions; in particular, its large antenna is made of mesh so that a small body just passes through. In addition most of the 18 on-board experiments were off for all or most of the time in the belt. We were conservative and considered only one Pioneer 10 component, its on-board propellant tank which was known not to have been penetrated. We assumed, after conversations with experts and consultation with the hypervelocity impact literature that a kilojoule of energy was needed to penetrate the tank (and the thin foil in front of part of it). This corresponds to a minimum mass of about 0.02 gm while Eq. (3) corresponds to a particle density, $n < 10^{-17}$ cm^{-3}. The result is a limit of less than about 0.06 $M_\oplus$ in 0.02 gm objects (Anderson et al., 1998b).

COBE DIRBE KUIPER BELT CONSTRAINTS

The Cosmic Background Explorer (COBE) satellite carried a Direct IR Background Experiment (DIRBE) with four relevant bands (60, 100, 140, and 240 microns). Measurements were taken in the plane of the ecliptic which can be used to give high upper bounds on the IR signal of the KB. If one subtracts known IR sources such as the Asteroid Belt one gets lower upper limits, although model-dependent ones, on the Kuiper Belt IR signal. These values, drawn from Backman et al. (1995) for the high (unsubtracted) and low (subtracted) limits for the four DIRBE bands are (17.0, 7.5, 6.0, 3.5) and (0.3, 1.0, 2.5, 2.0) respectively, where the figures in the parentheses for the four bands in the two cases are in Megajansky per steradian (MJy/sr = 10^{-20}Wm^{-2}Hz^{-1}sr^{-1}).

To calculate KB limits from the IR signal we need: (i) the IR signal from a body with radius a, albedo α, and heliocentric distance R; and (ii) a model for the dust distribution in the belt, we address these in turn.

(i) IR signal. The IR signal from N bodies with (a, α, R) spread over inclination $(-\theta, +\theta)$ is

$$I_N(\theta, a, R, \alpha, \lambda) = \frac{a^2\, N}{R^2\, 4\pi sin\theta}\, \epsilon_{PR}\, I_{BB} \tag{4}$$

where

$$I_{BB} = \frac{2\pi hc^2}{\lambda}\,[exp(hc/\lambda kT)\,-1]^{-1} \tag{5}$$

and

$$\begin{aligned}\epsilon_{IR} &= 1; \qquad \lambda < 0 \\ &= a/\lambda; \quad \lambda > 0\end{aligned} \tag{6}$$

The temperature T is found from energy balance

$$\pi a^2(1-\alpha)L_\odot/(4\pi R^2) = 4\pi a^2 \int_0^\infty \epsilon_{IR}(\lambda)I_{BB}(\lambda)\,d\lambda \tag{7}$$

Eq. (6) takes account of the fact that emission is inhibited for wavelength greater than the size of the emitting object. No ϵ factor is needed on the left hand side of Eq.(7) because sunlight is dominated by the sub micron optical region but sub micron particles don't remain long enough to play a role in the IR budget. Because of the ϵ-factor, Eq. (7) gives an integral equation for T. We found the following numerical approximation to the solution

$$y = z^B \tag{8}$$

$$B = \sum_{n=1}^{5} B_n(ln\,z)^{n-1} \tag{9}$$

$$T = hc/(k\,\alpha\,y) \tag{10}$$

$$z = (1-\alpha)L_\odot\,a^4/(32\pi^2\,bc^2R^2) \tag{11}$$

Equations (4-11) can be used to find "model independent" limits on the amount of KB (R, a, α) material and hence, by integration, model dependent limits on a distribution over these parameters (Anderson et al., 1998a). This work, which began as a project for an energetic SMU undergraduate, contains the numerical values for the B_n.

(ii) Dust model. We consider a simplified dust model in which, after solar GMC passage, dust is made afresh by collisions of $a_C = 5$ km radius comets. We assume such collisions result in complete fragmentation obeying the $a^{-7/2}$ rule ($dN_D/da = n_0a^{-7/2}$) discussed above. We assume that the 5 km comets are uniformly distributed over $(-\theta, +\theta)$ with $\theta = \bar{e}$ where $\bar{e}$ is the average KBO eccentricity (the relation is that of equipartition of energy of excitation from common circular orbits). Finally, we assume the number density of comets, $n_C(R)$ falls off as R^{-2} as discussed above. We can then write for the comet collision rate

$$\frac{d^2N_{col}(R)}{dR\,dt} = n_C^2(R)\,\pi a_C^2 v_C(R)\,4\pi R^2\sin\theta \tag{12}$$

with n_C, the number of comets given by

$$n_C(R) = \frac{M_T}{m_C}/[ln(R_2/R_1)4\pi\sin\theta R^3] \tag{13}$$

where R_1 andR_2 are the inner and outer sector radii and M_T and m_C total sector and individual comet masses respectively. We take the relative velocity between two colliding comets to be $v_C(R)$ to be

$$v_C(R) = 30\sqrt{2}\,\bar{e}/R^{1/2} \tag{14}$$

where $\bar{e}$ is the average eccentricity and velocity is given in km/s. The result of combining both contributions (i) and (ii) above for the net sector IR signal is

$$I(\lambda) = \int_{R_1}^{R_2} dR \int_{a_1}^{a_2} da \, \frac{d^2 N_{col}}{dR\, dt} \frac{dN_D}{da} \tau(R, a, t) I_1(\theta, R, a, T, \alpha, \lambda) \tag{15}$$

where I_1 is given by Eq.(4) and τ is the minimum of the time since last GMC passage and the Poynting-Robertson time.

Interesting preliminary numerical results from Eq.(15) include the two listed below. For these we assume that the subtraction of known IR sources to go from the high constraints to the low constrains is correct and draw consequences from the low constraints by treating them a first as upper bounds on the KB IR signal and then consider the possibility that they represent precisely the KB signal.

(i) In Figure 2, we show the limits on the inner (40–70 AU) and outer (70–120 AU) sector masses as a function of time since last GMC passage. The kinks at 2×10^8 years are related to the fact we have taken the albedo to be

$$\alpha(t) = 0.5 - 0.46[1 - exp(-t/t_D)] \tag{16}$$

with t_D, the characteristic time for radiation to lower albedo taken as 2×10^8years. The limit on the outer sector is about 5 times larger than that on the inner. The larger volume and lower temperatures in the outer sector give a larger mass bound for the outer sector. The magnitude of these effects is limited by the fact that larger Poynting-Robertson time increases the fraction of dust particles produced that are retained.

These limits can be compared with the Eq. (1) result of continuing the surface mass density $\Sigma(R)$ past Neptune. Using 48 $M_\oplus$ of solids between R=5 and 35 AU gives about $20M_\oplus$ between 35 and 70 AU and about 14 between 70 and 120. We see the outer sector at 2×10^8 years is bounded about a factor of 5 below the simple extrapolation of Eq. (1) while the inner is bounded by a factor of 40. The result is that there certainly must be a "Kuiper gap" in the inner sector compared to Eq. (1) extrapolation. Whether past 70 AU or so the density increases toward the extrapolation from R< 35 AU is still to be seen. Below is some indication that this may be the case.

(ii) Figures 3 and 4 show intensities in the 4 COBE DIRBE bands. Each assumes: (a) one sector (the inner in Figure 3 , the outer in Figure 4) dominates the IR signal; (b) enough sector mass to saturate the most constraining, among the 4 bands, of the low constraints in the first paragraph of this section; and (c) that the low constraints are IR signal values, not just limits. The band intensity limits, if interpreted as KB signals, imply relatively high ratios between the three longer wave length bands and the 60 micron one (3, 8, 7). We see first from both Figures 3 and 4 that we need the time since last GMC passage to be greater than 10^8 years to get the ratios ratios above one. We can, however go further. Assuming the signal comes from the inner belt implies ratios to the 60 micron band of roughly (2, 2.5, 1.5) while assuming it comes from the outer belt implies (3, 4.5, 3) for time of about 3.5×10^8 years or more in both cases. The outer belt ratios are closer to the "data" possibly providing some support for the prediction of an outer KB sector more massive than the inner one. An alternative to this conclusion would be that the 60 micron KB signal is significantly lower than even the low constraint.

CONCLUSIONS

A new component of the Solar System, the Kuiper Belt, has just been discovered and is now open for exploration by everyone's pet technique. Because it is made from original solar nebula material subjected only to limited weathering effects over the last 4.5×10^9

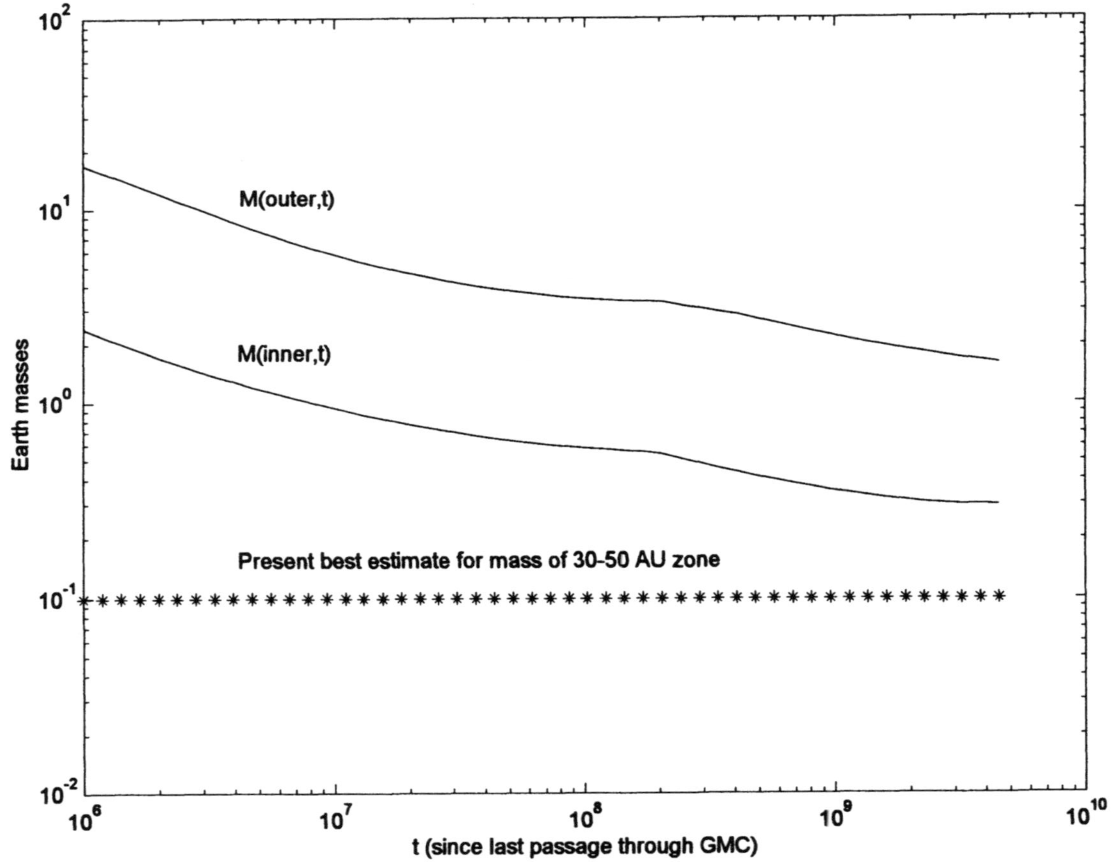

Figure 2. Upper mass bounds for inner and outer sectors (preliminary).

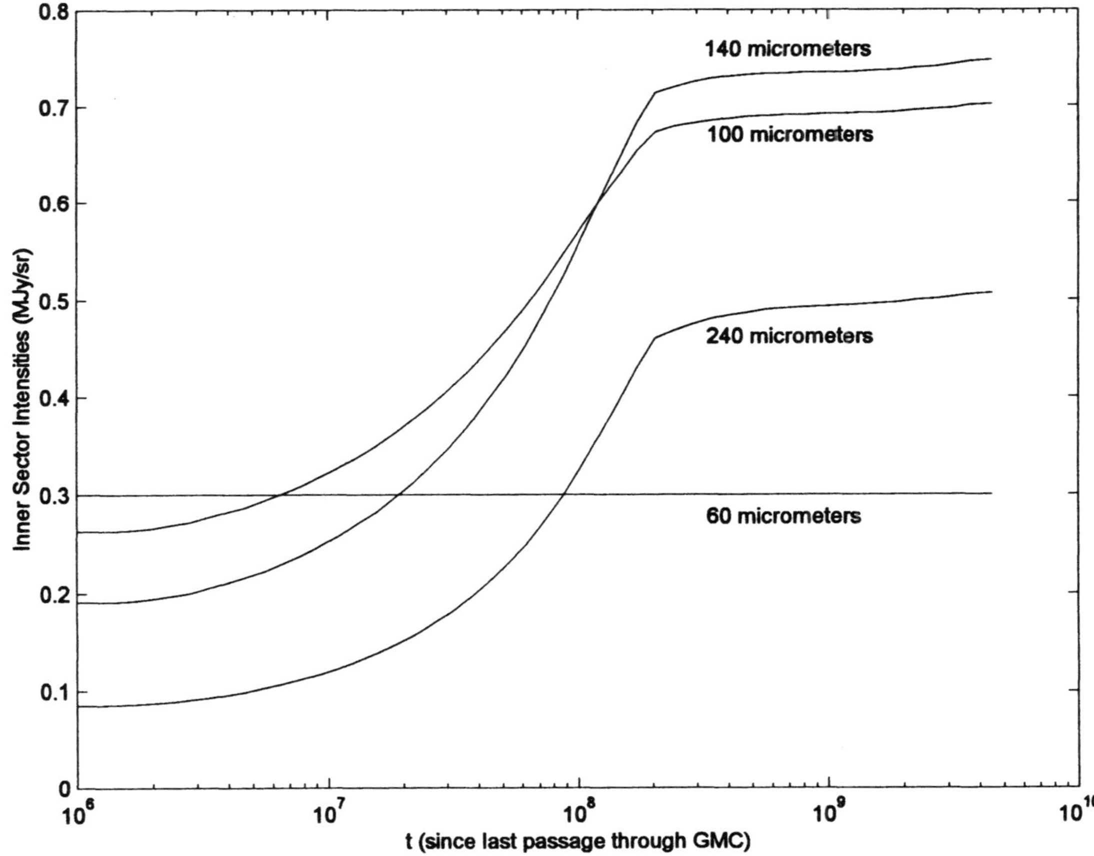

Figure 3. Band intensities for inner sector (preliminary).

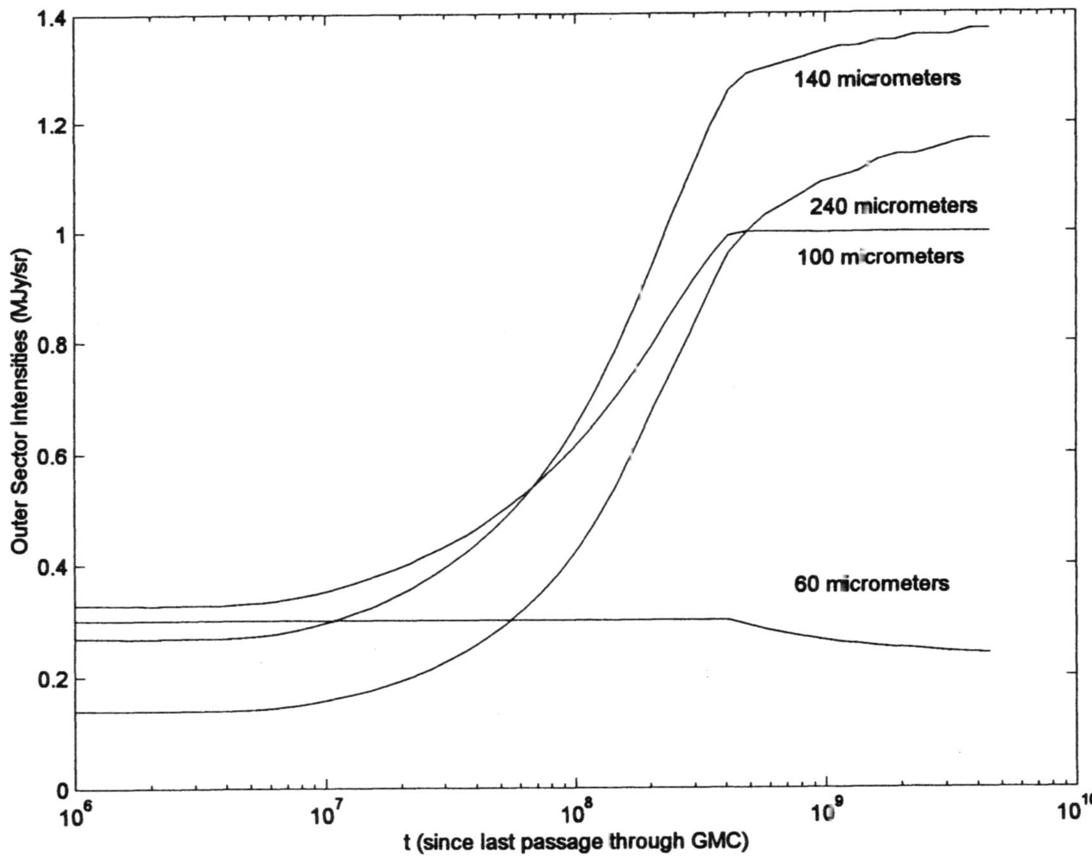

Figure 4. Band intensities for outer sector (preliminary).

years, its exploration is going to teach us a lot about the original nebula, the weathering, and whether, today, adhesive or erosive processes dominate. We have been able to place a limit on its mass, out to about 65 AU, in cm-sized objects (about 0.1 $M_{\oplus}$) from survival of the Pioneer 10 space craft. We have also studied limits on the sector masses in a two sector model of the KB as a function of time since the Sun last passed through a molecular cloud with preliminary results as given in Figure 2. We have some indication, from this preliminary study, that saturation of the constraints on the IR signal resulting from subtracting from the IR signal observed by COBE of the asteroid belt component is consistent with a picture in which the outer sector is more massive than the inner.

ACKNOWLEDGMENTS

We have benefited from working with several collaborators, E. Lau, T. Krisher, K. Scherer, and P. Wentzel, from conversations about the fragility of Pioneer's tank with several experts including E. Christiansen, G. Cleghorn, W. Dixon, D. Kessler, D. Lozier, and M. Wirth, and communications with E. Dwek.

NOTE ADDED

At the January, 1998, meeting of the American Astronomical Society, Hauser et al. (1997) presented a new analysis of the COBE DIRBE data in which they find evidence for a non-zero Cosmic Infrared Background (CIB) in the 140 micron and 240 micron bands as well as limits on the CIB at other DIRBE wavelengths. Because the KB is not included in the model used to subtract the Solar System contribution to the IR signal, the implications of these results for (limits on) the KB IR signal are not obvious. However the outcome of reanalysis with KB inclusion or other appropriate procedures is likely to be further lowering of bounds on the KB IR signal and hence of bounds on the mass of the Kuiper Belt such as those presented here.

REFERENCES

Anderson, J.D., et al., 1998a, (in preparation).
Anderson, J.D., et al., 1998b, *Icarus* 131:191.
Anderson, J.D., et al., 1995, *Ap. J.* 448:885.
Anderson, J. D. and Standish, E. M.,1986, in: *The Galaxy and the Solar System*, R. Smoluchowski, J. N. Bahcall, M.S. Matthews, Eds. University of Arizonia Press, Tucson, p. 286.
Backman, D. E., Dasgupta, A., and Stencel, R. E., 1995, *ApJ* 450:L35.
Burns, J.A., Lamy, P.L., and Soter, S., 1979, *Icarus* 40:1.
Cochran, A.L., Levison, A., and Stern, S.A., 1995, *ApJ* 455:342.
Donyhanyi, J.S., 1969, *J.Geophys. Res.* 74:2531.
Duncan, M., Levison, H. F,. Budd, S.M., 1995, *AJ* 110:3073.
Duncan, M. Quinn, T., Tremaine, S., 1989, *Icarus* 82:402.
Duncan, M., Quinn, T., Tremaine, S., 1988, *ApJ* 328:L69.
Dyson, F., 1994, *Astronomy*, Apr.
Edgeworth, K.E., 1949, *MNRAS* 109:600.
Edgeworth, K.E., 1943, *JBAA* 53:181.
Fernandez, J.A., 1980, *MNRAS* 192:481.
Flynn, G. J., 1994, *Lun. Planet. Sci. Conf.* 25:379 (abstract).
Hamid, S. E., Marsden, B.E., Whipple, F.L., 1968, *Astron. J.* 73:727.
Hauser, M.C. et al., 1997, in: *The COBE Diffuse Infrared Background Experiment*

Search for the Cosmic Infrared Background, I-IV, Bull. of the Amer. Astron. Soc. 29:1359.
Holman, M.J. and Wisdom, J., 1993, *AJ* 105:1987.
Ip, W. H. and Fernandez, J. A., 1991, *Icarus* 92:485.
Jewitt, D. and J. Luu, J.X., 1995, *Astron. J.* 109:1867.
Jewitt, D. and J. Luu, J.X., 1993, *Nature* 362:730.
Johnson, R.E., 1990, *Geophys. Research Lett.* 16:1233.
Kuiper, G.P., 1951, in: *Astrophysics*, J. A. Hyneck, Ed. McGraw Hill, New York, p. 357.
Lang, K.R., 1980, *Astrophysical Formulae*, Springer-Verlag, New York, NY .
Levison, H.F. and Duncan, M., 1993, *ApJ* 406:235.
Levy, E.H. and Lunine, J.I., 1993, *Protostars and Planets III*, D. C. Black and M. S. Matthews, eds. University of Arizona Press Tucson.
Malhotra, R., 1995, *AJ* 110: 420.
Mann, I., Grün, E., Wilck, M., 1996, *Icarus* 120:399.
Soberman, R.K., Neste, S.L. Lichtenfeld, K.,1974, *J. of Geophys. Res.* 79: 3685.
Stern, S.A. and Colwell, J.E., 1997, *ApJ* 490:879.
Stern, S.A., 1996a, *Astron. Astrophys.* 310: 999.
Stern, S.A., 1996b, *AJ* 112:1203.
Stern, S.A., 1995, *AJ* 110:856.
Stern, S.A., 1990, *Icarus* 84: 447.
Tanaka, H., Inaba, S., & Kiyoshi, N. 1995, *Icarus* 123,450.
Tremaine, S., 1990, in: *Baryonic Dark Matter*, D. Linden-Bell and G. Gilmore,eds., Kluwer, Boston, p. 37.
Whipple, F.L., 1964, *Proc. Natl. Acad. Sci.U.S.* 51:711-718.
Weissman, P. R., 1995, *Ann. Rev. Astron. and Astrophys.* 33:327.
Weissman P. R. and H. Levison, L., 1997, in *Pluto and Charon*, S.A. Stern, D.J. Tholen and A.R. Schumer, eds., Univ. of Ariz. Press; Tucson.
Yanagisiwa M., et al., 1996, *Icarus* 123:192.

NUCLEON STABILITY AND DARK MATTER CONSTRAINTS ON SUSY UNIFICATION

Pran Nath* and R. Arnowitt†
*Department of Physics, Northeastern University
Boston, MA 02115-5005
†Center for Theoretical Physics, Department of Physics
Texas A & M University, College Station, TX 77843-4242

March 31, 1998

Abstract

1. Introduction

Proton stability has played an important role in the development of unified models of particle interactions. Thus, for example, the ordinary SU(5) model[1] was ruled out by the data on the decay mode $p \rightarrow e + \pi^0$. In this talk we give a brief review of the constraints on SUSY GUT models generated by the proton stability constraint. In supersymmetric grand unification one has several sources of baryon and lepton number violation which can contribute to proton decay. These include lepto-quark mediated proton decay which is generic to all grand unified models, and proton decay generated by dimension 4 and dimension 5 baryon and lepton number violating operators in SUSY theories. We will also discuss the simultaneous constraints that proton stability and dark matter place on the SUSY spectrum and event rates[2]. We shall see that these constraints are remarkably strong and some of the predictions of these models can be tested in the near future at the upgraded Tevatron.

We begin by discussion of proton decay mediated by lepto-quark exchange. The dominant mode in this type of exchange in minimal SU(5) is $p \rightarrow e^+\pi^0$. A recent analysis gives for this mode in SUSY SU(5) the result[3]

$$\tau(p \rightarrow e^+\pi^0) \geq 1 \times 10^{35\pm1}\ yr \qquad (1)$$

while the current experimental limit on the decay mode is[4]

$$\tau(p \rightarrow e^+\pi^0) > 9 \times 10^{32} yr, (90\% CL) \quad (current) \qquad (2)$$

It is expected that in the future Super K will reach for this mode the limit[5]

$$\tau(p \rightarrow e^+\pi^0) > 1 \times 10^{34} yr, (90\% CL) \quad (Super K) \qquad (3)$$

Thus the $e^+\pi^0$ mode in SUSY SU(5) may be on the edge of detection if Super-K can reach its maximum sensitivity.

Physics of Mass
Edited by Kursunoglu *et al.*, Kluwer Academic / Plenum Publishers, New York, 1999

As already mentioned in SUSY GUTS one has in addition to the baryon and lepton number violation arising from lepto-quark exchange also baryon and lepton number violation arising from dimension 4 and dimension 5 operators. The dimension 4 interactions that violate B & L are interactions of the type

$$W = \lambda_u Q u^c H_2 + \lambda_d Q d^c H_1 + \lambda_e L e^c H_1 + \mu H_1 H_2 \tag{4}$$

$$+(\lambda'_B u^c d^c d^c + \lambda'_L Q d^c L + \lambda''_L L L e^c) \tag{5}$$

Here nucleon decay can occur at a rapid rate via squark exchange and suppression requires

$$\lambda'_B \lambda'_L < (\frac{m_{\tilde{d}}^2}{10^{16} GeV})^2 \sim O(10^{-26\pm1}) \tag{6}$$

In the MSSM one eliminates this type of fast p decay via the discrete R symmetry

$$R = (-1)^{3B+L+2S} \tag{7}$$

Such a symmetry could be a left over piece of a continuous global R symmetry. However, in this case the symmetry would not be preserved by gravitational interactions and worm holes could generate dangerous dimension 4 operators[6]. To guard against dangerous terms of this type one needs to promote global symmertries to gauge symmetries[7]. One hopes that the correct string model will contain the appropriate symmetry of this type.

We discuss next the B & L violating dimension 5 operators. In the minimal supersymmertic standard model (MSSM) one can write many dimension 5 operators that violate B & L number, such as $QQQL$, $u^c u^c d^c e^c$, $QQQH_1$, $Qu^c e^c H_1$, etc[8]. All of these operators contribute to p decay only at the loop order. Of these only the first two arise in the minimal SU(5) GUT models and generate observable p decay. We shall discuss the minimal model in detail shortly. Normally most SUSY/string models will exhibit p instability via dim 5 operators. It is possible to derive a simple condition that allows one one to test if p decay via dimension five operators will be suppressed. Thus in SUSY/string models one has in general that the Higgsino mediated p decay is governed by

$$\bar{H}_1 J + \bar{K} H_1 + \bar{H}_i M_{ij} H_j \tag{8}$$

$H_1, \bar{H}_1$ represent the Higgs triplets that couple with matter (J and $\bar{K}$), and M_{ij} is the Higgs triplet mass matrix. The condition for the suppression of p decay is then given by[9]

$$(M^{-1})_{11} = 0 \tag{9}$$

Now a suppression of this type can arise either via discrete symmetries or via non-standard embeddings. In general SUSY/string models (except for flipped $SU(5) \times U(1)$ models[10]) do not have a natural suppression, and suppression requires a doublet-triplet splitting. There are many mechanisms discussed in the literature for doublet-triplet splitting. These include the fine tuning mechanism to achieve Higgs doublet-Higgs triplet splitting, the sliding singlet mechanism which holds for SU(n) with n$\geq$ 6, the missing partner mechanism, method of VEV alignment, Higgs as pseudo-Goldstones, and use of more than one adjoint in the breaking of the GUT group. In our discussion below we shall assume that one of these mechanisms is operative to achieve the desired doublet- triplet splitting.

2. p Decay in Minimal SU(5)

We discuss now the case of minimal SU(5). The p decay interactions in minimal SU(5) are governed by[11, 12]

$$W_Y = -\frac{1}{8} f_{1ij} \epsilon_{uvwxy} H_1^u M_i^{vw} M_j^{xy} + f_{2ij} \bar{H}_{2u} \bar{M}_{iv} M_j^{uv} \tag{10}$$

After breakdown of the GUT symmetry and integration over the Higgs triplet fields the effective dimension five interaction below the GUT scale which governs p decay is given by

$$L_5^L = \frac{1}{M} \epsilon_{abc} (P f_1^u V)_{ij} (f_2^d)_{kl} (\tilde{u}_{Lbi} \tilde{d}_{Lcj} (\bar{e}_{Lk}^c (V u_L)_{al} - \nu_k^c d_{Lal}) + ...) + H.c. \tag{11}$$

$$L_5^R = -\frac{1}{M} \epsilon_{abc} (V^\dagger f^u)_{ij} (P V f^d)_{kl} (\bar{e}_{Ri}^c u_{Raj} \tilde{u}_{Rck} \tilde{d}_{Rbl} + ...) + H.c. \tag{12}$$

where L_5^L (LLLL) and L_5^R (RRRR) are dim 5 operators, V is the CKM matrix and f_i are related to quark masses by

$$m_i^u = f_i^u (sin 2\theta_W / e) M_Z sin\beta \tag{13}$$

$$m_i^d = f_i^d (sin 2\theta_W / e) M_Z sin\beta \tag{14}$$

P_i are the inter-generational phases

$$P_i = (e^{i\gamma_i}), \; \sum_i \gamma_i = 0; \; i = 1,2,3 \tag{15}$$

The dominant p decay modes involve pseudo-scalar bosons and leptons

$$\bar{\nu}_i K^+, \bar{\nu}_i \pi^+; i = e, \mu, \tau \tag{16}$$

$$e^+ K^0, \mu^+ K^0, e^+ \pi^0, \mu^+ \pi^0, e^+ \eta, \mu^+ \eta \tag{17}$$

Their relative strength depends on quark masses, CKM factors, and on the 3rd generation enhancement factors y^{tK} are discussed in ref.[12]. Using the analysis of this work we can deduce a rough hierarchy of the branching ratios

$$BR(\bar{\nu} K) > BR(\bar{\nu} \pi) > BR(lK) > BR(l\pi) \tag{18}$$

This hierarchy can be negated in special situations. For example, it is known that if there are large cancellations between the second and the third generations then the $\bar{\nu}\pi$ mode can become dominant[12]. It is also argued recently that in certain SO(10) scenarios one can achieve relatively large branching ratios for the $l\pi$ modes[13].

For the minimal SU(5) supergravity model[14] the decay width of $p \to \bar{\nu}_i K^+$ mode is given by

$$\Gamma(p \to \bar{\nu}_i K^+) = (\frac{\beta_p}{M_{H_3}})^2 |A|^2 |B_i| C \tag{19}$$

Here β_p is defined by

$$\beta_p U_L^\gamma = \epsilon_{abc} \epsilon_{\alpha\beta} < 0 | d_{aL}^\alpha u_{bL}^\beta u_{cL}^\gamma | p > \tag{20}$$

where lattice gauge analysis gives for this quantity the result[15]

$$\beta_p = (5.6 \pm 0.5) \times 10^{-3} GeV^3 \tag{21}$$

The remaining factors in the decay width formula are as follows: A contains the quark mass and CKM factors

$$A = \frac{\alpha_2^2}{2 M_W^2} m_s m_c V_{21}^\dagger V_{21} A_L A_S \tag{22}$$

where m_s, m_c are the quark mass factors, V_{ij} are the CKM factors, and A_L (A_R) are the long (short) RG suppression factor in evolution from the GUT scale down to the electro-weak scale. The quantities B_i are the dressing loop functions

$$B_i = \frac{1}{sin2\beta} \frac{m_i^d V_{i1}^\dagger}{m_s V_{21}^\dagger} [P_2 B_{2i} + \frac{m_t V_{31} V_{32}}{m_c V_{21} V_{22}} P_3 B_{3i}] \tag{23}$$

Here the first (second) term in the bracket is the contribution from the second (third) generation exchange in the dressing loops. Finally the factor C in the decay width formula is the chiral Lagrangian factor and is given by

$$C = \frac{m_N}{32\pi f_\pi^2} [(1 + \frac{m_N(D+F)}{m_B})(1 - \frac{m_K^2}{m_N^2})]^2 \tag{24}$$

where f_π is the pion decay constant with the value $f_\pi = 139MeV$, D=0.76, F= 0.48, $m_B = 1154$ MeV and m_K is the kaon mass and m_N is the nucleon mass. The result of the p lifetime analysis must be compared with the current limits on the p lifetime and the limits that one expects will be achievable in the future. For the $p \to \bar{\nu} K^+$ decay mode the current limit is[4]

$$\tau(p \to \bar{\nu} K^+) \geq 1 \times 10^{32} yr \tag{25}$$

One expects that at Super K[5] one will be sensitive to lifetime of

$$\tau(p \to \bar{\nu} K^+) \leq 2 \times 10^{33} yr \tag{26}$$

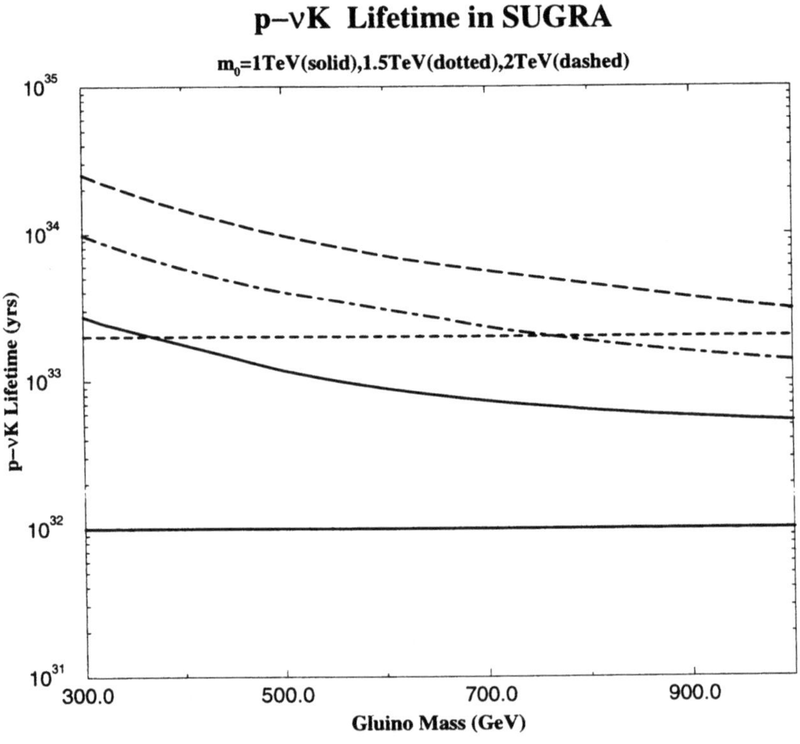

Figure 1. The maximum $\tau(p \to \bar{\nu} K)$ lifetime in Minimal SUGRA from ref.[2]. The curves are for naturalness assumption on m_0 of 1 TeV (solid), 1.5 TeV (dot-dashed) and 2 TeV (dashed). The horizontal solid line is the current experimental lower limit and the dashed horizontal line is the limit expected from Super K

and it is possible that at ICARUS[16] one will be sensitive up to $\tau(p \to \bar{\nu}K^+) = 10^{34}yr$.

A recent analysis of the proton life time in minimal SU(5) supergravity unification with and without p dark matter constraints was in given in ref.[2]. We discuss the results of that analysis here. We consider first the analysis without dark matter constraint. Here the maximum proton lifetime in minimal SU(5) is given in Fig. 1 for three different assumptions on the naturalness limit on m_0; i.e., of 1 TeV, 1.5 TeV, and 2 TeV. The analysis shows that Super K will exhaust a significant part of the parameter space for the naturalness assumption of $m_0 = 1$ TeV. For the case when the naturalness limit on m_0 is 1.5 TeV, Super K will exhaust the parameter space for gluino masses larger than 750 GeV while no restrictions on the gluino mass will result for a naturalness limit on m_0 of 2 TeV or larger[2].

3. Non-minimal Extensions

Similar generic results hold for SU(N)(N$\geq$ 5), $SU(3)^3$, etc. However, significant modifications can occur because of possible problems with the unification of Yukawa couplings vs proton stability. For example, in SO(10) models one needs typically a large tanβ, i.e., $tan\beta \sim 50$ to get b-t-τ unification. Under this constraint the mass scale necessary to suppress p decay to the current experimental value is

$$(M^{-1})_{11} > tan\beta(0.57 \times 10^{16})\ GeV \tag{27}$$

For tan$\beta \sim 50$ one requires a GUT mass of 2.5×10^{17} for the suppression of p decay to the current experimental lower limit. However, a mass scale this large upsets the consistency of the unification of gauge couplings with the LEP data and one needs large threshold corrections to get agreement with experiment[17, 18, 19].

Next we discuss the effects of quark-lepton textures on p decay. As is well known the usual GUT models give poor predictions for quark-lepton mass ratios and one needs textures to achieve the correct mass hierarchies[20, 21]. In the context of supergravity unified models textures can arise from higher dimensional operators given by Planck scale correcions. A minimal model of such textures leads to an upward correction to the $p \to \bar{\nu}K$ lifetime by a factor of $3 \sim 5$[22].

Cosmological Constraints

Next we discuss the effects of neutralino dark matter constraints[23] on the maximum proton life time. The quantity of interest in the computation of dark matter is $\Omega_{\chi_1^0} = \rho_{\chi_1^0}/\rho_c$ where $\rho_{\chi_1^0}$ is the neutralino matter density and ρ_c is the critical matter density needed to close the universe

$$\rho_c = 3H_0^2/8\pi G_N = 1.88h_0^2 \times 10^{-29} gm/cm^3 \tag{28}$$

where h_0 is the Hubble parameter H_0 in units of 100 km/sec.Mpc. The number density for χ_1 obeys

$$\frac{dn}{dt} = -3Hn - <\sigma v> (n^2 - n_0^2) \tag{29}$$

$$<\sigma v> = \int_0^\infty dv v^2 (\sigma v) e^{-v^2/4x} / \int_0^\infty dv v^2 e^{-v^2/4x}; \quad x = \frac{kT}{m_{\chi_i}} \tag{30}$$

At the "freeze-out" temperature T_f when the annihilation rate becomes smaller than the expansion rate the χ_1^0 decouple from the background and integration from T_f to the current temperature gives for $\Omega_{\chi_1^0}h^2$ ($\Omega_{\chi_1^0} = \rho_{\chi_1^0}/\rho_c$)

$$\Omega_{\chi_1^0}h^2 \cong 2.48 \times 10^{-11} \left(\frac{T_{\chi_1^0}}{T_\gamma}\right)^3 \left(\frac{T_\gamma}{2.73}\right)^3 \frac{N_f^{1/2}}{J(x_f)} \tag{31}$$

$$J\ (x_f) = \int_0^{x_f} dx\ \langle\ \sigma v\ \rangle\ (x) GeV^{-2} \tag{32}$$

T_γ is the current background temperature, $(T_{\chi_1^0}/T_\gamma)^3$ is the reheating factor, N_f is number of massless degrees of freedom at freezeout, and $x_f = kT_f/m_{\chi_1}$. Assuming inflationary scenario ($\Omega = 1$), a baryonic component $\Omega_B \leq 0.1$, hot (HDM) and cold (CDM) dark matter in the ratio $\Omega_{HDM} : \Omega_{CDM} = 1 : 2$, and $0.5 \leq h \leq 0.75$ one has

$$0.1 \leq \Omega_{\chi_1^0} h^2 \leq 0.4 \tag{33}$$

The effect of including the relic density constraint on the maximum p decay lifetime into the $\bar{\nu}K^+$ mode is given in Fig. 2[2]. Here one finds the remarkable result that the maximum p lifetime into the $\bar{\nu}K^+$ is reduced by a factor up to 10-30 with the inclusion of the dark matter constraint for gluino masses greater than about 500 GeV (see Table 1 taken from ref.[2]). Further, this reduction appears to be essentially independent of the naturalness limit on m_0. The reason for this independence of the naturalness limit on m_0 is due to the fact that the relic density constraint requires a small value of m_0, i.e., ($m_0 \leq 200$) GeV, for gluino masses in this mass range. However, an m_0 value this small does not lead to sufficient suppression for p decay to be consistent with experiment. This result is similar to the case of no-scale models which also has problems with p stabilty[24].

The analysis of Fig. 2 is for the case of universal boundary conditions for the soft SUSY breaking parameters at the GUT scale. An analysis of maximum p lifetime with non-universality boundary conditions has also been carried out. The non-universalities investigated were those in the Higgs sector and in the third generation sector. For the analysis here we shall discuss only the non-universalities in the Higgs sector. It is found convenient to parametrize the non-universalities in this sector by δ_1 and δ_2 such that[25]

$$m_{H_1}^2(M_G) = m_0^2(1+\delta_1), m_{H_1}^2(M_G) = m_0^2(1+\delta_1) \tag{34}$$

where a reasonable range for δ_1 and δ_2 is given by $|\delta_i| \leq 1$ (i=1,2). In Fig. 3 an analysis of the maximum p decay lifetime for the case $\delta_1 = 1 = -\delta_2$ is given[2]. One finds that the results in this case are similar to those for the case of Fig. 2. In both cases one finds that the gluino mass must lie below 500 GeV to be achieve consistency with the current constraints on p lifetime and dark matter. For the case $\delta_1 = -\delta_2 = 1$ one finds that the maximum p lifetime also falls below the current experimental limits except for a small gluino mass range.

An analysis of event rates expected in dark matter detectors under the combined relic density and p stability constraints has also been carried out[2]. Here one finds that the maximum and minimum event rates with the inclusion of proton stability constraints lie in a narrow range. The analysis shows that the inclusion of p decay constraint reduces the maximum event rates by a significant amount and one needs more sensitive detectors[26], more sensitive by a factor of 10^{3-4} than the detectors currently available[27] to detect dark matter in models with relic density and p stabilty constraints.

4. Unmasking Planck Effects

Proton decay can play an important role in disentangling Planck scale effects from GUT effects. As has been discussed extensively in the literature Planck corrections bring in a field dependence in the gauge kinetic energy function, i.e., the gauge kinetic energy has the form

$$f_{\alpha\beta} F^{\mu\nu\alpha} F_{\mu\nu}^{\beta} \tag{35}$$

where for SU(5) one may write[28]

$$f_{\alpha\beta} = (\delta_{\alpha\beta} + \frac{c}{2M_P} d_{\alpha\beta\gamma}\Sigma^{\gamma}) \tag{36}$$

Here Σ are the adjoint scalars and c parameterizes Planck physics [c=O(1)]. Now in the R-G analysis of the gauge coupling constants α_i one finds that the GUT threshold effects and the Planck scale effects are comparable[29]. One may see the comparable size of the effects by examining the R-G equations for the gauge coupling constants in the region $Q \sim M_G$ where the one loop evolution equations for the gauge coupling constants are given by

$$\alpha_i^{-1}(Q) = \alpha_G^{-1} + C_{ia} ln(\frac{M_a}{Q}) + \frac{cM}{2M_P} \alpha_G^{-1} n_i \tag{37}$$

Here the last term gives the contribution from the Planck scale corrections where $n_i = (-1, -3, 2)$ for the subgroups i=$(U(1), SU(2)_L, SU(3)_C)$ and $< \Sigma >= M$. It is easily seen that one may absorb the Planck effects into the GUT thresholds by rescaling and obtain an evolution equation in the region $Q \sim M_G$ of the form

$$\alpha_i^{-1}(Q) = \alpha_G^{eff-1} + C_{ia} ln(\frac{M_a^{eff}}{Q}) \tag{38}$$

Thus we see that the renormalization group analysis at this level cannot distinguish between the GUT thresholds and the Planck scale corrections. However, p decay depends on the unscaled GUT parameters. Thus p decay can unmask Planck effects and remove the GUT-Planck confusion. Thus one can use the renormalization group analysis along with p decay to compute the size of the Planck scale correction c. One finds[29]

$$p \to \bar{\nu} K^+, \quad c = \frac{\alpha_G}{10} \frac{M_P}{\pi M_V} ln \frac{M_{H_3}^{eff}}{M_{H_3}} \tag{39}$$

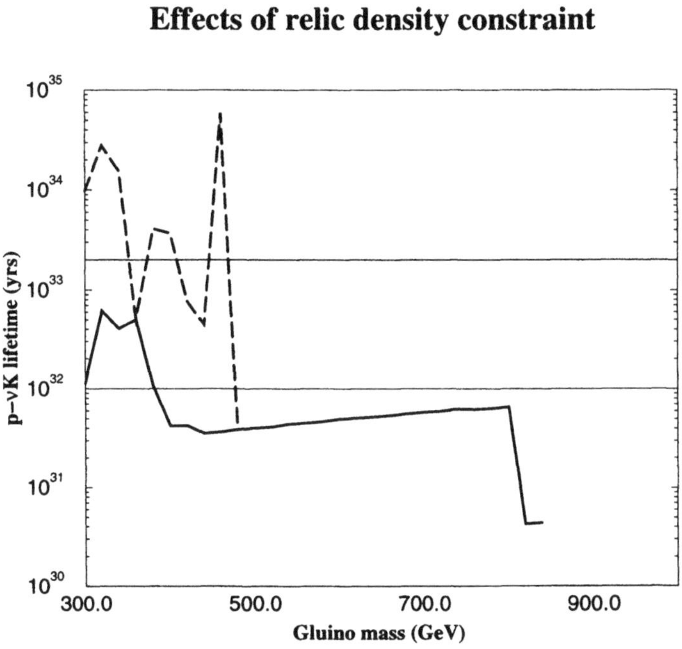

Figure 2. Analysis of the maximum $\tau(p \to \bar{\nu} K)$ life time with relic density constraints for minimal supergravity model. The solid curve is for the naturalness assumption of $m_0 \leq 1$ TeV and the dashed curve is for the naturalness assumption of $m_0 \leq 5$ TeV (from ref.[2]).

Effects of relic density constraint

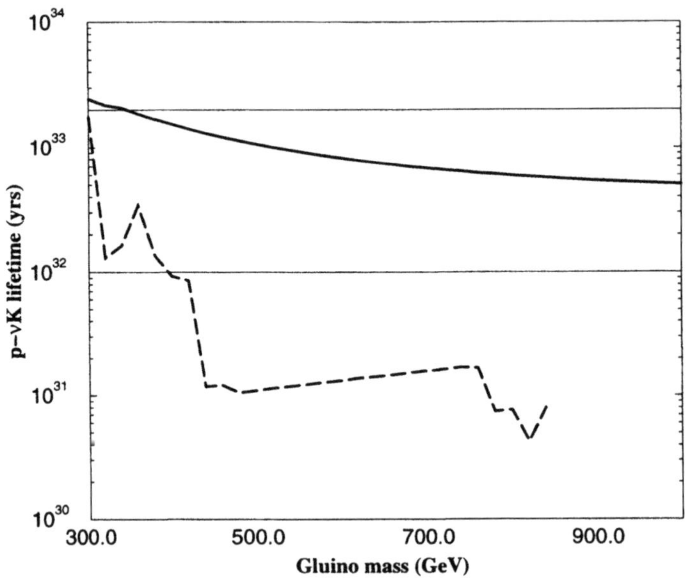

Figure 3. Analysis of the maximum p decay lifetime with (dashed) and without (solid) relic density constraint ($0.1 \leq \Omega h^2 \leq 0.4$) for the case $\delta_1 = 1 = -\delta_2$ for the case when the naturalness constraint on m_0 is 1 TeV. (from ref.[2])

and

$$p \to e^+\pi^0, \quad c = \frac{100}{3}\sqrt{\frac{2}{3}}\alpha_G^{3/2}\frac{M_P}{M_V}ln\frac{M_V}{M_V^{eff}} \tag{40}$$

5. Conclusions

One can probe a majority of the parameter space of the minimal SUGRA model within the naturalness constraint of $m_0 \leq 1$ TeV, and $m_{\tilde{g}} \leq 1$ TeV if the Super-K and Icarus experiments can reach the expected sensitivity of 2×10^{34} y for the $\bar{\nu}K^+$ mode[2]. With the inclusion of dark matter constraints one finds that the gluino mass must lie below 500 GeV within any reasonable naturalness constraint to satify the current lower limit on the $\bar{\nu}K^+$ mode. The simultaneous p stability and dark matter constraints will be tested in the near future in p decay experiments. The predictions on the constraints on the sparticle spectrum can be tested in large measure at the upgraded Tevatron using the trilepton and other signals[30, 31, 32]. Of course, the minimal model will be tested in full at the LHC.

Acknowledgements

This research was supported in part by NSF grant numbers PHY-9602074 and PHY-9722090.

Table 1. Reduction of $\tau(p \to \bar{\nu}K)_{max}$ from dark matter constraint

gluino mass (GeV)	reduction factor when $0.1 < \Omega h^2 < 0.4$
500	29.5
600	18.6
700	12.9
800	9.8

References

1. H. Georgi and S.L. Glashow, Phys. Rev. Lett. **32**, 438 (1974).
2. R. Arnowitt and P. Nath, hep-ph/9801246.
3. W.J. Marciano, talk at SUSY-97, University of Pennsylvania, Philadelphia, May 27-31, 1997.
4. Particle Data Group, Phys.Rev. **D50**,1173(1994).
5. Y.Totsuka, Proc. XXIV Conf. on High Energy Physics, Munich, 1988,Eds. R.Kotthaus and J.H. Kuhn (Springer Verlag, Berlin, Heidelberg,1989).
6. G. Gilbert, Nucl. Phys. **B328**, 159 (1989).
7. L. Krauss and F. Wilczek, Phys. Rev. Lett. **62**, 1221 (1989).
8. L. Ibanez and G.G. Ross, Nucl Phys. **368**, 4 (1992).
9. R. Arnowitt and P. Nath, Phys. Rev. **D49**, 1479 (1994).
10. I.Antoniadis, J.Ellis, J.S.Hagelin and D.V.Nanopoulos, Phys.Lett.**B231**,65 (1987); ibid, **B205**, 459(1988).
11. S. Weinberg, Phys. Rev. **D26**, 287 (1982); N. Sakai and T. Yanagida, Nucl. Phys.**B197**, 533 (1982); S. Dimopoulos, S. Raby and F. Wilczek, Phys.Lett. **112B**, 133 (1982); J. Ellis, D.V. Nanopoulos and S. Rudaz, Nucl. Phys. **B202**, 43 (1982); B.A. Campbell, J. Ellis and D.V. Nanopoulos, Phys. Lett. **141B**, 299 (1984); S. Chadha, G.D. Coughlan, M. Daniel and G.G. Ross, Phys. Lett.**149B**, 47 (1984).
12. R. Arnowitt, A.H. Chamseddine and P. Nath, Phys. Lett. **156B**, 215 (1985); P. Nath, R. Arnowitt and A.H. Chamseddine, Phys. Rev. **32D**, 2348 (1985); J. Hisano, H. Murayama and T. Yanagida, Nucl. Phys. **B402**, 46 (1993); R. Arnowitt and P. Nath, Phys. Rev. **49**, 1479 (1994).
13. K.S.Babu, J.C. Pati, and F. Wilczek, hep-ph/9712307.
14. A.H. Chamseddine, R. Arnowitt and P. Nath, Phys. Rev. Lett **29**. 970 (1982); For review see, P.Nath,Arnowitt and A.H.Chamseddine , "Applied N=1 Supergravity" (World Scientific, Singapore, 1984); H.P. Nilles, Phys. Rep. **110**, 1 (1984); R. Arnowitt and P. Nath, Proc of VII J.A. Swieca Summer School (World Scientific, Singapore 1994).
15. M.B.Gavela et al, Nucl.Phys.**B312**,269(1989).
16. The ICARUS Collaboration: "ICARUS II, a Second -Generation Proton Decay Experiment and Neutrino Observatory at the Gran Sasso Laboratory", Proposal Vol I II, LNGS-94/99-III, May 1994 and "A first 600 ton ICARUS Detector Installed at the Gran Sasso Laboratory", Addendum to proposal LNGS-94/99 III, May 19, 1995.
17. K.S. Babu and S.M. Barr, Phys. Rev. **D51**(1995)2463.
18. V. Lucas and S. Raby, Phys. Rev. **D54** (1996) 2261; ibid. **D55** (1997) 6986.
19. S. Urano and R. Arnowitt, hep-ph/9611389.
20. H. Georgi and C. Jarlskog, Phys. Lett. **B86**,297 (1979); J. Harvey, P. Ramon and D. Reiss, Phys. Lett. **B92**,309 (1980); P. Ramond, R.G. Roberts, G.G. Ross, Nucl. Phys. **B406**, 19 (1993); K.S. Babu and R. N. Mohapatra, Phys. Lett.**74**, 2418 (1995).
21. G. Anderson, S. Raby, S. Dimopoulos, L. Hall, and G.D. Starkman, Phys. Rev. **D49**, 3660 (1994).
22. P. Nath, Phys. Rev. Lett. **76**, 2218 (1996).

23. For a review see G. Jungman, M. Kamionkowski and K. Greist, Phys. Rep. **267**,195(1995); E.W. Kolb and M.S. Turner, "The Early Universe" (Addison-Wesley, Redwood City, 1989); P. Nath and R. Arnowitt, Proc. of the Workshop on Aspects of Dark Matter in Astrophysics and Particle Physics, Heidelberg, Germany 16-20 September, 1996.

24. P. Nath and R. Arnowitt, Phys. Lett. **B289**, 308 (1992).

25. R. Arnowitt and P. Nath, Phys. Rev.**D56**, 2833(1997); P. Nath and R. Arnowitt, Phys. Rev. **D56**, 2820(1997).

26. For recent developments in detector technology in dark matter detectors see D. Cline, Nucl. Phys. B (Proc. Suppl.) **51B**, 304 (1996); P. Benetti et al, Nucl. Inst. and Method for Particle Physics Research, **A307**,203 (1993).

27. R. Bernabei, et.al., Phys. Lett. **B389**, 757(1996).

28. C.T. Hill, Phys. Lett. **B135**, 47 (1984); Q. Shafi and C. Wetterich, Phys. Rev. Lett. **52**, 875 (1984).

29. T. Dasgupta, P. Mamales and P. Nath, Phys. Rev. **D52**, 5366 (1995); D. Ring, S. Urano and R. Arnowitt, Phys. Rev. **D52**, 6623 (1995); S. Urano, D. Ring and R. Arnowitt, Phys. Rev. Lett. **76**, 3663 (1996); P. Nath, Phys. Rev. Lett. **76**, 2218 (1996).

30. P. Nath and R. Arnowitt, Mod. Phys. Lett. **A2**, 331(1987); R. Barbieri, F. Caravaglio, M. Frigeni, and M. Mangano, Nucl. Phys. **B367**, 28(1991); H. Baer and X. Tata, Phys. Rev. **D47**, 2739(1992); J.L. Lopez, D.V. Nanopoulos, X. Wang and A. Zichichi, Phys. Rev. **D48**, 2062(1993); H. Baer, C. Kao and X. Tata, Phys. Rev. **D48**, 5175(1993).

31. T. Kamon, J. Lopez, P. McIntyre and J.J. White, Phys. Rev.**D50**,5676 (1994); H. Baer, C-H. Chen, C. Kao and X. Tata, Phys. Rev. **D52**, 1565 (1995); S. Mrenna, G.L. Kane, G.D. Kribbs, and T.D. Wells, Phys. Rev. **D53**, 1168 (1996).

32. D. Amidie and R. Brock, " Report of the tev-2000 Study Group", FERMILAB-PUB-96/082.

LARGE N DUALITY OF YANG-MILLS THEORY ON A TORUS

Zachary Guralnik

Joseph Henry Laboratories
Princeton university
Princeton, NJ 08544

1. Introduction

The purpose of this talk is to present some evidence suggesting the existence of a $GL(2,Z)$ duality of confining large N gauge theory which resembles T-duality of a string description. We will study $SU(N)/Z_N$ Yang-Mills theory on a spacelike torus $T^2 \times R^{d-2}$, with magnetic flux m through the torus. Supersymmetry will play no role in our discussion. Unlike Olive Montonen duality, too much supersymmetry certainly ruins the conjecture, which depends critically on confinement. We will consider the t'Hooft large N limit with g^2N and m/N fixed, and focus almost entirely on the planar (free string) limit.

Two generators of the conjectured $GL(2,Z)$ are trivial and manifest even at finite N. One of these takes m to $m+N$, and is a symmetry because the magnetic flux m is only defined modulo N [1]. The other trivial generator corresponds to parity and takes m to $-m$. If N and m could also be exchanged, one would obtain $GL(2,Z)$. Such an exchange duality, often referred to as Nahm duality [2][3], has been studied in the context of maximally supersymmetric Yang-Mills theory [4]. However for exact duality of the theory, it appears one must consider a non-commutative deformation of the Yang-Mills theory [5][6], in part because exchange of N and m does not make sense in the case $m=0$. This difficulty does not arise in our conjecture, for which the modular parameter is not $\tau = m/N$, but instead

$$\tau = \frac{m}{N} + i\Lambda^2 A, \tag{1.1}$$

where A is the area of the torus and Λ is the string tension. $GL(2,Z)$ is generated by

Physics of Mass
Edited by Kursunoglu *et al.*, Kluwer Academic / Plenum Publishers, New York, 1998

$$\begin{aligned} \tau &\to \tau + 1, \\ \tau &\to -\bar{\tau}, \\ \tau &\to -\frac{1}{\tau} \end{aligned} \tag{1.2}$$

Since m and N are integers, the last generator makes sense for non-zero $\Lambda^2 A$ only if N is taken to infinity. In this limit $\frac{m}{N}$ becomes a continuous parameter.

In section 2 we will attempt to motivate the conjecture by discussing some qualitative properties of confining Yang Mills which are consistent with T-duality of a string description. The $GL(2, Z)$ we propose resembles string T-duality after identifying $\frac{m}{N}$ with a two form modulus in the string theory.

If this duality exists, then large N pure $4d$ QCD on R^4 is dual to a large N $2d$ QCD on R^2 with two adjoint scalars. There is some qualitative evidence for this which we discuss in section 3. The two dimensional description makes possible some rigorous tests of duality which can me made in the future.

In section 4 we discuss the status of the conjecture for pure two dimensional QCD on T^2, for which the large N partition function is calculable. In this case the partition function is indeed a function of the modular parameter $\tau = \frac{m}{N} + i\frac{\lambda}{2\pi}A$, with $\lambda = g^2 N$. The partition function is almost but not exactly modular invariant. A very simple modification removes the anomaly. It may be that T-duality relates theories in the same universality class as QCD, but is not a self duality of conventional QCD.

There are domains in which the proposed $GL(2, Z)$ takes weak coupling to weak coupling, and should therefore be visible perturbatively. This includes the limit in which $\Lambda^2 A N^2/m^2 << 1$. In section 4 we construct an explicit map which relates theories with the same greatest common divisor of N and m in the limit of vanishing coupling. This map treats $\frac{m}{N}$ as a modular parameter. Under $\tau \to -\frac{1}{\tau}$ the area is mapped linearly instead of being inverted. It is in this sense that our proposed duality differs from Nahm duality, which exchanges rank and flux but inverts the area.

2. Duality and confinement

It has been suspected for some time that confining Yang Mills theories may have a string description [7]. Pure two dimensional QCD is known to be a string theory [8][9][10][11]. In higher dimensions the situation is less clear, although some progress has been made [12]. In this section we will assume that a string description exists, and discuss some qualitative features of a confining large N Yang-Mills theory on $T^2 \times R^n$ which suggest that this string theory may have a T-duality. For a review of T-duality in the context of critical strings, the reader is referred to [13].

The T-duality group for compactifications on T^2 includes a $GL(2, Z)$ subgroup for which the modular parameter is $\tau = B + i\Lambda^2 A$. B is a flat two form background on T^2, A

is the area of T^2, and Λ is the string tension. It is convenient to use complex coordinates and define $w = w^1 + iw^2$ where w^i are the string winding numbers, and $P = P_1 + iP_2$ where P_i are the integer string momenta. The generator $\tau \to -\frac{1}{\tau}$ is accompanied by the exchange $w \leftrightarrow P$. Parity takes $\tau \to -\bar{\tau}$, $w \to \bar{w}$ and $P \to \bar{P}$. The remaining generator, $\tau \to \tau + 1$, is accompanied by $P \to P + iw$. This momentum shift arises because the canonical string momentum P has a contribution iBw arising from the two form term in the world sheet action, $2\pi \int_\Sigma B$. The $GL(2, Z)$ invariant energy of a multiplet of mass M is given by

$$E = \sqrt{M^2 + \frac{\Lambda^2}{Im\tau}|P - i\tau w|^2}. \tag{2.1}$$

Now consider an electrically confining Yang-Mills theory on $T^2 \times R^n$ with the time direction lying in R^n. The energy of an t'Hooft electric flux [1] on a large (square) torus is given by the confining potential $e\Lambda^2\sqrt{A}$. This energy is equal to that of a Kaluza Klein mode on a torus with the area inverted, $A' = 1/(\Lambda^4 A)$. This is consistent with T-duality for vanishing B if one identifies the electric flux with the string winding number. A more difficult question is whether the energy of an electric flux on a small torus is equal to the energy of a kaluza klein mode on a large dual torus. This question will be discussed in the next section.

One must also identify the quantity in the Yang Mills theory corresponding to the two form modulus B. This quantity should be continuous and periodic. A natural candidate is $\frac{m}{N}$, where m is the $SU(N)$ magnetic flux, which is defined only modulo N. $\frac{m}{N}$ becomes a continuous parameter in the $N \to \infty$ limit. With this identification, the standard two form contribution to the canonical string momentum $P_i = B_{ij}w^j + \ldots$ has the form one expects when written in terms of Yang-Mills variables; $P_i = \frac{m_{ij}}{N}e^j + \ldots$.

This last point is actually somewhat subtle since the $SU(N)$ fluxes are defined in terms of twisted bounbdary conditions rather than integrals of local operators. Therefore the fluxes do not contribute to the Yang Mills momentum in any direct way. The $U(N)$ theory does not have this difficulty, but we will not consider it because it has free massless photon having nothing to do with the QCD string. Nevertheless it would seem natural, albeit somewhat arbitrary, to divide the Yang Mills momentum P_i^{YM} into a contribution coming from global structure (ie fluxes), and a contribution p_i from everything else; $P_i^{YM} = \frac{m_{ij}}{N}e^j + p_i$, for $-N/2 < m < N/2$ so that p_i is well defined. One can then define a momentum $P_i = \frac{m_{ij}}{N}e^j + p_i$ having the same transformation properties as the string momentum under $\tau \to \tau + 1$ and parity.

3. $D = 4 \leftrightarrow D = 2$ duality

If the conjectured duality exists, there is a remarkable consequence for electrically confining large N Yang Mills theories on R^d. One can obtain R^d from $R^{d-2} \times T^2$ by making the torus very large. Under inversion of the area of the torus (for $m = 0$), one obtains a dimensionally reduced theory. Thus duality would imply that large N pure QCD in 4 dimensions is equivalent to Large N QCD with two adjoint scalars in 2 dimensions. In fact this two dimensional model has been used to approximate the dynamics of pure QCD in 4 dimensions QCD [14][15][16]. The adjoint scalars in this model play the role of transversely polarized gluons. In [16] the spectrum of this two dimensional model, computed by discrete light cone quantization, was compared to the glueball spectrum of pure 4-d QCD computed using Monte-Carlo simulation. The degree of numerical accuracy allows only crude comparison, however the spectra have some qualitatively agreement. Perhaps in the $N \to \infty$ limit the agreement is more than just qualitative.

If such an equivalence exists, duality must map the QCD scale nontrivially. The mass gap is only proportional to the QCD scale, defined in terms of the running coupling, if the torus is very large. In general the mass gap, or string tension Λ, depends on both the QCD scale Λ_{QCD4} and the area of the torus. Let us fix the mass gap and take the size of the torus to infinity. Λ_{QCD4} on the small dual torus with area A can be found by matching the dimensionless running coupling of the two dimensional reduced theory to the 4 dimensional running coupling at the Kaluza Klein scale;

$$\sqrt{A}g_{2D} = \sqrt{A}\Lambda = g_{4D}(\frac{1}{\sqrt{A}}, \Lambda_{QCD4}). \tag{3.1}$$

The notion that large N can generate extra dimensions is not novel. The 2-d, 4-d equivalence we suggest here is similar to Eguchi Kawai reduction [17], although our discussion pertains to the continuum theory. Assuming the proposed duality exists, the momentum in the two hidden dimensions of the reduced theory must correspond to the electric flux on the vanishingly small torus in the unreduced theory. More precisely, the hidden momentum should be related to the length of a string wrapped around the small torus, $p_i = e_i\sqrt{A}\Lambda^2$, which is held fixed as $A \to 0$. For $SU(N)/Z_N$. The flux e_i is defined [1] by considering certain large gauge transformations $\hat{T}_i$ which leave the boundary conditions on the torus invariant. Such gauge transformations correspond to elements of $SU(N)$ satisfying

$$g_i(x + 2\pi R_s) = g_i(x)e^{\frac{2\pi i}{N}}, \tag{3.2}$$

where R_s is the radius of a cycle of the small (square) torus. $\hat{T}_i^{\,N}$ is a "small" gauge transformation leaving physical states invariant. The electric fluxes are defined by the $\hat{T}_i$ eigenvalues, $\exp(2\pi i \frac{e_i}{N})$. To see what e_i becomes in the reduced theory, consider a Wilson loop around the i'th cycle of T^2. Under a large gauge transformation

$$Pe^{i\oint A} \to e^{\frac{2\pi i}{N}} Pe^{i\oint A}, \tag{3.3}$$

In the reduced theory, this transformation becomes a a global symmetry,

$$e^{iR_sX^i} \to e^{\frac{2\pi i}{N}} e^{iR_sX^i}. \tag{3.4}$$

Here X^i are the adjoint scalars of the reduced theory. If hidden momenta exist in the 2-d theory, they should be the generators of the transformation (3.4). Since N such transformations give the identity map, we wish ultimately to interpret them as translations around a large discretized hidden torus which becomes continuous as $N \to \infty$.

A problem arises because the naive reduced $SU(N)/Z_N$ theory,

$$S = \frac{N}{\lambda_{2d}} \int d^2x Tr(F_{\mu\nu}^2 + D_\mu X^i D^\mu X^i + [X^2, X^3]^2), \tag{3.5}$$

is not a finite theory and requires a mass counterterm. It also does not have a symmetry corresponding to (3.4). It is tempting to consider a $U(N)$ theory instead, in which case the naive reduced theory has a continuous $U(1) \times U(1)$ symmetry. However in this case there is a free photon and no mass gap. States charged under the $U(1) \times U(1)$ have energies inconsistent with (2.1). Therefore we will only consider the $SU(N)$ theory. For R_s finite but small compared to $\frac{1}{\Lambda}$, an effective action which is $Z_N \times Z_N$ symmetric may be written in terms of the $SU(N)$ Wilson loops $h_i = \exp(iR_s X_i)$;

$$S_{SU(N)} = \frac{N}{\lambda_{2d}} \int d^2x Tr \left(F_{\mu\nu}^2 + \frac{1}{R_s^2}(h_i D_\mu h_i^\dagger)^2 + \frac{1}{R_s^4}[h_2, h_3][h_2^\dagger, h_3^\dagger] \right). \tag{3.6}$$

The naive reduced action is recovered by writing $h_i = 1 + \exp(iR_s X^i) + \ldots$ and taking $R_s \to 0$ with X_i fixed. The metric of this sigma model is proportional to $1/R_s^2$, which corresponds to the area of the hidden torus on which $Z_N \times Z_N$ acts as a translation. Note that a mass counterterm proportional to $\frac{1}{R_s^2} \sum_i Trh_i$ is prohibited by the $Z_N \times Z_N$ symmetry.

The truth of our conjecture depends on whether the $Z_N \times Z_N$ symmetry is spontaneously broken below a critical R_s. The two dimensional theory can only generate extra dimensions as $N \to \infty$ if $R_s > R_s^c$. The $Z_N \times Z_N$ symmetry is proposed to correspond to translations symmetry on the dual torus, which should not be spontaneously broken. ¿From experience with finite temperature deconfinement transitions, one might expect that $Z_N \times Z_N$ should be broken for sufficiently small R_s. In our case, symmetry breaking would not be interpreted as deconfinement, since the order parameters are spacelike wilson loops, rather than a timelike Wilson loop*.

The question of whether symmetry breaking occurs at finite R_s is closely tied to the question of whether the limit we wish to take exists. This limit is $N \to \infty$ followed by $R_s \to 0$, while tuning Λ_{QCD4} to keep the mass gap fixed. If Z_N can be thought of as a continuous $U(1)$ in the $N \to \infty$ limit, then the broken phase would have a goldstone boson. If the reduced theory has a gap in the large N limit, then there can not be any symmetry breaking. Furthermore if a string description remains valid for small R_s, one would not expect symmetry breaking in the $N \to \infty$ limit. If the symmetry were broken, states with different $Z_N \times Z_N$ charges, or "hidden momenta," would become degenerate and condense. The hidden momenta are identified with $e_i R_s \Lambda^2$. $e_i R_s$ is the minimum length of a QCD string wrapping e_i times around a cycle of the small torus. Since e_i is defined modulo N, the minimum length can be at most $(N-1)R_s$. If N is infinite, then the minimum length is unbounded and can be held fixed as $R_s \to 0$ by scaling e_i like $1/R_s$. For fixed string

* In the finite temperature theory, the Wilson loop wrapped around the Euclidean time direction is an order parameter of the deconfining phase transition [18]. In this case, the Z_N symmetry is unbroken in the confining phase.

tension Λ one would expect wrapped strings with arbitrarily large minimum lengths to be heavy, in which case they could not condense.

The absence of symmetry breaking at infinite N would mean that the vacuum has zero electric flux. Introducing even the minimum amount of flux, or a singly wrapped string, would lead to an energy in excess of the mass gap. A possible explanation for this behavior is that the string spreads in the non-compact direction as $R_s \to 0$.

Showing that the $Z_N \times Z_N$ symmetry is unbroken in the large N limit is essential for demonstrating that a dual torus exists. One must also show that the spectrum of the $R_s \to 0$ theory has four dimensional Lorentz invariance. This may be testable numerically using the two dimensional description. Having found candidates for momentum in the hidden directions, it may also be possible to see if a four dimensional Lorentz algebra exists. However we leave these tests for the future.

4. Modular invariance in two dimensions

In this section we consider pure Euclidean $SU(N)$ Yang-Mills theory on T^2 in the t'Hooft large N limit. While this theory has no dynamics, we can use it to test the conjecture that the t'Hooft twist and the area combine into a modular parameter (1.1). The partition function of this theory on a surface of arbitrary genus is known in terms of a sum over representations of $SU(N)$ [19][20]. On a two torus, the partition function on a lattice with twisted boundary conditions may be easily computed using the heat kernal action. The result is

$$Z = \sum_R e^{g^2 A C_2(R)} \frac{\mathcal{X}_R(D_m)}{d_R}. \tag{4.1}$$

$C_2(R)$ is the quadratic casimir in the representation R. $\mathcal{X}_R(D_m)$ is the trace of the element D_m in the center of $SU(N)$ corresponding to t'Hooft twist m. In a representation whose Young Tableaux has n_R boxes,

$$D_m = e^{2\pi i \frac{m}{N} n_R} \tag{4.2}$$

To compute the large N expansion of the partition function, we repeat the calculations of [9], for the case of nonvanishing twist. The $\frac{1}{N}$ expansion is obtained by considering composite representations $\bar{S}R$ obtained by gluing the Young Tableaux of a representation R with a finite number of boxes onto the right of the complex conjugate of a representation S with a finite number of boxes. The quadratic Casimir of such a representation in the $\frac{1}{N}$ expansion is

$$C_2(\bar{S}R) = n_R N + n_S N + \ldots \tag{4.3}$$

Furthermore,

$$D_m = e^{2\pi i \frac{m}{N}(n_R - n_S)} \tag{4.4}$$

Therefore the free energy at leading order is

$$F = ln \left| \sum_n^{\infty} \rho(n) e^{g^2 N A n + 2\pi i \frac{m}{N} n} \right|^2 \tag{4.5}$$

where $\rho(n)$ counts the number of representations with n boxes. This sum is computed just as in [8], giving

$$F = ln \left| \frac{e^{2\pi i \frac{\tau}{24}}}{\eta(\tau)} \right|^2 \tag{4.6}$$

where η is a Dedekind eta function, and

$$\tau = \frac{m}{N} - \frac{\lambda A}{2\pi i}, \tag{4.7}$$

with $g^2 N = \lambda$. Thus the complexification of the area generated by modular transformations corresponds to a non-zero t'Hooft twist. Upon completing this work we became aware that M. Douglas has also made this observation using a Jevicki-Sakita boson description of 2-d QCD on the torus [21]. As noted in [22][21] the free energy is almost, but not exactly invariant under inversion of the area. The eta function has the modular properties

$$\begin{aligned} \eta(\tau + 1) &= \eta(\tau) \\ \eta(-\frac{1}{\tau}) &= \sqrt{i\tau}\eta(\tau). \end{aligned} \tag{4.8}$$

A simple modification of the partition function,

$$\mathcal{Z} = Z \frac{e^{\frac{\lambda A}{24}}}{\sqrt{\lambda A}} \tag{4.9}$$

is modular invariant. The extra term exponential in the area has the form of a local counterterm $\int \sqrt{\det g}$. Although the theory is superrenormalizable, such a counterterm is known to arise [23]. On the other hand, the $\sqrt{\lambda A}$ factor poses a problem for modular invariance, since it is non local. Nonetheless the deviation from modular invariance is very simple and we feel deserves better understanding. The two dimensional QCD string has at least an "approximate" T-duality.

5. Exchange of rank and twist

The $GL(2, Z)$ we have proposed treats $\frac{m}{N}$ as a modular parameter as long as one only acts with $\tau \to -1/\tau$ in the region $\Lambda^2 A N^2/m^2 << 1$. In this domain the area is mapped linearly and is never inverted. If this duality exists, it should be possible to see it perturbatively in this region. In this section we discuss an attempt to construct a classical $GL(2, Z)$ under which N and m transform as a doublet. We are only able to prove the validity of the map we construct for the limit of vanishing coupling. If it does extend to finite coupling, it has the correct qualitative property that the area of the torus is mapped linearly rather than being inverted. While the results of this section are very far from quantitative rigor, we feel they are at least suggestive. The map we propose will take an $SU(N)/Z_N$ theory with flux m, or an (N, m) theory, into a $(p, 0)$ theory where p is the greatest common divisor of N and m. $GL(2, Z)$ is then generated by inverting this map to get other theories with the same greatest common divisor.

We begin by reviewing the definition of magnetic flux for $SU(N)/Z_N$ Yang-Mills on a torus. Following t'Hooft [1], translations around a cycle of the torus are equivalent to gauge transformations:

$$A_i(x+a_j) = U_j(x)A_i(x)U_j^\dagger(x) + U_j(x)i\partial_i U_j^\dagger(x) \tag{5.1}$$

where a_i are the periodicities of the torus. For adjoint matter, one has the following constraint:

$$U_i(x)U_j(x+a_i)U_i^\dagger(x+a_j)U_j^\dagger(x) = e^{2\pi i\frac{m_{ij}}{N}} \tag{5.2}$$

where m_{ij} is an integer, and is taken as the definition of nonabelian magnetic flux. We consider the case of Yang Mills on $T^2 \times R^n$. We will drop the indices ij, and the magnetic flux m will often be referred to as the twist. Note that m is defined modulo N, so that the generator of $SL(2,Z)$ which takes $m \to m + jN$ is a manifest symmetry. We shall assume the timelike direction lies in R^n. In $A_0 = 0$ gauge we choose the following twisted boundary conditions

$$\begin{aligned} U_1(x) &= Q, \\ U_2(x) &= P^m, \end{aligned} \tag{5.3}$$

where

$$Q = e^{i\theta}\begin{pmatrix} 1 & & & \\ & e^{\frac{2\pi i}{N}} & & \\ & & \ddots & \\ & & & e^{\frac{2\pi i(N-1)}{N}} \end{pmatrix}, \qquad P = e^{i\theta'}\begin{pmatrix} & 1 & & \\ & & 1 & \\ & & & \ddots \\ 1 & & & \end{pmatrix} \tag{5.4}$$

The phases θ and θ' are chosen so that Q and P have determinant 1. Q and P satisfy

$$PQ = QPe^{\frac{2\pi i}{N}}. \tag{5.5}$$

The constraints imposed by twisted boundary conditions can be solved to find the independent perturbative degrees of freedom. To this end we shall work in momentum space. Since $Q^N = 1$, the gauge potentials are periodic in x^1 on the interval $[0, Na^1]$. To find the periodicity in x^2, one looks for the minimal power to which one must raise P^m to get 1. Writing the pair (N,m) as $(p\alpha, p\beta)$ where α and β are relatively prime, one finds that this power is α. Therefore the gauge potentials are periodic in x^2 on the interval $[0, \alpha a_2]$. The fourier modes of the gauge field strength F_{n_1,n_2} satisfy the twisted boundary conditions,

$$e^{2\pi i\frac{n_1}{N}}F_{n_1,n_2} = QF_{n_1,n_2}Q^\dagger \tag{5.6}$$

and

$$e^{2\pi i\frac{n_2}{\alpha}}F_{n_1,n_2} = P^mF_{n_1,n_2}P^{m\dagger}. \tag{5.7}$$

Making use of the algebra (5.5) a general solution of (5.6) is

$$F_{n_1,n_2} = M_{n_1,n_2}P^{n_1}, \tag{5.8}$$

where M_{n_1,n_2} is a diagonal $N \times N$ matrix which is traceless when $n_1 = 0 mod N$. Note that if m vanished there would be $N-1$ degrees of freedom for each fourier mode with

$n_1 = 0 mod N$ and N degrees of freedom for all the others, rather than $N^2 - 1$ degrees of freedom for every fourier mode. This is because the (non-gauge invariant) momentum in the x^1 direction of the torus is fractional in units of $\frac{1}{N}$. Heuristically, in going to a more conventional gauge with $U_1 = U_2 = I$, the fractional modes become integral and fill out the Lie algebra.

Now consider arbitrary m. The second constraint (5.7) gives

$$P^m M_{n_1,n_2} P^{-m} = M_{n_1,n_2} e^{2\pi i \frac{n_2}{\alpha}}, \tag{5.9}$$

Conjugating M by P^m shifts the elements of M cyclically by an amount m. Thus we find that the number of independent elements of M_{n_1,n_2} is p, the greatest common divisor of N and m;

$$M_n = \begin{pmatrix} M'_n & & \\ & M'_n e^{2\pi i \frac{n_2}{\alpha\beta}} & \\ & & \cdots \end{pmatrix} \tag{5.10}$$

where M'_n is a diagonal $p \times p$ matrix. It is now easy to construct a candidate for the gauge field of the dual theory with rank p and vanishing magnetic flux, or a $(p, 0)$ theory. We define the dual field strength as

$$F'_n = e^{i\phi(n)} M'_n P'^{n_1} \tag{5.11}$$

where P' is the $p \times p$ shift matrix. The diagonal phase factor ϕ is chosen so that the dual field strength is real, $F'_{-n} = F'^{\dagger}_n$ This is general solution of the constraint

$$e^{2\pi i \frac{n_1}{p}} F'_n = Q' F'_{n_1,n_2} Q'^{\dagger}, \tag{5.12}$$

where Q' is the $p \times p$ matrix

$$Q' = \begin{pmatrix} 1 & & & \\ & e^{\frac{2\pi i}{p}} & & \\ & & \ddots & \\ & & & e^{\frac{2\pi i(p-1)}{p}} \end{pmatrix}. \tag{5.13}$$

Therefore F' is a candidate for a field strength on a dual torus with the twisted boundary conditions given by $U'_1 = Q'$ and $U'_2 = I$, which corresponds to vanishing magnetic flux.

The action of the (N, m) theory is

$$S = \int_{R^n} \sum_n Tr_{N\times N} F_{\mu\nu,n} F^{\dagger}_{\mu\nu,n} = \alpha \int_{\mathcal{M}} \sum_n \tag{5.14}$$

Written in terms of the proposed dual field strength this becomes

$$S = \int_{R_n} \alpha Tr_{p\times p} F'_{\mu\nu,n} F'^{\dagger}_{\mu\nu,n} \tag{5.15}$$

However, for duality to hold, we must be able to define a dual gauge potential which solves the Jacobi identity and gives the correct measure in the path integral. Let us define the

The dual gauge potential the same way we defined the the dual field strength. Then the map between the $(N.m)$ gauge potential and the dual $(p,0)$ gauge potential is linear, so one might expect that the measure maps correctly. We will scale the fields such that the coupling constant appears only in the interaction terms. If one neglects the the $[A_\mu, F^{\alpha\beta}]$ terms, it is easy to check that the Jacobi identity is preserved by the map, provided that that the periodicities of fields on the original torus are the same as those on the dual torus. In other words $Na^1 = N'a'^1 = pa'^1$ and $\alpha a^2 = \alpha' a'^2 = a'^2$ Working in a Hamiltonian formulation, one can easily check that the Hamiltonian, commutation relations, and Gauss law constraint of the $(N.m)$ theory at zero coupling map to those of the $(p,0)$ theory at zero coupling.

At finite coupling however, the Jacobi identity is no longer satisfied. It is violated by terms involving the difference between the $N \times N$ shift matrix P and α copies of the $p \times p$ shift matrix,

$$P - \begin{pmatrix} P' & & \\ & P' & \\ & & \cdots \end{pmatrix}. \tag{5.16}$$

These matrices differ by a finite number of ones in the limit that $p \to \infty$. Thus it is very tempting to neglect the difference. However when these matrices are raised to a power of order p, the difference is not always negligible. This occurs when n_1 is of order p. We can not discard such modes from the action, since they may correspond to a finite gauge invariant physical momenta. It may be possible that this discrepancy is negligible at leading order in some weak coupling expansion, however we will not attempt to prove this.

6. Conclusion

We have given evidence that large N confining Yang Mills theories on tori may have an $SL(2,Z)$ duality which appears to be T-duality of a string description. The existence of such a duality would be quite useful, since it relates non compact 4 dimensional theories to more numerically tractable 2 dimensional theories. It should be interesting to study the two dimensional gauged sigma model of (3.6)İt may be possible to explicitly test whether this model generates two extra dimensions in the large N limit. The extent to which the duality we propose is exact is not known, however at least in two dimensions, it seems to have a very mild anomaly. Perhaps duality only relates theories in the same universality class as QCD, and does not act on conventional QCD.

Acknowledgments

I am especially grateful to J. Nunes, S. Ramgoolam, and W. Taylor for very helpful discussions. I am also indebted to C. Callan, A, Cohen, A. Jevicki, I. Klebanov, and M. Schmaltz. This work was supported in part by NSF grant PHY96-00258.

References

[1] G. 't Hooft, *"A property of electric and magnetic flux in non-abelian gauge theories,"* Nucl. Phys. B153 (1979) 141-160.

[2] W. Nahm, *"The construction of all self-dual multimonopoles by the ADHM method,"* Phys. Lett.90B (1980) 413.

[3] P. J. Braam and P. van Baal, *"Nahms transformation for Instantons,"* Commun. Math. Phys. 122 (1989) 267; P. van Baal, *"Instanton moduli for $T^3 \times R$,"* Nucl. Phys. Proc. Suppl. 49 (1996) 238, hep-th/9512223.

[4] H. Verlinde and F. Hacquebord, *"Duality symmetry of $N = 4$ Yang-Mills theory on T^3,"* hep-th/9707179.

[5] M. Douglas and C. Hull, *"D-branes and the noncommutative torus,"* hep-th/9711165.

[6] A. Connes, M. Douglas, A. Schwarz, *"Noncommutative geometry and Matrix theory: compactification on tori,"* hep-th/9711162.

[7] A. Polyakov, *"String representations and hidden symmetries for gauge fields,"* Phys. Lett. 82B (1979) 247-250. *"Gauge fields as rings of glue,"* Nucl. Phys.B164 (1980) 171-188.

[8] D. Gross, *"Two dimensional QCD as a string theory,"* Nucl. Phys. B400 (1993) 161, hep-th/9212149.

[9] D. Gross and W. Taylor, *"Two dimensional QCD is a string theory,"* Nucl. Phys. B400 (1993) 181, hep-th/9301068.

[10] S. Cordes, G. Moore, S. Ramgoolam, *"Large N 2-D Yang-Mills theory and topological string theory,"* Commun. Math. Phys.185 (1997) 543-619, hep-th/9402107.

[11] P. Horava, *"Topological rigid string theory and two-dimensional QCD,"* Nucl. Phys.B463 (1996) 238-286, hep-th/9507060.

[12] A. Polyakov, *"String theory and quark confinement,"* hep-th/9711002. *"Confining strings,"* Nucl.Phys.B486 (1997) 23-33, hep-th/9607049.

[13] A. Giveon, M. Porrati, and E. Rabinovici, *"Target space duality in string theory,"* Phys. Rept. 244 (1994) 77-202, hep-th/9401139.

[14] S. Dalley, I. Klebanov, *String spectrum of* (1 + 1) *dimensional large N QCD with adjoint matter,"* Phys. Rev. D47 (1993) 2517-2527.

[15] K. Demeterfi, I. Klebanov, Gyan Bhanot, *"Glueball spectrum in a* $(1+1)$ *dimensional model for QCD,"* Nucl. Phys.B418 (1994) 15-29, hep-th/9311015.

[16] F. Antonuccio and S. Dalley, *" Glueballs from (1+1) dimensional gauge theories with transverse degrees of freedom,"* Nucl. Phys. B461 (1996) 275-304, hep-ph/9506456.

[17] T. Eguchi and H. Kawai, *"Reduction of dynamical degrees of freedom in the large N gauge theory,"* Phys. Rev. Lett.48 (1982) 1063.

[18] B. Svetitsky and L. Yaffe, *"Critical behavior at finite temperature confinement transitions,"* Nucl.Phys.B210 (1982) 423.

[19] B. Rusakov, *"Loop averages and partition functions in U(N) gauge theory on two-dimensional manifolds,"* Mod.Phys.Lett.A5 (1990) 693-703.

[20] A. Migdal, *"Recursion equations in gauge theories,"* Sov.Phys.JETP42 (1975) 413, Zh.Eksp.Teor.Fiz.69 (1975) 810-822.

[21] M. Douglas, *"Conformal field theory techniques in large N Yang-Mills theory,"* RU-93-57, Presented at Cargese Workshop on Strings, Conformal Models and Topological

Field Theories, Cargese, France, May 12-26, 1993, hep-th/9311130.
[22] R. Rudd, *"The string partition function for QCD on the torus,"* hep-th/9407176.
[23] M. Blau and G. Thompson, *"Lectures on 2-D gauge theories: topological aspects and path integral techniques,"* Trieste HEP Cosmol. 1993: 175-244, hep-th/9310144.

SUSY MASSES WITH NON-UNIVERSAL SOFT BREAKING

R. Arnowitt[1] and Pran Nath[2]

[1]Center for Theoretical Physics, Department of Physics,
Texas A&M University, College Station, TX 77843-4242,
[2]Department of Physics, Northeastern University,
Boston, MA 02115-5005

Abstract

The effects of non-universal scalar soft breaking masses are examined within the framework of gravity mediated supergravity GUT models with R-parity invariance. For the domain $\tan\beta \lesssim 25$, cosmological constraints on the amount of cold dark matter predicted require that $m_0 \lesssim 200\ GeV$ for a gluino mass of $m_{\tilde{g}} \gtrsim 450\ GeV$. Thus sleptons and squarks will generally be light for $m_{\tilde{g}} \gtrsim 450\ GeV$. Also, significant corrections to the gaugino mass scaling relations can occur for $m_{\tilde{g}} \lesssim 450\ GeV$. Using estimates of the accuracy expected for measurements of the cosmological parameters by the Planck sattelite, two models are examined. For the ΛCDM model, one finds that $m_{\tilde{g}} \lesssim 540\ GeV$, and that terrestial CDM detector event rates can be significantly reduced or enhanced depending on the sign of the non-universal corrections. For the νCDM model, we find $m_{\tilde{g}} \lesssim 720\ GeV$ and gaps (forbidden regions) can occur at $m_{\tilde{g}} \simeq 500\ GeV$ and $m_{\tilde{g}} \simeq 600\ GeV$ for one sign of the non-universal corrections.

INTRODUCTION-SUPERGRAVITY MODELS

The matter that we see in the universe [i.e. quarks and leptons] or speculate to exist [Higgs, massive neutrinos (possible hot dark matter, HDM), neutralinos (possible cold dark matter, CDM)] appear to exist at relatively low energies, i.e. below 1 TeV. One line of theoretical thought is that the principles that determine the existence and numerical values of these masses resides, however, at a much higher mass scale, perhaps the Planck scale. We consider here supergravity grand unified models with R-parity invariance (GUT models) where supersymmetry is broken in a hidden sector by gravity with gravity the messinger field transmitting this breaking to the physical sectors[1,2]. While these models do not predict the value of the Yukawa coupling constants, they do relate the supersymmetry (SUSY) breaking scale with the electroweak breaking scale, and hence produce relations between the masses of SUSY particles.

Physics of Mass
Edited by Kursunoglu *et al.*, Kluwer Academic / Plenum Publishers, New York, 1999

The simplest GUT model involves four soft breaking parameters: m_0 (the scalar soft breaking mass), $m_{1/2}$ (the gaugino soft breaking mass), A_0 (the cubic soft breaking parameter) and $\tan\beta = \langle H_2\rangle/\langle H_1\rangle$ (where $\langle H_{1,2}\rangle$ are the VEVs of the two Higgs doublets required by SUSY). In addition, there is a Higgs mixing parameter μ, which enters into superpotential as $\mu H_1 H_2$. However, the renormalization group equations[3] (RGE) lead to the spontaneous breaking of $SU(2)\times U(1)$ at the electroweak scale and determine μ up to its sign in terms of M_Z and the other parameters. Thus one finds at scale $Q \simeq M_Z$

$$\mu^2 = \frac{m_{H_1}^2 - \tan^2\beta\, m_{H_2}^2}{\tan^2\beta - 1} - \frac{1}{2}M_Z^2 \tag{1}$$

where $m_{H_{1,2}}$ are the running Higgs masses with loop corrections. Over most of the parameter space, μ^2/M_Z^2 is large, leading to scaling relations for the gauginos χ_i^0 (neutralinos), $\chi_i^\pm$ (charginos) and $\tilde g$ (gluino): $2m_{\chi_1^0} \cong \chi_1^\pm \cong \chi_2^0 \cong (\frac{1}{3}-\frac{1}{4})m_{\tilde g}$ (where $m_{\tilde g}$ $\cong$ $(\alpha_3/\alpha_G)m_{1,2}$, and $\alpha_G \cong 1/24$ is the GUT coupling constant). Also $m_{\chi_2^\pm} \cong m_{\chi_{3,4}^0} \gg m_{\chi_1^0}$. Corrections to these relations are $O(M_Z^2/\mu)$.

However, one may have non-universal soft breaking at M_G. We will assume in the following that the first two generations of squarks and sleptons have a universal mass m_0 (to suppress flavor changing neutral currents) and also assume as above that the gaugino mass, $m_{1/2}$, is universal. Non-universality may occur, however, in the Higgs and third generation masses which we parametrize at M_G as follows: $m_{H_{1,2}}^2 = m_0^2(1+\delta_{1,2})$, $m_{q_L}^2 = m_0^2(1+\delta_3)$, $m_{u_R}^2 = m_0^2(1+\delta_4)$, $m_{e_R}^2 = m_0^2(1+\delta_5)$, $m_{d_R}^2 = m_0^2(1+\delta_6)$ and $m_{l_L}^2 = m_0^2(1+\delta_7)$. Here q_L is the squark doublet (t_L, b_L), l_L the lepton doublet, etc. In addition there may be separate cubic soft breaking parameters $A_{0t}, A_{0b}, A_{0\tau}$. (We note that for any GUT group that contains an $SU(5)$ subgroup with matter in the usual $10+\bar 5$ representations, $\delta_3=\delta_4=\delta_5$, $\delta_6=\delta_7$ and $A_{0b}=A_{0\tau}$.)

In the following we will assume $|\,\delta_i\,| \geq 1$, $m_0, m_{\tilde g} \leq 1\,TeV$, $|\,A_t/m_0\,| \leq 7$ (A_t is the t-quark cubic parameter at $Q=M_Z$) and $\tan\beta \leq 25$. The last condition means that to a good approximation $\delta_5, \delta_6, \delta_7, A_{0b}, A_{0\tau}$ may be neglected (though these parameters would be important for larger $\tan\beta$). The RGE then give [4]

$$\begin{aligned}\mu^2 = \;& \frac{t^2}{t^2-1}\left[\left\{\frac{1-3D_0}{2}+\frac{1}{t^2}\right\}+\left\{\frac{1-D_0}{2}(\delta_3+\delta_4)-\frac{1+D_0}{2}\delta_2+\frac{1}{t^2}\delta_1\right\}\right]m_0^2\\ &+\frac{t^2}{t^2-1}\left[\frac{1}{2}(1-D_0)\frac{A_R^2}{D_0}+C_{\tilde g}m_{\tilde g}^2\right]-\frac{1}{2}M_Z^2+\frac{1}{22}\frac{t^2+1}{t^2-1}S_0\left(1-\frac{\alpha_1}{\alpha_G}\right)\end{aligned} \tag{2}$$

where $D_0 \cong 1 - m_t^2/(200\sin\beta)^2$, $A_R \cong A_t - 0.613\, m_{\tilde g}$, $C_{\tilde g}$ is given in Ibañez at al.[3] and $S_0 = TrYm^2$ ($Y=$ hypercharge and m^2 are the scalar (mass)^2 at $Q=M_G$). $D_0=0$ is the t-quark Landau pole (A_R is the residue). For $m_t = 175\,GeV$, one has $D_0 \leq 0.23$. Since D_0 is small, and t^2 is mostly large, we see that μ^2 depends approximately on the combination $\delta \equiv \delta_2 - (\delta_3+\delta_4)$. Thus $\delta > 0$ decreases μ^2, and $\delta < 0$ increases μ^2. Accelerator data now has eliminated a considerable amount of the parameter space, i. e. from $m_t \cong 175\,GeV$ one has $A_t/m_0 \gtrsim -0.5$ and the $b\to s+\gamma$ data eliminates most of the $A_t/\mu < 0$ domain[5].

COSMOLOGICAL CONSTRAINTS

Supergravity models with R-parity invariance automatically predict the existence of dark matter i.e. the lightest supersymmetric particle (LSP) which is absolutely stable. Over almost all the parameter space, the LSP is the χ_1^0. One of the successes of supergravity GUTs is that for a significant part of this parameter space, it predicts a relic density for the χ_1^0 in accord with current astronomical CDM measurements: $0.1 \lesssim \Omega_{CDM}h^2 \lesssim 0.4$, where $H = (100h)\ km/s\ Mpc$ is the Hubble constant, $\Omega_i = \rho_i/\rho_c$, ρ_i = density of matter of type "i", and $\rho_c = 3H^2/8\pi G_N$. Future astronomical measurements by the MAP and Planck sattelites (and many balloon and ground based experiments) will greatly narrow this $\Omega_{CDM}h^2$ window. We consider first, as an example, the ΛCDM model and assume for the CDM $\Omega_{\chi_1^0} = 0.4$, a baryonic (B) part $\Omega_B = 0.05$, a vacuum energy (cosmological constant) of $\Omega_\Lambda = 0.55$ and a Hubble constant of $h = 0.62$. (The above numbers are consistent with current astronomical measurements.) The errors with which the Planck sattelite can measure the above quantities have been estimated [6], and from this we find [7] $\Omega_{\chi_1^0}h^2 = 0.154 \pm 0.017$, which shows the accuracy of future determinations of the amount of cold dark matter. In calculating $\Omega_{\chi_1^0}h^2$, one finds two domains:

(i) $m_{\chi_1^0} \lesssim 60\ GeV$ ($m_{\tilde{g}} \lesssim 450\ GeV$). Here rapid annihilation of χ_1^0 in the early universe can occur through s-channel Z and h-poles allowing m_0 to get large and still satisfy the above astronomical bounds on $\Omega_{\chi_1^0}h^2$. Thus this regions can have heavy sfermions.

(ii) $m_{\chi_1^0} \gtrsim 60\ GeV$ ($m_{\tilde{g}} \gtrsim 450\ GeV$). Here the t-channel sfermion pole diagrams dominate the annihilation requiring m_0 to be small ($m_0 < 200\ GeV$) to get sufficient annihilation to satisfy the astronomical bounds. Thus sfermions will in general be light.

The above ideas are illustrated in Fig. 1 which is a scatter plot of the $\tilde{e}_R$ as a function of $m_{\tilde{g}}$, as one scans the other SUSY parameters (with $\delta_i = 0$). One sees that for $m_{\tilde{g}} \leq 450\ GeV$, $m_{\tilde{e}_R}$ can get quite large (since m_0 can be large), but for $m_{\tilde{g}} > 460\ GeV$, $m_{\tilde{e}_R}$ should lie below $100\ GeV$. Thus in this model, the $\tilde{e}_R$ should be accesible to LEP200 if $m_{\tilde{g}}$ is large, but for $m_{\tilde{g}} < 450\ GeV$ the $\tilde{e}_R$ would not be observable even at the LHC (though the gluino would then be accessible to the upgraded Tevatron). Results similar to this hold with non-universal soft breaking. Deviations

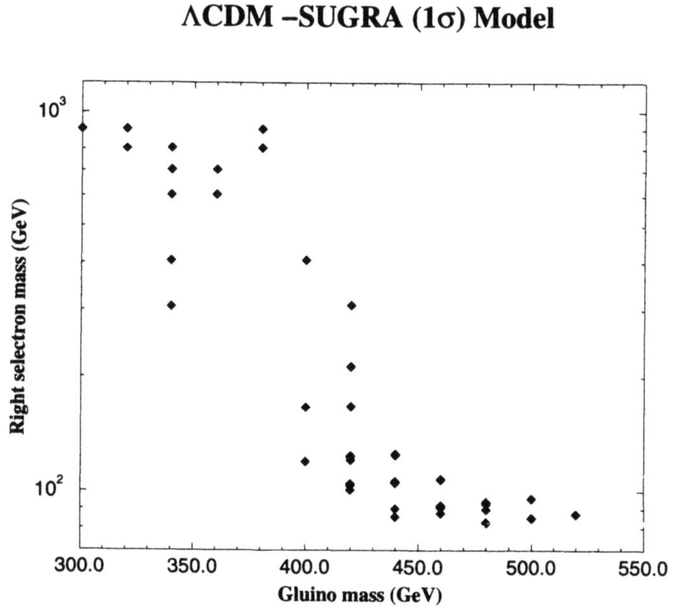

Figure1. Scatter plot for $m_{\tilde{e}_R}$ vs. $m_{\tilde{g}}$ for $\delta_i = 0$ as other parameters are varied for ΛCDM model.

from scaling of the gaugino masses are $O(M_Z^2/\mu)$. For $\delta < 0$, where by Eq.(2) μ^2 is reduced, one may have significant breakdown of scaling when $m_{\tilde{g}} \lesssim 450\ GeV$. [Above 450 GeV these effects are suppressed since then m_0 is small and non-universal terms are scaled by m_0^2 in Eq.(2).] Similarly, for $\delta > 0$ the scaling relations are better obeyed, since μ^2 is increased.

Terrestial dark matter detector event rates generally increase (decrease) as μ^2 decreases (increases). This effect is shown in Fig.2 where maximum and minimum detector event rates for a Xe detector are plotted as a function of $m_{\tilde{g}}$ for the case $\delta_i = 0$ (solid), $\delta_2 = -1 = -\delta_1$ [and hence $\delta < 0$] (dotted) and $\delta_2 = 1 = -1$ [or $\delta > 0$] (dashed). One sees there is a significant decrease particulary in the minimum event rates, for $\delta > 0$, and an increase for the $\delta < 0$ case. Also, all the curves show a maximum allowed gluino mass of about 540 GeV (arising from the maximum value of $\Omega_{\chi_1^0} h^2$ of 0.188 at the 2 sigma level). For the $\delta < 0$ case, one also has minimum gluino mass of 400 GeV.

As a second model, we consider a mixed dark matter νCDM with hot dark matter (HDM) arising from possible massive neutrinos. We assume here $\Omega_\nu = 0.2$, $\Omega_{\chi_1^0} = 0.75$, $\Omega_B = 0.05$ and $h = 0.62$. Using [6,8] to estimate the expected Planck sattelite accuracy for these quantities, we obtain [7] $\Omega_{\chi_1^0} h^2 = 0.288 \pm 0.013$. Here one finds a larger range for $m_{\tilde{g}}$ i.e. $m_{\tilde{g}} \leq 720\ GeV$ (since $(\Omega_{\chi_1^0} h^2)_{max}$ is larger), with $m_{\tilde{e}_R} \simeq 120\ GeV$ for $m_{\tilde{g}} > 500\ GeV$ (though the $\tilde{e}_R$ can be quite heavy for $m_{\tilde{g}} \lesssim 450\ GeV$). A heavy gluino would imply for this model that the selectrons would be observable at the LHC, though they would not necessarily be observable if $m_{\tilde{g}} \lesssim 450\ GeV$. The most striking effect for this model, however, is the appearance of forbidden regions of $m_{\tilde{g}}$ (or $m_{\chi_1^0}$) appearing at $m_{\tilde{g}} \simeq 500\ GeV$ and $m_{\tilde{g}} \simeq 600\ GeV$. Fig.3 shows the appearance of these gaps in the Xe detector event rates for $\delta_i = 0$ and Fig.4 for $\delta_2 = -1 = -\delta_1$. Note that the gap at $m_{\tilde{g}} \simeq 500\ GeV$ widens for the non-universal case of Fig.4 (but in fact disappears eventually when δ_2 becomes positive).

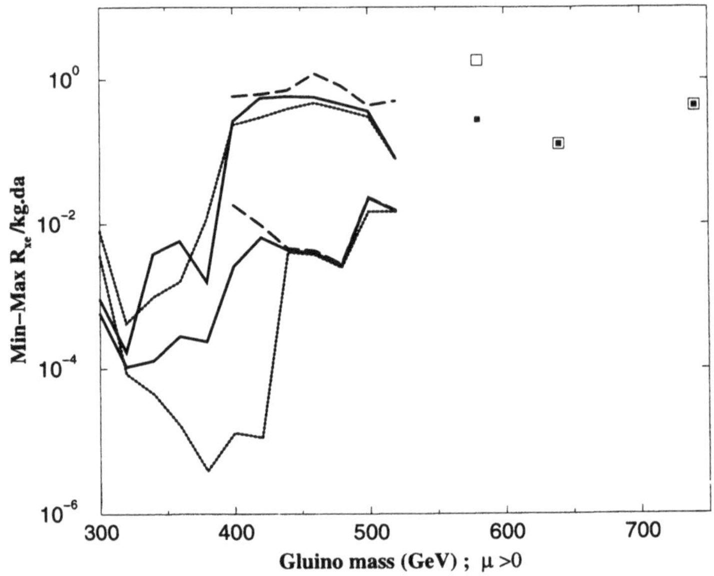

Figure 2. Maximum and minimum event rates for a Xe detector as a function of $m_{\tilde{g}}$ with $\mu > 0$ for 1 std band of the ΛCDM model with $\delta_1 = 0 = \delta_2$ (solid), $\delta_2 = -1 = -\delta_1$ (dotted), $\delta_2 = 1 = -\delta_1$ (dashed) [7].

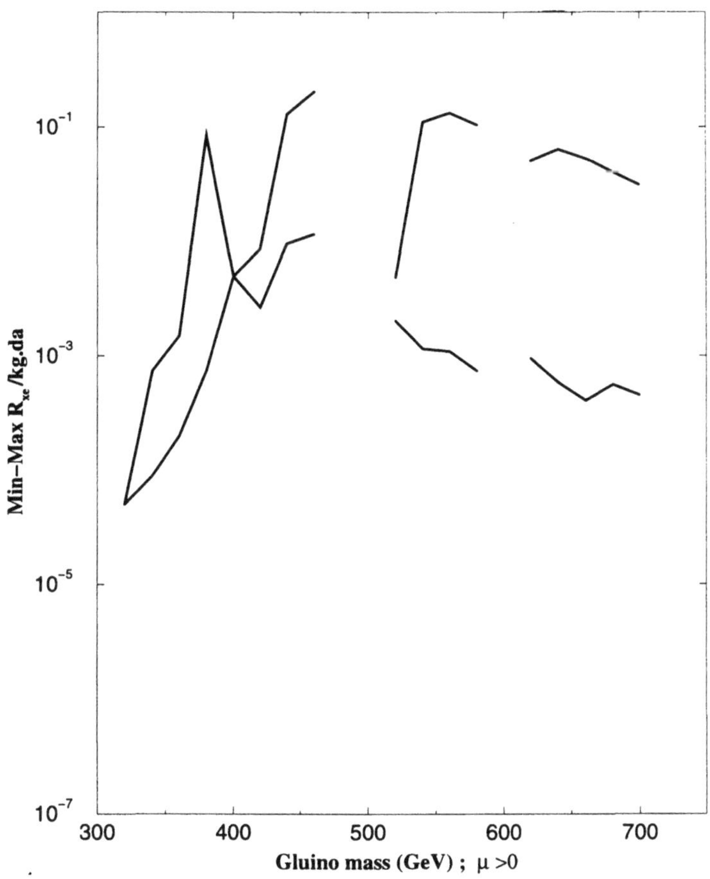

Figure 3. Maximum and minimum event rates for a Xe detector as a function of $m_{\tilde{g}}$ with $\mu > 0$ for the 1 std band of the νCDM model with $\delta_i = 0$ [7].

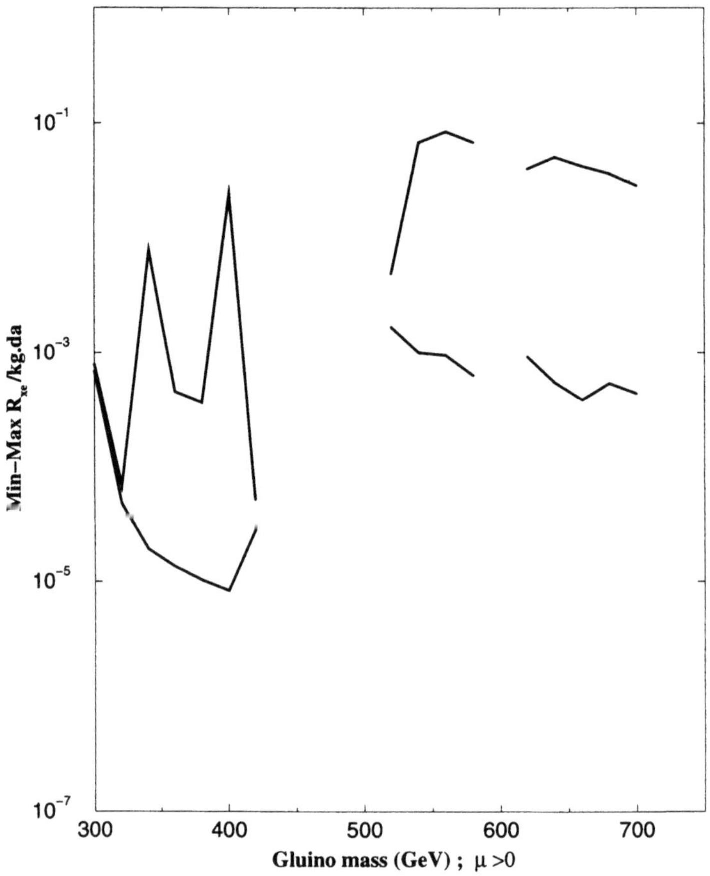

Figure 4. Same as Fig. 3. with $\delta_2 = -1 = -\delta_1$ [7].

CONCLUSIONS

The possibility of non-universal soft breaking masses tends to complicate the predictions of supergravity GUT models. Fortunately, we've seen that some of these predictions are sensitive only to a combination of soft breaking parameters $\delta = \delta_2 - (\delta_3 + \delta_4)$. Cosmological data concerning the existence of cold dark matter produce further constraints on the parameter space. Thus in order to obtain the amount of CDM seen, m_0 must be small for $m_{\tilde{g}} \gtrsim 450\ GeV$, but can be large for $m_{\tilde{g}} \lesssim 450\ GeV$. Thus one expects sleptons and squarks to be relatively light for $m_{\tilde{g}} \gtrsim 450\ GeV$, but may be heavy for light gluinos, and further non-universal effects arising from the μ^2 parameter will be small for $m_{\tilde{g}} \gtrsim 450\ GeV$ since they are scaled by m_0^2 there.

Future sattelite and balloon experiments are expected to determine the cosmological parameters to good accuracy. Using the expected accuracy, we've exemined two models, the ΛCDM model and the νCDM model. In the former, we found that $m_{\tilde{g}}$ is expected to lie below about 540 GeV. In the latter, remarkable gaps (forbiden regions) can occur in $m_{\tilde{g}}$ at $m_{\tilde{g}} \simeq 500\ GeV$ and $m_{\tilde{g}} \simeq 600\ GeV$. Both models are sensitive to non-universal soft breaking. Thus cosmological constraints should be an important tool for disentangling the nature of SUSY breaking.

1. A. H. Chamseddine, R. Arnowitt and P. Nath, Phys. Rev. Lett. **49**, 970 (1982).

2. For reviews see P. Nath, R. Arnowitt and A. H. Chamseddine, ***Applied N=1 Supergravity*** (World Scientific, Singapore, 1984); H. P. Nilles, Phys. Rep. **110**, 1 (1984); R. Arnowitt and P. Nath, Proc. of VII J.A. Swieca Summer School ed. E. Eboli (World Scientific, Singapore, 1994).

3. K. Inoue et al. Prog. Theor. Phys.**68**, 927 (1982); L. Ibañez and G. G. Ross, Phys. Lett. B **110**, 227 (1982); L. Alvarez-Gaumé, J. Polchinski and M. B. Wise, Nucl. Phys. B **221**, 495 (1983); J. Ellis, J. Hagelin, D. V. Nanopoulos and K. Tamvakis, Phys. Lett. B **125**, 2275 (1983); L. E. Ibañez and C. Lopez, Nucl. Phys. B **233**, 545 (1984); L.E. Ibañez, C. Lopez and C. Muños, Nucl. Phys. B **256**, 218 (1985).

4. P. Nath and R. Arnowitt, Phys. Rev. D **56**, 2820 (1997).

5. P. Nath and R. Arnowitt, Phys. Lett. B **336**, 395 (1994); F. Borzumati, M. Drees and M. Nojiri, Phys. Rev. D **51**, 341 (1995).

6. A. Kosowsky, M. Kamionkowski, G. Jungman and D. Spergel, Nucl. Phys. Proc. Suppl. **51B**, 49 (1996).

7. P. Nath and R. Arnowitt, hep-ph/9801454.

8. S. Dodelson, E. Gates and A. Stebbins, ApJ. **467**, 10 (1996).

INDEX

MIX
Papier aus verantwortungsvollen Quellen
Paper from responsible sources
FSC® C105338

If you have any concerns about our products,
you can contact us on
ProductSafety@springernature.com

In case Publisher is established outside the EU,
the EU authorized representative is:
Springer Nature Customer Service Center GmbH
Europaplatz 3, 69115 Heidelberg, Germany

Printed by Libri Plureos GmbH
in Hamburg, Germany